An Overview of String Theory

Contents

Chapter 1

String theory

For the study of strings of characters, see Concatenation theory.
For a more accessible and less technical introduction to this topic, see Introduction to M-theory.

In physics, **string theory** is a theoretical framework in which the point-like particles of particle physics are replaced by one-dimensional objects called strings. String theory describes how these strings propagate through space and interact with each other. On distance scales larger than the string scale, a string looks just like an ordinary particle, with its mass, charge, and other properties determined by the vibrational state of the string. In string theory, one of the vibrational states of the string corresponds to the graviton, a quantum mechanical particle that carries gravitational force. Thus string theory is a theory of quantum gravity.

String theory is a broad and varied subject that attempts to address a number of deep questions of fundamental physics. String theory has been applied to a variety of problems in black hole physics, early universe cosmology, nuclear physics, and condensed matter physics, and it has stimulated a number of major developments in pure mathematics. Because string theory potentially provides a unified description of gravity and particle physics, it is a candidate for a theory of everything, a self-contained mathematical model that describes all fundamental forces and forms of matter. Despite much work on these problems, it is not known to what extent string theory describes the real world or how much freedom the theory allows to choose the details.

String theory was first studied in the late 1960s as a theory of the strong nuclear force, before being abandoned in favor of quantum chromodynamics. Subsequently, it was realized that the very properties that made string theory unsuitable as a theory of nuclear physics made it a promising candidate for a quantum theory of gravity. The earliest version of string theory, bosonic string theory, incorporated only the class of particles known as bosons. It later developed into superstring theory, which posits a connection called supersymmetry between bosons and the class of particles called fermions. Five consistent versions of superstring theory were developed before it was conjectured in the mid-1990s that they were all different limiting cases of a single theory in eleven dimensions known as M-theory. In late 1997, theorists discovered an important relationship called the AdS/CFT correspondence, which relates string theory to another type of physical theory called a quantum field theory.

One of the challenges of string theory is that the full theory does not yet have a satisfactory definition in all circumstances. Another issue is that the theory is thought to describe an enormous landscape of possible universes, and this has complicated efforts to develop theories of particle physics based on string theory. These issues have led some in the community to criticize these approaches to physics and question the value of continued research on string theory unification.

1.1 Fundamentals

In the twentieth century, two theoretical frameworks emerged for formulating the laws of physics. One of these frameworks was Albert Einstein's general theory of relativity, a theory that explains the force of gravity and the structure of space and time. The other was quantum mechanics, a radically different formalism for describing physical phenomena using probability. By the late 1970s, these two frameworks had proven to be sufficient to explain most of the observed features of the universe, from elementary particles to atoms to the evolution of stars and the universe as a whole.[1]

The fundamental objects of string theory are open and closed strings.

In spite of these successes, there are still many problems that remain to be solved. One of the deepest problems in modern physics is the problem of quantum gravity.[2] The general theory of relativity is formulated within the framework of classical physics, whereas the other fundamental forces are described within the framework of quantum mechanics. A quantum theory of gravity is needed in order to reconcile general relativity with the principles of quantum mechanics, but difficulties arise when one attempts to apply the usual prescriptions of quantum theory to the force of gravity.[3] In addition to the problem of developing a consistent theory of quantum gravity, there are many other fundamental problems in the physics of atomic nuclei, black holes, and the early universe.[lower-alpha 1]

String theory is a theoretical framework that attempts to address these questions and many others. The starting point for string theory is the idea that the point-like particles of particle physics can also be modeled as one-dimensional objects called strings. String theory describes how strings propagate through space and interact with each other. In a given version of string theory, there is only one kind of string, which may look like a small loop or segment of ordinary string, and it can vibrate in different ways. On distance scales larger than the string scale, a string will look just like an ordinary particle, with its mass, charge, and other properties determined by the vibrational state of the string. In this way, all of the different elementary particles may be viewed as vibrating strings. In string theory, one of the vibrational states of the string gives rise to the graviton, a quantum mechanical particle that carries gravitational force. Thus string theory is a theory of quantum gravity.[4]

One of the main developments of the past several decades in string theory was the discovery of certain "dualities", mathematical transformations that identify one physical theory with another. Physicists studying string theory have discovered a number of these dualities between different versions of string theory, and this has led to the conjecture that all consistent versions of string theory are subsumed in a single framework known as M-theory.[5]

Studies of string theory have also yielded a number of results on the nature of black holes and the gravitational interaction. There are certain paradoxes that arise when one attempts to understand the quantum aspects of black holes, and work on string theory has attempted to clarify these issues. In late 1997 this line of work culminated in the discovery of the anti-de Sitter/conformal field theory correspondence or AdS/CFT.[6] This is a theoretical result which relates string theory to other physical theories which are better understood theoretically. The AdS/CFT correspondence has implications for the study of black holes and quantum gravity, and it has been applied to other subjects, including nuclear[7] and condensed matter physics.[8][9]

Since string theory incorporates all of the fundamental interactions, including gravity, many physicists hope that it fully describes our universe, making it a theory of everything. One of the goals of current research in string theory is to find a solution of the theory that reproduces the observed spectrum of elementary particles, with a small cosmological constant, containing dark matter and a plausible mechanism for cosmic inflation. While there has been progress toward these goals, it is not known to what extent string theory describes the real world or how much freedom the theory allows to choose the details.[10]

One of the challenges of string theory is that the full theory does not yet have a satisfactory definition in all circumstances. The scattering of strings is most straightforwardly defined using the techniques of perturbation theory, but it is not known in general how to define string theory nonperturbatively.[11] It is also not clear whether there is any principle by which string theory selects its vacuum state, the physical state that determines the properties of our universe.[12] These problems have led some in the community to criticize these approaches to the unification of physics and question the value of continued research on these problems.[13]

1.1.1 Strings

Main article: String (physics)

The application of quantum mechanics to physical objects such as the electromagnetic field, which are extended in

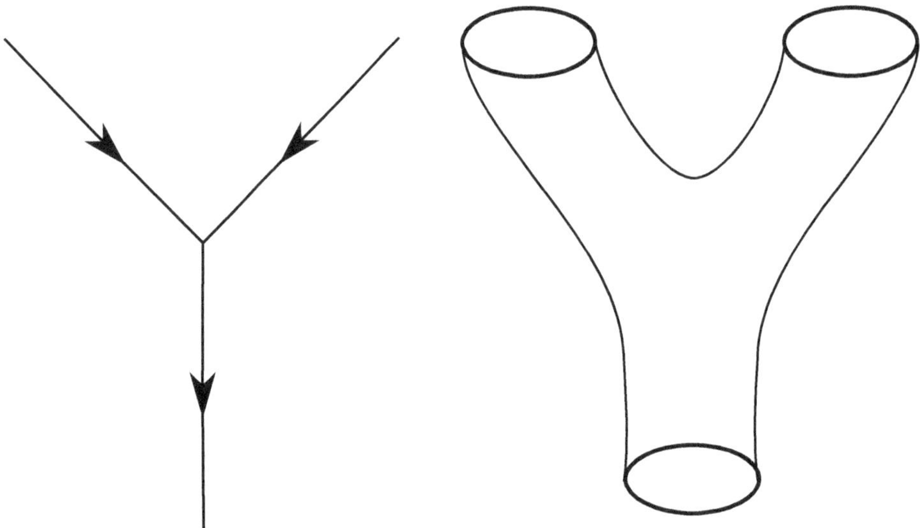

Interaction in the quantum world: worldlines of point-like particles or a worldsheet swept up by closed strings in string theory.

space and time, is known as quantum field theory. In particle physics, quantum field theories form the basis for our understanding of elementary particles, which are modeled as excitations in the fundamental fields.[14]

In quantum field theory, one typically computes the probabilities of various physical events using the techniques of perturbation theory. Developed by Richard Feynman and others in the first half of the twentieth century, perturbative quantum field theory uses special diagrams called Feynman diagrams to organize computations. One imagines that these diagrams depict the paths of point-like particles and their interactions.[15]

The starting point for string theory is the idea that the point-like particles of quantum field theory can also be modeled as one-dimensional objects called strings.[16] The interaction of strings is most straightforwardly defined by generalizing the perturbation theory used in ordinary quantum field theory. At the level of Feynman diagrams, this means replacing the one-dimensional diagram representing the path of a point particle by a two-dimensional surface representing the motion of a string.[17] Unlike in quantum field theory, string theory does not yet have a full non-perturbative definition, so many of the theoretical questions that physicists would like to answer remain out of reach.[18]

In theories of particle physics based on string theory, the characteristic length scale of strings is assumed to be on the order of the Planck length, or 10^{-35} meters, the scale at which the effects of quantum gravity are believed to become significant.[19] On much larger length scales, such as the scales visible in physics laboratories, such objects would be indistinguishable from zero-dimensional point particles, and the vibrational state of the string would determine the type of particle. One of the vibrational states of a string corresponds to the graviton, a quantum mechanical particle that carries the gravitational force.[20]

The original version of string theory was bosonic string theory, but this version described only bosons, a class of particles which transmit forces between the matter particles, or fermions. Bosonic string theory was eventually superseded by theories called superstring theories. These theories describe both bosons and fermions, and they incorporate a theoretical idea called supersymmetry. This is a mathematical relation that exists in certain physical theories between the bosons and fermions. In theories with supersymmetry, each boson has a counterpart which is a fermion, and vice versa.[21]

There are several versions of superstring theory: type I, type IIA, type IIB, and two flavors of heterotic string theory ($SO(32)$ and $E_8 \times E_8$). The different theories allow different types of strings, and the particles that arise at low energies exhibit different symmetries. For example, the type I theory includes both open strings (which are segments with endpoints) and closed strings (which form closed loops), while types IIA and IIB include only closed strings.[22]

1.1.2 Extra dimensions

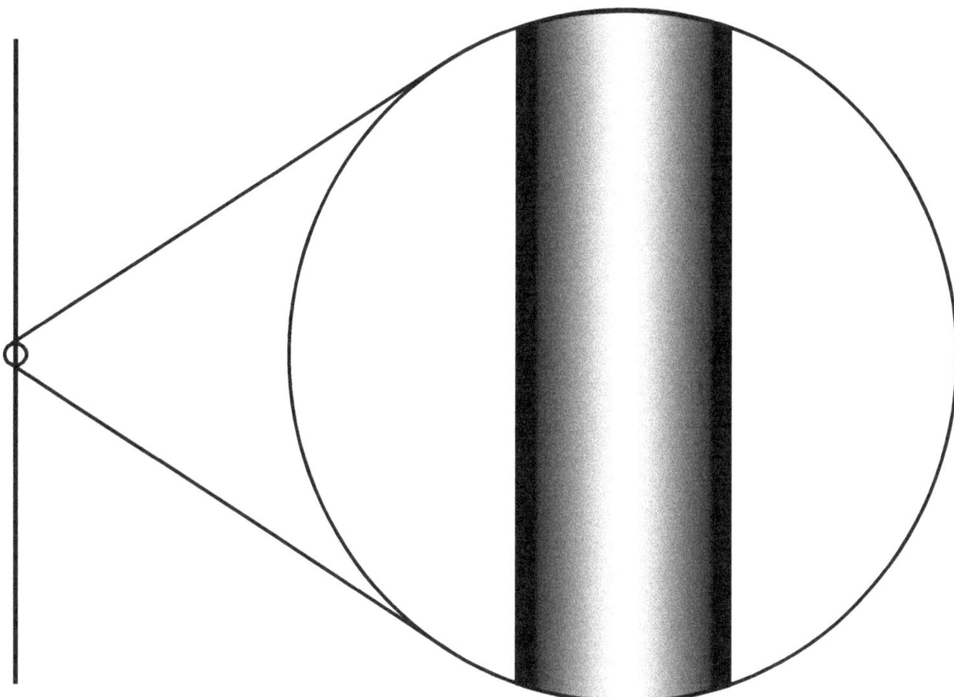

An example of compactification: At large distances, a two dimensional surface with one circular dimension looks one-dimensional.

In everyday life, there are three familiar dimensions of space: height, width and length. Einstein's general theory of relativity treats time as a dimension on par with the three spatial dimensions; in general relativity, space and time are not modeled as separate entities but are instead unified to a four-dimensional spacetime. In this framework, the phenomenon of gravity is viewed as a consequence of the geometry of spacetime.[23]

In spite of the fact that the universe is well described by four-dimensional spacetime, there are several reasons why physicists consider theories in other dimensions. In some cases, by modeling spacetime in a different number of dimensions, a theory becomes more mathematically tractable, and one can perform calculations and gain general insights more easily.[lower-alpha 2] There are also situations where theories in two or three spacetime dimensions are useful for describing phenomena in condensed matter physics.[24] Finally, there exist scenarios in which there could actually be more than four dimensions of spacetime which have nonetheless managed to escape detection.[25]

One notable feature of string theories is that these theories require extra dimensions of spacetime for their mathematical consistency. In bosonic string theory, spacetime is 26-dimensional, while in superstring theory it is ten-dimensional. In order to describe real physical phenomena using string theory, one must therefore imagine scenarios in which these extra dimensions would not be observed in experiments.[26]

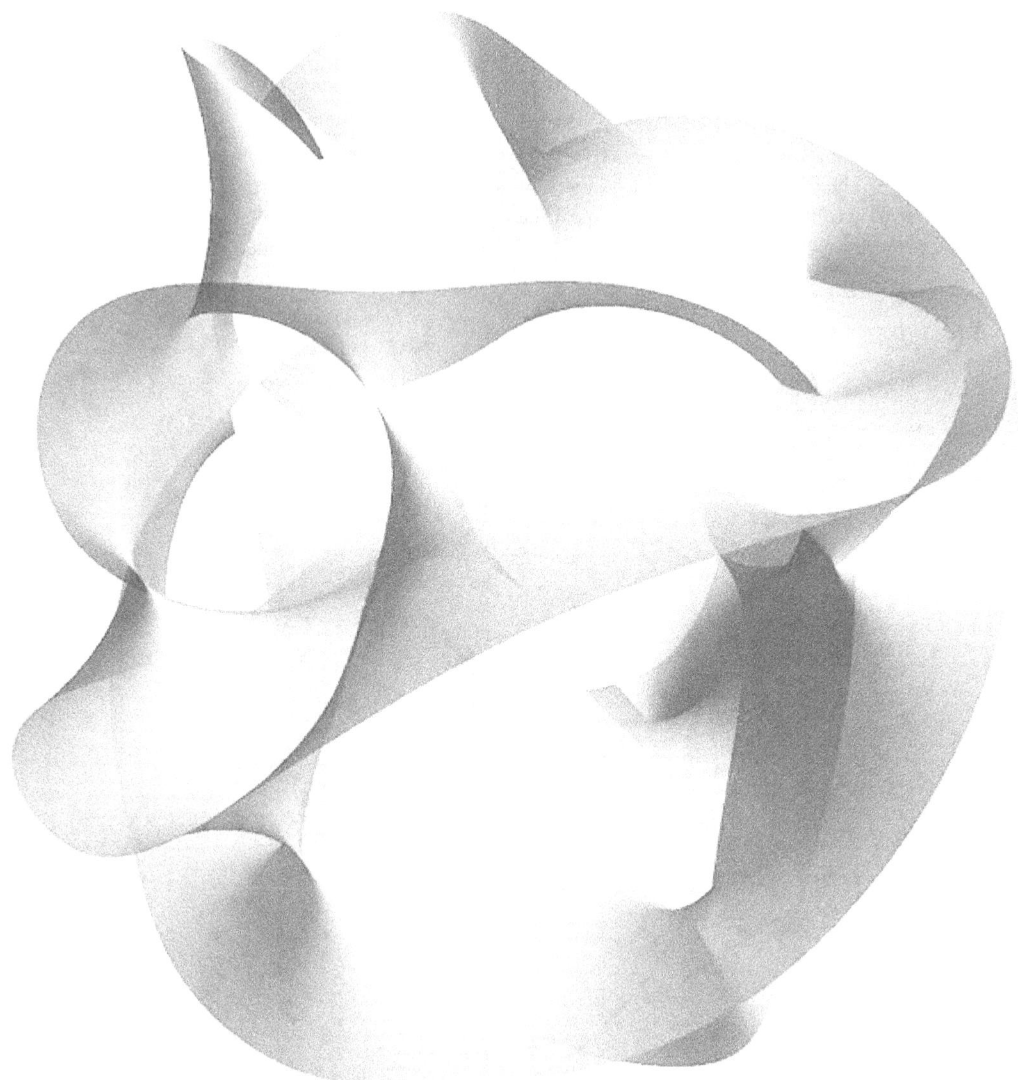

A cross section of a quintic Calabi–Yau manifold

Compactification is one way of modifying the number of dimensions in a physical theory. In compactification, some of the extra dimensions are assumed to "close up" on themselves to form circles.[27] In the limit where these curled up dimensions become very small, one obtains a theory in which spacetime has effectively a lower number of dimensions. A standard analogy for this is to consider a multidimensional object such as a garden hose. If the hose is viewed from a sufficient distance, it appears to have only one dimension, its length. However, as one approaches the hose, one discovers that it contains a second dimension, its circumference. Thus, an ant crawling on the surface of the hose would move in two dimensions.[28]

Compactification can be used to construct models in which spacetime is effectively four-dimensional. However, not every way of compactifying the extra dimensions produces a model with the right properties to describe nature. In a viable model of particle physics, the compact extra dimensions must be shaped like a Calabi–Yau manifold.[29] A Calabi–Yau manifold is a special space which is typically taken to be six-dimensional in applications to string theory. It is named after mathematicians Eugenio Calabi and Shing-Tung Yau.[30]

Another approach to reducing the number of dimensions is the so called brane-world scenario. In this approach, physicists assume that the observable universe is a four-dimensional subspace of a higher dimensional space. In such models, the force-carrying bosons of particle physics arise from open strings with endpoints attached to the four-dimensional subspace, while gravity arises from closed strings propagating through the larger ambient space. This idea plays an important role in attempts to develop models of real world physics based on string theory, and it provides a natural explanation for the weakness of gravity compared to the other fundamental forces.[31]

1.1.3 Dualities

Main articles: S-duality and T-duality

One notable fact about string theory is that the different versions of the theory all turn out to be related in highly nontrivial ways. One of the relationships that can exist between different string theories is called S-duality. This is a relationship which says that a collection of strongly interacting particles in one theory can, in some cases, be viewed as a collection of weakly interacting particles in a completely different theory. Roughly speaking, a collection of particles is said to be strongly interacting if they combine and decay often and weakly interacting if they do so infrequently. Type I string theory turns out to be equivalent by S-duality to the $SO(32)$ heterotic string theory. Similarly, type IIB string theory is related to itself in a nontrivial way by S-duality.[32]

Another relationship between different string theories is T-duality. Here one considers strings propagating around a circular extra dimension. T-duality states that a string propagating around a circle of radius R is equivalent to a string propagating around a circle of radius $1/R$ in the sense that all observable quantities in one description are identified with quantities in the dual description. For example, a string has momentum as it propagates around a circle, and it can also wind around the circle one or more times. The number of times the string winds around a circle is called the winding number. If a string has momentum p and winding number n in one description, it will have momentum n and winding number p in the dual description. For example, type IIA string theory is equivalent to type IIB string theory via T-duality, and the two versions of heterotic string theory are also related by T-duality.[33]

In general, the term *duality* refers to a situation where two seemingly different physical systems turn out to be equivalent in a nontrivial way. Two theories related by a duality need not be string theories. For example, Montonen–Olive duality is example of an S-duality relationship between quantum field theories. The AdS/CFT correspondence is example of a duality which relates string theory to a quantum field theory. If two theories are related by a duality, it means that one theory can be transformed in some way so that it ends up looking just like the other theory. The two theories are then said to be *dual* to one another under the transformation. Put differently, the two theories are mathematically different descriptions of the same phenomena.[34]

1.1.4 Branes

Main article: Brane

In string theory and related theories, a brane is a physical object that generalizes the notion of a point particle to higher dimensions. For example, a point particle can be viewed as a brane of dimension zero, while a string can be viewed as a brane of dimension one. It is also possible to consider higher-dimensional branes. In dimension p, these are called p-branes. The word brane comes from the word "membrane" which refers to a two-dimensional brane.[35]

Branes are dynamical objects which can propagate through spacetime according to the rules of quantum mechanics. They have mass and can have other attributes such as charge. A p-brane sweeps out a $(p+1)$-dimensional volume in spacetime called its *worldvolume*. Physicists often study fields analogous to the electromagnetic field which live on the worldvolume of a brane.[36]

In string theory, D-branes are an important class of branes that arise when one considers open strings. As an open string propagates through spacetime, its endpoints are required to lie on a D-brane. The letter "D" in D-brane refers to a certain mathematical condition on the system known as the Dirichlet boundary condition. The study of D-branes in string theory has led to important results such as the AdS/CFT correspondence, which has shed light on many problems in quantum field theory.[37]

Branes are also frequently studied from a purely mathematical point of view. Mathematically, branes can be described as objects of certain categories, such as the derived category of coherent sheaves on a complex algebraic variety, or the Fukaya category of a symplectic manifold.[38] The connection between the physical notion of a brane and the mathematical notion of a category has led to important mathematical insights in the fields of algebraic and symplectic geometry[39] and representation theory.[40]

Type I

SO(32) heterotic

E_8xE_8 heterotic

M-theory

Type IIA

Type IIB

A diagram of string theory dualities. Yellow arrows indicate S-duality. Blue arrows indicate T-duality.

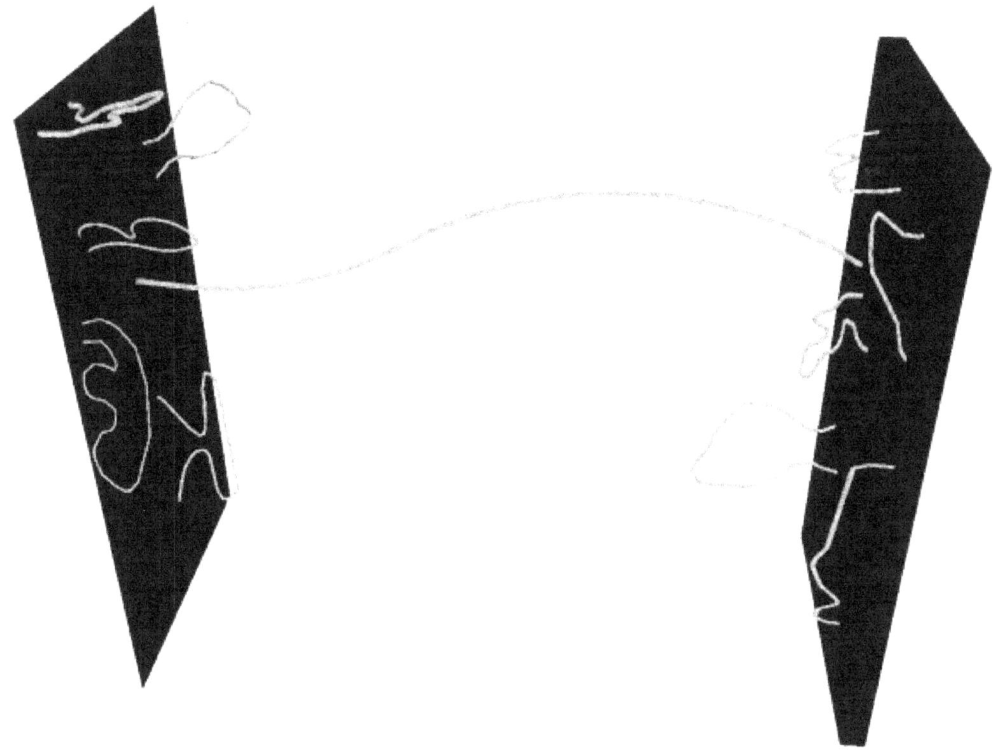

Open strings attached to a pair of D-branes

1.2 M-theory

Main article: M-theory

Prior to 1995, theorists believed that there were five consistent versions of superstring theory (type I, type IIA, type IIB, and two versions of heterotic string theory). This understanding changed in 1995 when Edward Witten suggested that the five theories were just special limiting cases of an eleven-dimensional theory called M-theory. Witten's conjecture was based on the work of a number of other physicists, including Ashoke Sen, Chris Hull, Paul Townsend, and Michael Duff. His announcement led to a flurry of research activity now known as the second superstring revolution.[41]

1.2.1 Unification of superstring theories

In the 1970s, many physicists became interested in supergravity theories, which combine general relativity with supersymmetry. Whereas general relativity makes sense in any number of dimensions, supergravity places an upper limit on the number of dimensions.[42] In 1978, work by Werner Nahm showed that the maximum spacetime dimension in which one can formulate a consistent supersymmetric theory is eleven.[43] In the same year, Eugene Cremmer, Bernard Julia, and Joel Scherk of the École Normale Supérieure showed that supergravity not only permits up to eleven dimensions but is in fact most elegant in this maximal number of dimensions.[44][45]

Initially, many physicists hoped that by compactifying eleven-dimensional supergravity, it might be possible to construct realistic models of our four-dimensional world. The hope was that such models would provide a unified description of the four fundamental forces of nature: electromagnetism, the strong and weak nuclear forces, and gravity. Interest in eleven-dimensional supergravity soon waned as various flaws in this scheme were discovered. One of the problems was that the laws of physics appear to distinguish between clockwise and counterclockwise, a phenomenon known as chirality. Edward Witten and others observed this chirality property cannot be readily derived by compactifying from eleven dimensions.[46]

In the first superstring revolution in 1984, many physicists turned to string theory as a unified theory of particle

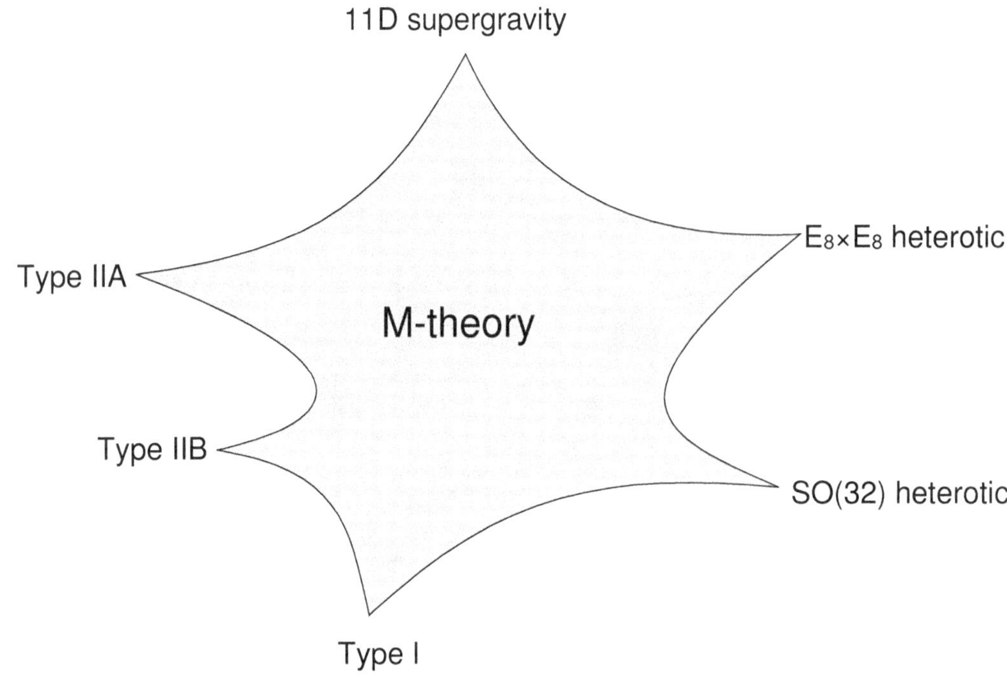

A schematic illustration of the relationship between M-theory, the five superstring theories, and eleven-dimensional supergravity. The shaded region represents a family of different physical scenarios that are possible in M-theory. In certain limiting cases corresponding to the cusps, it is natural to describe the physics using one of the six theories labeled there.

physics and quantum gravity. Unlike supergravity theory, string theory was able to accommodate the chirality of the standard model, and it provided a theory of gravity consistent with quantum effects.[47] Another feature of string theory that many physicists were drawn to in the 1980s and 1990s was its high degree of uniqueness. In ordinary particle theories, one can consider any collection of elementary particles whose classical behavior is described by an arbitrary Lagrangian. In string theory, the possibilities are much more constrained: by the 1990s, physicists had argued that there were only five consistent supersymmetric versions of the theory.[48]

Although there were only a handful of consistent superstring theories, it remained a mystery why there was not just one consistent formulation.[49] However, as physicists began to examine string theory more closely, they realized that these theories are related in intricate and nontrivial ways. They found that a system of strongly interacting strings can, in some cases, be viewed as a system of weakly interacting strings. This phenomenon is known as S-duality. It was studied by Ashoke Sen in the context of heterotic strings in four dimensions[50][51] and by Chris Hull and Paul Townsend in the context of the type IIB theory.[52] Theorists also found that different string theories may be related by T-duality. This duality implies that strings propagating on completely different spacetime geometries may be physically equivalent.[53]

At around the same time, as many physicists were studying the properties of strings, a small group of physicists was examining the possible applications of higher dimensional objects. In 1987, Eric Bergshoeff, Ergin Sezgin, and Paul Townsend showed that eleven-dimensional supergravity includes two-dimensional branes.[54] Intuitively, these objects look like sheets or membranes propagating through the eleven-dimensional spacetime. Shortly after this discovery, Michael Duff, Paul Howe, Takeo Inami, and Kellogg Stelle considered a particular compactification of eleven-dimensional supergravity with one of the dimensions curled up into a circle.[55] In this setting, one can imagine the membrane wrapping around the circular dimension. If the radius of the circle is sufficiently small, then this membrane looks just like a string in ten-dimensional spacetime. In fact, Duff and his collaborators showed that this construction reproduces exactly the strings appearing in type IIA superstring theory.[56]

Speaking at a string theory conference in 1995, Edward Witten made the surprising suggestion that all five superstring theories were in fact just different limiting cases of a single theory in eleven spacetime dimensions. Witten's announcement drew together all of the previous results on S- and T-duality and the appearance of higher dimensional branes in string theory.[57] In the months following Witten's announcement, hundreds of new papers appeared on the Internet confirming different parts of his proposal.[58] Today this flurry of work is known as the second superstring revolution.[59]

Initially, some physicists suggested that the new theory was a fundamental theory of membranes, but Witten was skeptical of the role of membranes in the theory. In a paper from 1996, Hořava and Witten wrote "As it has been proposed that the eleven-dimensional theory is a supermembrane theory but there are some reasons to doubt that interpretation, we will non-committally call it the M-theory, leaving to the future the relation of M to membranes."[60] In the absence of an understanding of the true meaning and structure of M-theory, Witten has suggested that the *M* should stand for "magic", "mystery", or "membrane" according to taste, and the true meaning of the title should be decided when a more fundamental formulation of the theory is known.[61]

1.2.2 Matrix theory

Main article: Matrix theory (physics)

In mathematics, a matrix is a rectangular array of numbers or other data. In physics, a matrix model is a particular kind of physical theory whose mathematical formulation involves the notion of a matrix in an important way. A matrix model describes the behavior of a set of matrices within the framework of quantum mechanics.[62]

One important example of a matrix model is the BFSS matrix model proposed by Tom Banks, Willy Fischler, Stephen Shenker, and Leonard Susskind in 1997. This theory describes the behavior of a set of nine large matrices. In their original paper, these authors showed, among other things, that the low energy limit of this matrix model is described by eleven-dimensional supergravity. These calculations led them to propose that the BFSS matrix model is exactly equivalent to M-theory. The BFSS matrix model can therefore be used as a prototype for a correct formulation of M-theory and a tool for investigating the properties of M-theory in a relatively simple setting.[63]

The development of the matrix model formulation of M-theory has led physicists to consider various connections between string theory and a branch of mathematics called noncommutative geometry. This subject is a generalization of ordinary geometry in which mathematicians define new geometric notions using tools from noncommutative algebra.[64] In a paper from 1998, Alain Connes, Michael R. Douglas, and Albert Schwarz showed that some aspects of matrix models and M-theory are described by a noncommutative quantum field theory, a special kind of physical theory in which spacetime is described mathematically using noncommutative geometry.[65] This established a link between matrix models and M-theory on the one hand, and noncommutative geometry on the other hand. It quickly led to the discovery of other important links between noncommutative geometry and various physical theories.[66][67]

1.3 Black holes

In general relativity, a black hole is defined as a region of spacetime in which the gravitational field is so strong that no particle or radiation can escape. In the currently accepted models of stellar evolution, black holes are thought to arise when massive stars undergo gravitational collapse, and many galaxies are thought to contain supermassive black holes at their centers. Black holes are also important for theoretical reasons, as they present profound challenges for theorists attempting to understand the quantum aspects of gravity. String theory has proved to be an important tool for investigating the theoretical properties of black holes because it provides a framework in which theorists can study their thermodynamics.[68]

1.3.1 Bekenstein–Hawking formula

In the branch of physics called statistical mechanics, entropy is a measure of the randomness or disorder of a physical system. This concept was studied in the 1870s by the Austrian physicist Ludwig Boltzmann, who showed that the thermodynamic properties of a gas could be derived from the combined properties of its many constituent molecules. Boltzmann argued that by averaging the behaviors of all the different molecules in a gas, one can understand macroscopic properties such as volume, temperature, and pressure. In addition, this perspective led him to give a precise definition of entropy as the natural logarithm of the number of different states of the molecules (also called *microstates*) that give rise to the same macroscopic features.[69]

In the twentieth century, physicists began to apply the same concepts to black holes. In most systems such as gases, the entropy scales with the volume. In the 1970s, the physicist Jacob Bekenstein suggested that the entropy of a black hole is instead proportional to the *surface area* of its event horizon, the boundary beyond which matter and radiation is lost to its gravitational attraction.[70] When combined with ideas of the physicist Stephen Hawking,[71] Bekenstein's work yielded a precise formula for the entropy of a black hole. The formula expresses the entropy S as

$$S = \frac{c^3 kA}{4\hbar G}$$

where c is the speed of light, k is Boltzmann's constant, $\hbar$ is the reduced Planck constant, G is Newton's constant, and A is the surface area of the event horizon.[72]

Like any physical system, a black hole has an entropy defined in terms of the number of different microstates that lead to the same macroscopic features. The Bekenstein–Hawking entropy formula gives the expected value of the entropy of a black hole, but by the 1990s, physicists still lacked a derivation of this formula by counting microstates in a theory of quantum gravity. Finding such a derivation of this formula was considered an important test of the viability of any theory of quantum gravity such as string theory.[73]

1.3.2 Derivation within string theory

In a paper from 1996, Andrew Strominger and Cumrun Vafa showed how to derive the Beckenstein–Hawking formula for certain black holes in string theory.[74] Their calculation was based on the observation that D-branes—which look like fluctuating membranes when they are weakly interacting—become dense, massive objects with event horizons when the interactions are strong. In other words, a system of strongly interacting D-branes in string theory is indistinguishable from a black hole. Strominger and Vafa analyzed such D-brane systems and calculated the number of different ways of placing D-branes in spacetime so that their combined mass and charge is equal to a given mass and charge for the resulting black hole. Their calculation reproduced the Bekenstein–Hawking formula exactly, including the factor of 1/4.[75] Subsequent work by Strominger, Vafa, and others refined the original calculations and gave the precise values of the "quantum corrections" needed to describe very small black holes.[76][77]

The black holes that Strominger and Vafa considered in their original work were quite different from real astrophysical black holes. One difference was that Strominger and Vafa considered only extremal black holes in order to make the calculation tractable. These are defined as black holes with the lowest possible mass compatible with a given charge.[78] Strominger and Vafa also restricted attention to black holes in five-dimensional spacetime with unphysical supersymmetry.[79]

Although it was originally developed in this very particular and physically unrealistic context in string theory, the entropy calculation of Strominger and Vafa has led to a qualitative understanding of how black hole entropy can be accounted for in any theory of quantum gravity. Indeed, in 1998, Strominger argued that the original result could be generalized to an arbitrary consistent theory of quantum gravity without relying on strings or supersymmetry.[80] In collaboration with several other authors in 2010, he showed that some results on black hole entropy could be extended to non-extremal astrophysical black holes.[81][82]

1.4 AdS/CFT correspondence

Main article: AdS/CFT correspondence

One approach to formulating string theory and studying its properties is provided by the anti-de Sitter/conformal field theory (AdS/CFT) correspondence. This is a theoretical result which implies that string theory is in some cases equivalent to a quantum field theory. In addition to providing insights into the mathematical structure of string theory, the AdS/CFT correspondence has shed light on many aspects of quantum field theory in regimes where traditional calculational techniques are ineffective.[83] The AdS/CFT correspondence was first proposed by Juan Maldacena in late 1997.[84] Important aspects of the correspondence were elaborated in articles by Steven Gubser, Igor Klebanov, and Alexander Markovich Polyakov,[85] and by Edward Witten.[86] By 2010, Maldacena's article had over 7000 citations, becoming the most highly cited article in the field of high energy physics.[lower-alpha 3]

1.4.1 Overview of the correspondence

In the AdS/CFT correspondence, the geometry of spacetime is described in terms of a certain vacuum solution of Einstein's equation called anti-de Sitter space.[87] In very elementary terms, anti-de Sitter space is a mathematical model of spacetime in which the notion of distance between points (the metric) is different from the notion of distance in ordinary Euclidean geometry. It is closely related to hyperbolic space, which can be viewed as a disk as illustrated on the left.[88] This image shows a tessellation of a disk by triangles and squares. One can define the distance between points of this disk in such a way that all the triangles and squares are the same size and the circular outer boundary is infinitely far from any point in the interior.[89]

One can imagine a stack of hyperbolic disks where each disk represents the state of the universe at a given time. The resulting geometric object is three-dimensional anti-de Sitter space.[90] It looks like a solid cylinder in which any cross section is a copy of the hyperbolic disk. Time runs along the vertical direction in this picture. The surface of this cylinder plays an important role in the AdS/CFT correspondence. As with the hyperbolic plane, anti-de Sitter space is curved in such a way that any point in the interior is actually infinitely far from this boundary surface.[91]

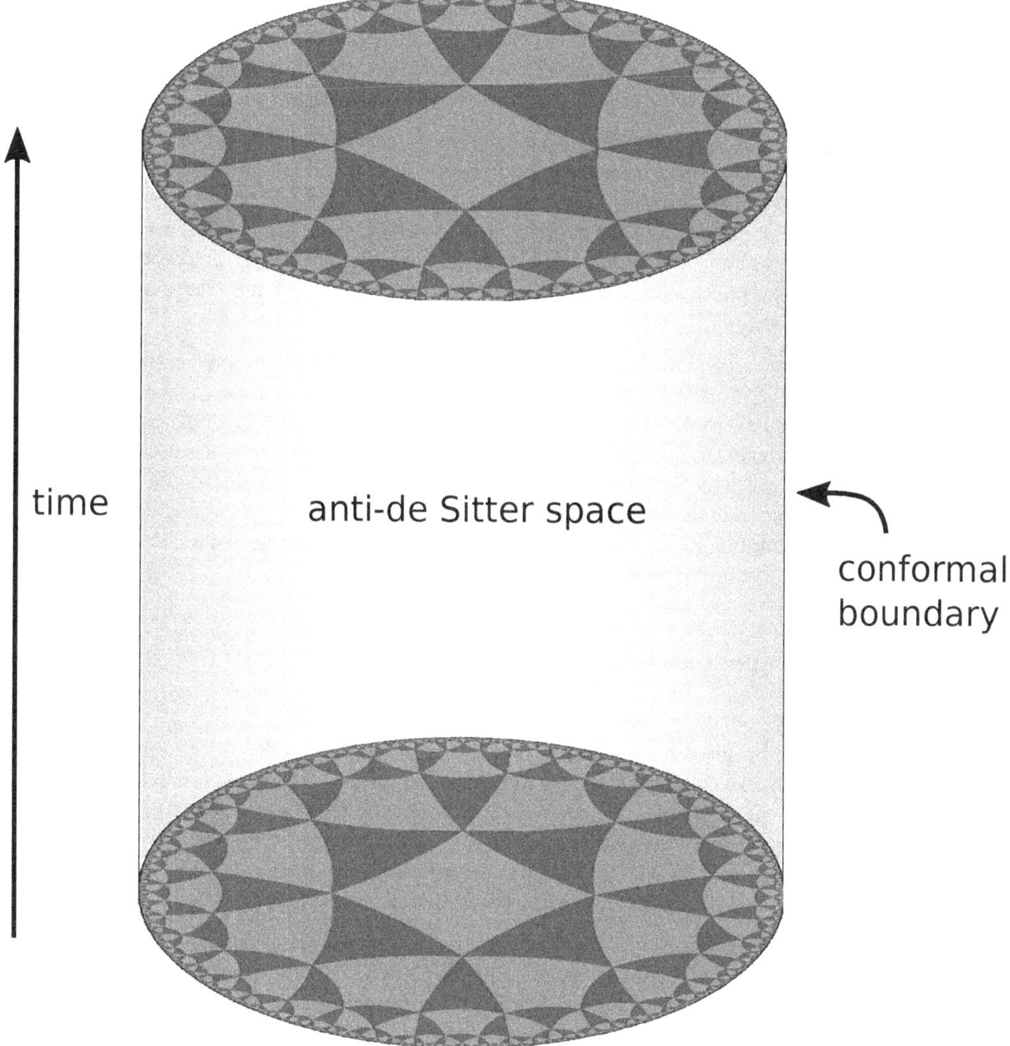

Three-dimensional anti-de Sitter space is like a stack of hyperbolic disks, each one representing the state of the universe at a given time. The resulting spacetime looks like a solid cylinder.

This construction describes a hypothetical universe with only two space dimensions and one time dimension, but it can be generalized to any number of dimensions. Indeed, hyperbolic space can have more than two dimensions and one can "stack up" copies of hyperbolic space to get higher-dimensional models of anti-de Sitter space.[92]

An important feature of anti-de Sitter space is its boundary (which looks like a cylinder in the case of three-dimensional anti-de Sitter space). One property of this boundary is that, within a small region on the surface around any given point, it looks just like Minkowski space, the model of spacetime used in nongravitational physics.[93] One can therefore consider an auxiliary theory in which "spacetime" is given by the boundary of anti-de Sitter space. This observation is the starting point for AdS/CFT correspondence, which states that the boundary of anti-de Sitter space can be regarded as the "spacetime" for a quantum field theory. The claim is that this quantum field theory is equivalent to a gravitational theory, such as string theory, in the bulk anti-de Sitter space in the sense that there is a "dictionary" for translating entities and calculations in one theory into their counterparts in the other theory. For example, a single particle in the gravitational theory might correspond to some collection of particles in the boundary theory. In addition, the predictions in the two theories are quantitatively identical so that if two particles have a 40 percent chance of colliding in the gravitational theory, then the corresponding collections in the boundary theory would also have a 40 percent chance of colliding.[94]

1.4.2 Applications to quantum gravity

The discovery of the AdS/CFT correspondence was a major advance in physicists' understanding of string theory and quantum gravity. One reason for this is that the correspondence provides a formulation of string theory in terms of quantum field theory, which is well understood by comparison. Another reason is that it provides a general framework in which physicists can study and attempt to resolve the paradoxes of black holes.[95]

In 1975, Stephen Hawking published a calculation which suggested that black holes are not completely black but emit a dim radiation due to quantum effects near the event horizon.[96] At first, Hawking's result posed a problem for theorists because it suggested that black holes destroy information. More precisely, Hawking's calculation seemed to conflict with one of the basic postulates of quantum mechanics, which states that physical systems evolve in time according to the Schrödinger equation. This property is usually referred to as unitarity of time evolution. The apparent contradiction between Hawking's calculation and the unitarity postulate of quantum mechanics came to be known as the black hole information paradox.[97]

The AdS/CFT correspondence resolves the black hole information paradox, at least to some extent, because it shows how a black hole can evolve in a manner consistent with quantum mechanics in some contexts. Indeed, one can consider black holes in the context of the AdS/CFT correspondence, and any such black hole corresponds to a configuration of particles on the boundary of anti-de Sitter space.[98] These particles obey the usual rules of quantum mechanics and in particular evolve in a unitary fashion, so the black hole must also evolve in a unitary fashion, respecting the principles of quantum mechanics.[99] In 2005, Hawking announced that the paradox had been settled in favor of information conservation by the AdS/CFT correspondence, and he suggested a concrete mechanism by which black holes might preserve information.[100]

1.4.3 Applications to quantum field theory

Main articles: AdS/QCD correspondence and AdS/CMT correspondence
In addition to its applications to theoretical problems in quantum gravity, the AdS/CFT correspondence has been applied to a variety of problems in quantum field theory. One physical system that has been studied using the AdS/CFT correspondence is the quark–gluon plasma, an exotic state of matter produced in particle accelerators. This state of matter arises for brief instants when heavy ions such as gold or lead nuclei are collided at high energies. Such collisions cause the quarks that make up atomic nuclei to deconfine at temperatures of approximately two trillion kelvins, conditions similar to those present at around 10^{-11} seconds after the Big Bang.[102]

The physics of the quark–gluon plasma is governed by a theory called quantum chromodynamics, but this theory is mathematically intractable in problems involving the quark–gluon plasma.[lower-alpha 4] In an article appearing in 2005, Đàm Thanh Sơn and his collaborators showed that the AdS/CFT correspondence could be used to understand some aspects of the quark–gluon plasma by describing it in the language of string theory.[103] By applying the AdS/CFT correspondence, Sơn and his collaborators were able to describe the quark gluon plasma in terms of black holes in five-dimensional spacetime. The calculation showed that the ratio of two quantities associated with the quark–gluon plasma, the shear viscosity and volume density of entropy, should be approximately equal to a certain universal constant. In 2008, the predicted value of this ratio for the quark–gluon plasma was confirmed at the Relativistic Heavy Ion Collider at Brookhaven National Laboratory.[104][105]

The AdS/CFT correspondence has also been used to study aspects of condensed matter physics. Over the decades, experimental condensed matter physicists have discovered a number of exotic states of matter, including superconductors

A magnet levitating above a high-temperature superconductor. Today some physicists are working to understand high-temperature superconductivity using the AdS/CFT correspondence.[101]

and superfluids. These states are described using the formalism of quantum field theory, but some phenomena are difficult to explain using standard field theoretic techniques. Some condensed matter theorists including Subir Sachdev hope that the AdS/CFT correspondence will make it possible to describe these systems in the language of string theory and learn more about their behavior.[106]

So far some success has been achieved in using string theory methods to describe the transition of a superfluid to an insulator. A superfluid is a system of electrically neutral atoms that flows without any friction. Such systems are often produced in the laboratory using liquid helium, but recently experimentalists have developed new ways of producing artificial superfluids by pouring trillions of cold atoms into a lattice of criss-crossing lasers. These atoms initially behave as a superfluid, but as experimentalists increase the intensity of the lasers, they become less mobile and then suddenly transition to an insulating state. During the transition, the atoms behave in an unusual way. For example, the atoms slow to a halt at a rate that depends on the temperature and on Planck's constant, the fundamental parameter of quantum mechanics, which does not enter into the description of the other phases. This behavior has recently been understood by considering a dual description where properties of the fluid are described in terms of a higher dimensional black hole.[107]

1.5 Phenomenology

Main article: String phenomenology

In addition to being an idea of considerable theoretical interest, string theory provides a framework for constructing models of real world physics that combine general relativity and particle physics. Phenomenology is the branch of theoretical physics in which physicists construct realistic models of nature from more abstract theoretical ideas. String phenomenology is the part of string theory that attempts to construct realistic models based on string theory.

Partly because of theoretical and mathematical difficulties and partly because of the extremely high energies needed to test these theories experimentally, there is so far no experimental evidence that would unambiguously point to any of these models being a correct fundamental description of nature. This has led some in the community to criticize these approaches to unification and question the value of continued research on these problems.[108]

1.5.1 Particle physics

The currently accepted theory describing elementary particles and their interactions is known as the standard model of particle physics. This theory provides a unified description of three of the fundamental forces of nature: electromagnetism and the strong and weak nuclear forces. Despite its remarkable success in explaining a wide range of physical phenomena, the standard model cannot be a complete description of reality. This is because the standard model fails to incorporate the force of gravity and because of problems such as the hierarchy problem and the inability to explain the structure of fermion masses or dark matter.

String theory has been used to construct a variety of models of particle physics going beyond the standard model. Typically, such models are based on the idea of compactification. Starting with the ten- or eleven-dimensional spacetime of string or M-theory, physicists postulate a shape for the extra dimensions. By choosing this shape appropriately, they can construct models roughly similar to the standard model of particle physics, together with additional undiscovered particles.[109] One popular way of deriving realistic physics from string theory is to start with the heterotic theory in ten dimensions and assume that the six extra dimensions of spacetime are shaped like a six-dimensional Calabi–Yau manifold. Such compactifications offer many ways of extracting realistic physics from string theory. Other similar methods can be used to construct realistic models of our four-dimensional world based on M-theory.[110]

1.5.2 Cosmology

Main article: String cosmology

The Big Bang theory is the prevailing cosmological model for the universe from the earliest known periods through its

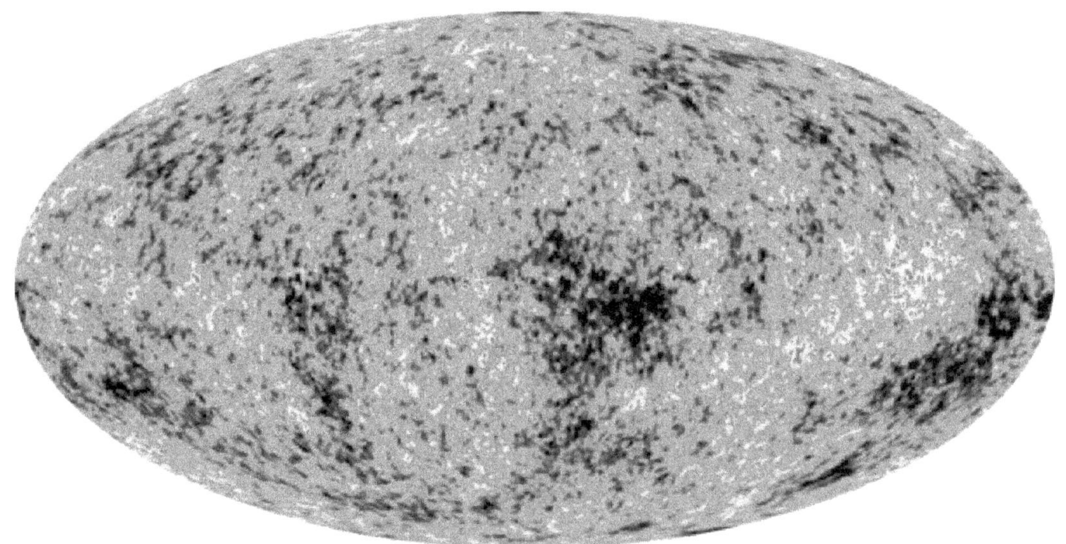

A map of the cosmic microwave background produced by the Wilkinson Microwave Anisotropy Probe

subsequent large-scale evolution. Despite its success in explaining many observed features of the universe including galactic redshifts, the relative abundance of light elements such as hydrogen and helium, and the existence of a cosmic microwave background, there are several questions that remain unanswered. For example, the standard Big Bang model does not explain why the universe appears to be same in all directions, why it appears flat on very large distance scales, or why certain hypothesized particles such as magnetic monopoles are not observed in experiments.[111]

Currently, the leading candidate for a theory going beyond the Big Bang is the theory of cosmic inflation. Developed by Alan Guth and others in the 1980s, inflation postulates a period of extremely rapid accelerated expansion of the universe prior to the expansion described by the standard Big Bang theory. The theory of cosmic inflation preserves the successes of the Big Bang while providing a natural explanation for some of the mysterious features of the universe.[112] The theory has also received striking support from observations of the cosmic microwave background, the radiation that has filled the sky since around 380,000 years after the Big Bang.[113]

In the theory of inflation, the rapid initial expansion of the universe is caused by a hypothetical particle called the inflaton. The exact properties of this particle are not fixed by the theory but should ultimately be derived from a more fundamental theory such as string theory.[114] Indeed, there have been a number of attempts to identify an inflation within the spectrum of particles described by string theory and to study inflation using string theory. While these approaches might eventually find support in observational data such as measurements of the cosmic microwave background, the application of string theory to cosmology is still in its early stages.[115]

1.6 Connections to mathematics

In addition to influencing research in theoretical physics, string theory has stimulated a number of major developments in pure mathematics. Like many developing ideas in theoretical physics, string theory does not at present have a mathematically rigorous formulation in which all of its concepts can be defined precisely. As a result, physicists who study string theory are often guided by physical intuition to conjecture relationships between the seemingly different mathematical structures that are used to formalize different parts of the theory. These conjectures are later proved by mathematicians, and in this way, string theory serves as a source of new ideas in pure mathematics.[116]

1.6.1 Mirror symmetry

Main article: Mirror symmetry (string theory)

After Calabi–Yau manifolds had entered physics as a way to compactify extra dimensions in string theory, many physicists began studying these manifolds. In the late 1980s, several physicists noticed that given such a compactification of string theory, it is not possible to reconstruct uniquely a corresponding Calabi–Yau manifold.[117] Instead, two different versions of string theory, type IIA and type IIB, can be compactified on completely different Calabi–Yau manifolds giving rise to the same physics. In this situation, the manifolds are called mirror manifolds, and the relationship between the two physical theories is called mirror symmetry.[118]

Regardless of whether Calabi–Yau compactifications of string theory provide a correct description of nature, the existence of the mirror duality between different string theories has significant mathematical consequences. The Calabi–Yau manifolds used in string theory are of interest in pure mathematics, and mirror symmetry allows mathematicians to solve problems in enumerative geometry, a branch of mathematics concerned with counting the numbers of solutions to geometric questions.[119][120]

Enumerative geometry studies a class of geometric objects called algebraic varieties which are defined by the vanishing of polynomials. For example, the Clebsch cubic illustrated on the right is an algebraic variety defined using a certain polynomial of degree three in four variables. A celebrated result of nineteenth-century mathematicians Arthur Cayley and George Salmon states that there are exactly 27 straight lines that lie entirely on such a surface.[121]

Generalizing this problem, one can ask how many lines can be drawn on a quintic Calabi–Yau manifold, such as the one illustrated above, which is defined by a polynomial of degree five. This problem was solved by the nineteenth-century German mathematician Hermann Schubert, who found that there are exactly 2,875 such lines. In 1986, geometer Sheldon Katz proved that the number of curves, such as circles, that are defined by polynomials of degree two and lie entirely in the quintic is 609,250.[122]

By the year 1991, most of the classical problems of enumerative geometry had been solved and interest in enumerative geometry had begun to diminish.[123] The field was reinvigorated in May 1991 when physicists Philip Candelas, Xenia de la Ossa, Paul Green, and Linda Parks showed that mirror symmetry could be used to translate difficult mathematical questions about one Calabi–Yau manifold into easier questions about its mirror.[124] In particular, they used mirror symmetry to show that a six-dimensional Calabi–Yau manifold can contain exactly 317,206,375 curves of degree three.[125] In addition to counting degree-three curves, Candelas and his collaborators obtained a number of more general results for counting rational curves which went far beyond the results obtained by mathematicians.[126]

Originally, these results of Candelas were justified on physical grounds. However, mathematicians generally prefer rigorous proofs that do not require an appeal to physical intuition. Inspired by physicists' work on mirror symmetry, mathematicians have therefore constructed their own arguments proving the enumerative predictions of mirror symmetry.[lower-alpha 5] Today mirror symmetry is an active area of research in mathematics, and mathematicians are working to develop a more complete mathematical understanding of mirror symmetry based on physicists'

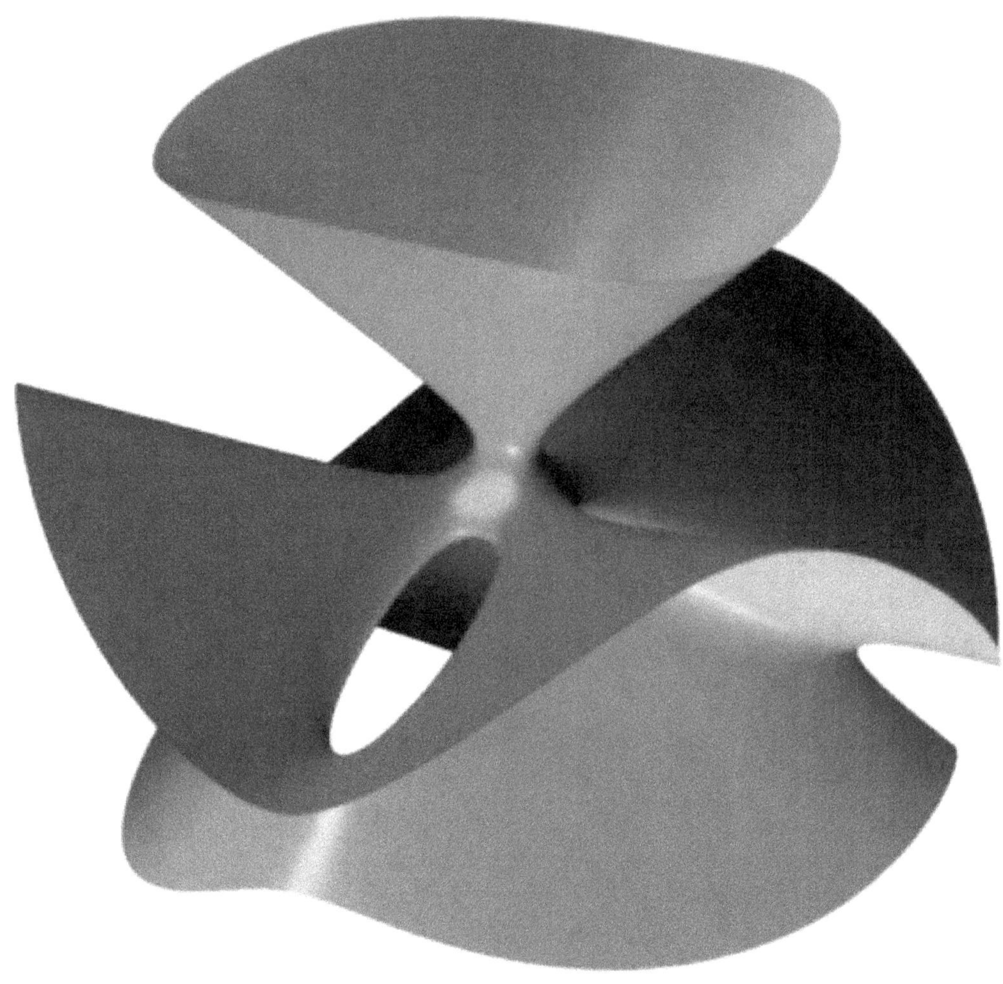

The Clebsch cubic is an example of a kind of geometric object called an algebraic variety. A classical result of enumerative geometry states that there are exactly 27 straight lines that lie entirely on this surface.

intuition.[127] Major approaches to mirror symmetry include the homological mirror symmetry program of Maxim Kontsevich[128] and the SYZ conjecture of Andrew Strominger, Shing-Tung Yau, and Eric Zaslow.[129]

1.6.2 Monstrous moonshine

Main article: Monstrous moonshine

Group theory is the branch of mathematics that studies the concept of symmetry. For example, one can consider a geometric shape such as an equilateral triangle. There are various operations that one can perform on this triangle without changing its shape. One can rotate it through 120°, 240°, or 360°, or one can reflect in any of the lines labeled S_0, S_1, or S_2 in the picture. Each of these operations is called a *symmetry*, and the collection of these symmetries satisfies certain technical properties making it into what mathematicians call a group. In this particular example, the group is known as the dihedral group of order 6 because it has six elements. A general group may describe finitely many or infinitely many symmetries; if there are only finitely many symmetries, it is called a finite group.[130]

Mathematicians often strive for a classification (or list) of all mathematical objects of a given type. It is generally believed that finite groups are too diverse to admit a useful classification. A more modest but still challenging problem is to classify all finite *simple* groups. These are finite groups which may be used as building blocks for constructing arbitrary finite groups in the same way that prime numbers can be used to construct arbitrary whole numbers by taking products.[lower-alpha 6] One of the major achievements of contemporary group theory is the classification of finite simple groups, a mathematical theorem which provides a list of all possible finite simple groups.[131]

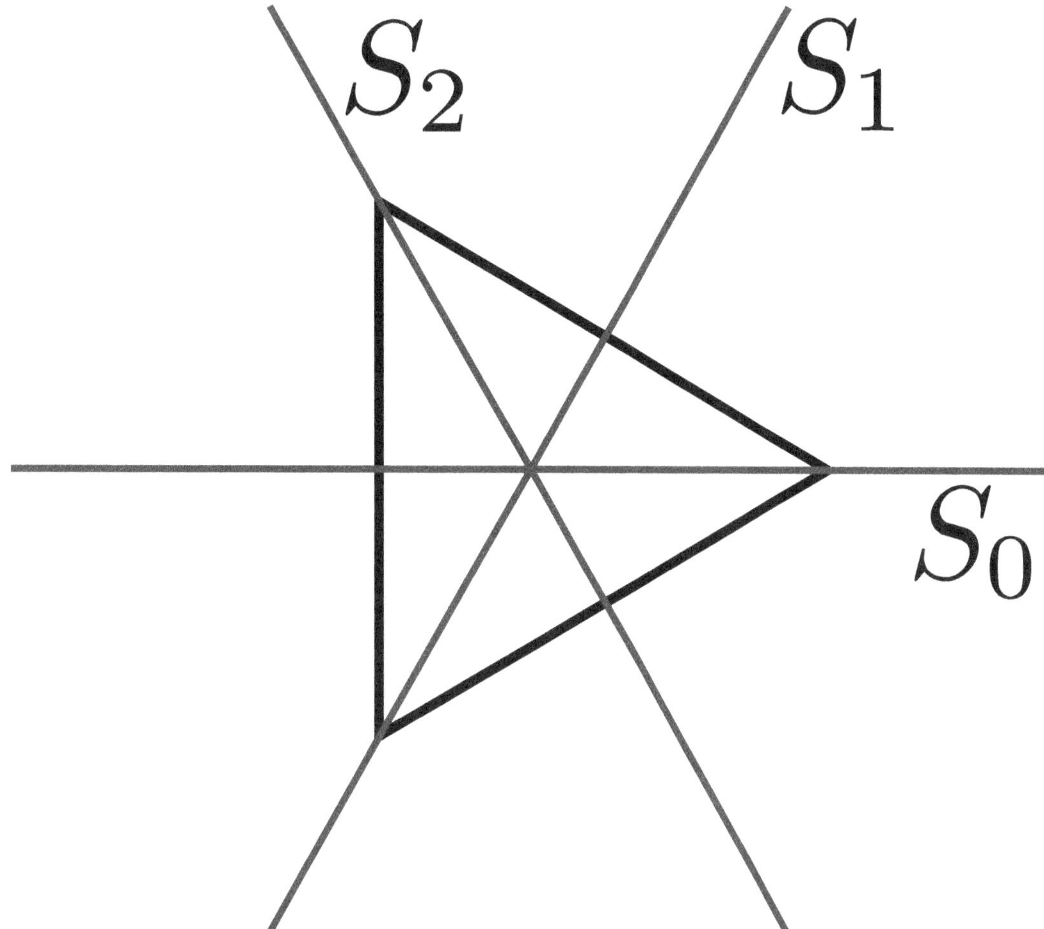

An equilateral triangle can be rotated through through 120°, 240°, or 360°, or reflected in any of the three lines pictured without changing its shape.

This classification theorem identifies several infinite families of groups as well as 26 additional groups which do not fit into any family. The latter groups are called the "sporadic" groups, and each one owes its existence to a remarkable combination of circumstances. The largest sporadic group, the so called monster group, has over 10^{53} elements, more than a thousand times the number of atoms in the Earth.[132]

A seemingly unrelated construction is the *j*-function of number theory. This object belongs to a special class of functions called modular functions, whose graphs form a certain kind of repeating pattern.[133] Although this function appears in a branch of mathematics which seems very different from the theory of finite groups, the two subjects turn out to be intimately related. In the late 1970s, mathematicians John McKay and John Thompson noticed that certain numbers arising in the analysis of the monster group (namely, the dimensions of its irreducible representations) are related to numbers that appear in a formula for the *j*-function (namely, the coefficients of its Fourier series).[134] This relationship was further developed by John Horton Conway and Simon Norton[135] who called it monstrous moonshine because it seemed so far fetched.[136]

In 1992, Richard Borcherds constructed a bridge between the theory of modular functions and finite groups and, in the process, explained the observations of McKay and Thompson.[137][138] Borcherds' work used ideas from string theory in an essential way, extending earlier results of Igor Frenkel, James Lepowsky, and Arne Meurman, who had realized the monster group as the symmetries of a particular version of string theory.[139] In 1998, Borcherds was awarded the Fields medal for his work.[140]

Since the 1990s, the connection between string theory and moonshine has led to further results in mathematics and physics.[141] In 2010, physicists Tohru Eguchi, Hirosi Ooguri, and Yuji Tachikawa discovered connections between a different sporadic group, the Mathieu group M_{24}, and a certain version of string theory.[142] Miranda Cheng, John Duncan, and Jeffrey A. Harvey proposed a generalization of this moonshine phenomenon called umbral moon-

A graph of the j*-function in the complex plane*

shine,[143] and their conjecture was proved mathematically by Duncan, Michael Griffin, and Ken Ono.[144] Witten has also speculated that the version of string theory appearing in monstrous moonshine might be related to a certain simplified model of gravity in three spacetime dimensions.[145]

1.7 History

Main article: History of string theory

1.7.1 Early results

Some of the structures reintroduced by string theory arose for the first time much earlier as part of the program of classical unification started by Albert Einstein. The first person to add a fifth dimension to a theory of gravity was Gunnar Nordström in 1914, who noted that gravity in five dimensions describes both gravity and electromagnetism in four. Nordström attempted to unify electromagnetism with his theory of gravitation, which was however superseded by Einstein's general relativity in 1919. Thereafter, German mathematician Theodor Kaluza combined the fifth dimension with general relativity, and only Kaluza is usually credited with the idea. In 1926, the Swedish physicist Oskar Klein gave a physical interpretation of the unobservable extra dimension—it is wrapped into a small circle. Einstein introduced a non-symmetric metric tensor, while much later Brans and Dicke added a scalar component to gravity. These ideas would be revived within string theory, where they are demanded by consistency conditions.

String theory was originally developed during the late 1960s and early 1970s as a never completely successful theory of hadrons, the subatomic particles like the proton and neutron that feel the strong interaction. In the 1960s, Geoffrey Chew and Steven Frautschi discovered that the mesons make families called Regge trajectories with masses related to spins in a way that was later understood by Yoichiro Nambu, Holger Bech Nielsen and Leonard Susskind to be the relationship expected from rotating strings. Chew advocated making a theory for the interactions of these trajectories

Leonard Susskind

that did not presume that they were composed of any fundamental particles, but would construct their interactions from self-consistency conditions on the S-matrix. The S-matrix approach was started by Werner Heisenberg in the

1940s as a way of constructing a theory that did not rely on the local notions of space and time, which Heisenberg believed break down at the nuclear scale. While the scale was off by many orders of magnitude, the approach he advocated was ideally suited for a theory of quantum gravity.

Working with experimental data, R. Dolen, D. Horn and C. Schmid developed some sum rules for hadron exchange. When a particle and antiparticle scatter, virtual particles can be exchanged in two qualitatively different ways. In the s-channel, the two particles annihilate to make temporary intermediate states that fall apart into the final state particles. In the t-channel, the particles exchange intermediate states by emission and absorption. In field theory, the two contributions add together, one giving a continuous background contribution, the other giving peaks at certain energies. In the data, it was clear that the peaks were stealing from the background—the authors interpreted this as saying that the t-channel contribution was dual to the s-channel one, meaning both described the whole amplitude and included the other.

Gabriele Veneziano

The result was widely advertised by Murray Gell-Mann, leading Gabriele Veneziano to construct a scattering amplitude that had the property of Dolen-Horn-Schmid duality, later renamed world-sheet duality. The amplitude needed poles where the particles appear, on straight line trajectories, and there is a special mathematical function whose poles are evenly spaced on half the real line— the Gamma function— which was widely used in Regge theory. By manipulating combinations of Gamma functions, Veneziano was able to find a consistent scattering amplitude with poles on straight lines, with mostly positive residues, which obeyed duality and had the appropriate Regge scaling at high energy. The amplitude could fit near-beam scattering data as well as other Regge type fits, and had a suggestive integral representation that could be used for generalization.

Over the next years, hundreds of physicists worked to complete the bootstrap program for this model, with many surprises. Veneziano himself discovered that for the scattering amplitude to describe the scattering of a particle that appears in the theory, an obvious self-consistency condition, the lightest particle must be a tachyon. Miguel Virasoro and Joel Shapiro found a different amplitude now understood to be that of closed strings, while Ziro Koba and Holger Nielsen generalized Veneziano's integral representation to multiparticle scattering. Veneziano and Sergio Fubini introduced an operator formalism for computing the scattering amplitudes that was a forerunner of world-sheet conformal theory, while Virasoro understood how to remove the poles with wrong-sign residues using a constraint on the states. Claud Lovelace calculated a loop amplitude, and noted that there is an inconsistency unless the dimension of the theory is 26. Charles Thorn, Peter Goddard and Richard Brower went on to prove that there are no wrong-sign propagating states in dimensions less than or equal to 26.

In 1969, Yoichiro Nambu, Holger Bech Nielsen, and Leonard Susskind recognized that the theory could be given a description in space and time in terms of strings. The scattering amplitudes were derived systematically from the action principle by Peter Goddard, Jeffrey Goldstone, Claudio Rebbi, and Charles Thorn, giving a space-time picture to the vertex operators introduced by Veneziano and Fubini and a geometrical interpretation to the Virasoro conditions.

In 1970, Pierre Ramond added fermions to the model, which led him to formulate a two-dimensional supersymmetry to cancel the wrong-sign states. John Schwarz and André Neveu added another sector to the fermi theory a short time later. In the fermion theories, the critical dimension was 10. Stanley Mandelstam formulated a world sheet conformal theory for both the bose and fermi case, giving a two-dimensional field theoretic path-integral to generate the operator formalism. Michio Kaku and Keiji Kikkawa gave a different formulation of the bosonic string, as a string field theory, with infinitely many particle types and with fields taking values not on points, but on loops and curves.

In 1974, Tamiaki Yoneya discovered that all the known string theories included a massless spin-two particle that obeyed the correct Ward identities to be a graviton. John Schwarz and Joel Scherk came to the same conclusion and made the bold leap to suggest that string theory was a theory of gravity, not a theory of hadrons. They reintroduced Kaluza–Klein theory as a way of making sense of the extra dimensions. At the same time, quantum chromodynamics was recognized as the correct theory of hadrons, shifting the attention of physicists and apparently leaving the bootstrap program in the dustbin of history.

String theory eventually made it out of the dustbin, but for the following decade all work on the theory was completely ignored. Still, the theory continued to develop at a steady pace thanks to the work of a handful of devotees. Ferdinando Gliozzi, Joel Scherk, and David Olive realized in 1976 that the original Ramond and Neveu Schwarz-strings were separately inconsistent and needed to be combined. The resulting theory did not have a tachyon, and was proven to have space-time supersymmetry by John Schwarz and Michael Green in 1981. The same year, Alexander Polyakov gave the theory a modern path integral formulation, and went on to develop conformal field theory extensively. In 1979, Daniel Friedan showed that the equations of motions of string theory, which are generalizations of the Einstein equations of General Relativity, emerge from the Renormalization group equations for the two-dimensional field theory. Schwarz and Green discovered T-duality, and constructed two superstring theories—IIA and IIB related by T-duality, and type I theories with open strings. The consistency conditions had been so strong, that the entire theory was nearly uniquely determined, with only a few discrete choices.

1.7.2 First superstring revolution

In the early 1980s, Edward Witten discovered that most theories of quantum gravity could not accommodate chiral fermions like the neutrino. This led him, in collaboration with Luis Alvarez-Gaumé to study violations of the conservation laws in gravity theories with anomalies, concluding that type I string theories were inconsistent. Green and Schwarz discovered a contribution to the anomaly that Witten and Alvarez-Gaumé had missed, which restricted the gauge group of the type I string theory to be SO(32). In coming to understand this calculation, Edward Witten became convinced that string theory was truly a consistent theory of gravity, and he became a high-profile advocate.

John Schwarz

Following Witten's lead, between 1984 and 1986, hundreds of physicists started to work in this field, and this is sometimes called the first superstring revolution.

During this period, David Gross, Jeffrey Harvey, Emil Martinec, and Ryan Rohm discovered heterotic strings. The gauge group of these closed strings was two copies of E8, and either copy could easily and naturally include the standard model. Philip Candelas, Gary Horowitz, Andrew Strominger and Edward Witten found that the Calabi–Yau manifolds are the compactifications that preserve a realistic amount of supersymmetry, while Lance Dixon and others worked out the physical properties of orbifolds, distinctive geometrical singularities allowed in string theory. Cumrun Vafa generalized T-duality from circles to arbitrary manifolds, creating the mathematical field of mirror symmetry. Daniel Friedan, Emil Martinec and Stephen Shenker further developed the covariant quantization of the superstring

Edward Witten

using conformal field theory techniques. David Gross and Vipul Periwal discovered that string perturbation theory was divergent. Stephen Shenker showed it diverged much faster than in field theory suggesting that new non-perturbative objects were missing.

In the 1990s, Joseph Polchinski discovered that the theory requires higher-dimensional objects, called D-branes and identified these with the black-hole solutions of supergravity. These were understood to be the new objects suggested by the perturbative divergences, and they opened up a new field with rich mathematical structure. It quickly became clear that D-branes and other p-branes, not just strings, formed the matter content of the string theories, and the physical interpretation of the strings and branes was revealed—they are a type of black hole. Leonard Susskind had incorporated the holographic principle of Gerardus 't Hooft into string theory, identifying the long highly excited string states with ordinary thermal black hole states. As suggested by 't Hooft, the fluctuations of the black hole horizon, the world-sheet or world-volume theory, describes not only the degrees of freedom of the black hole, but all nearby objects too.

1.7.3 Second superstring revolution

In 1995, at the annual conference of string theorists at the University of Southern California (USC), Edward Witten gave a speech on string theory that in essence united the five string theories that existed at the time, and giving birth to a new 11-dimensional theory called M-theory. M-theory was also foreshadowed in the work of Paul Townsend at

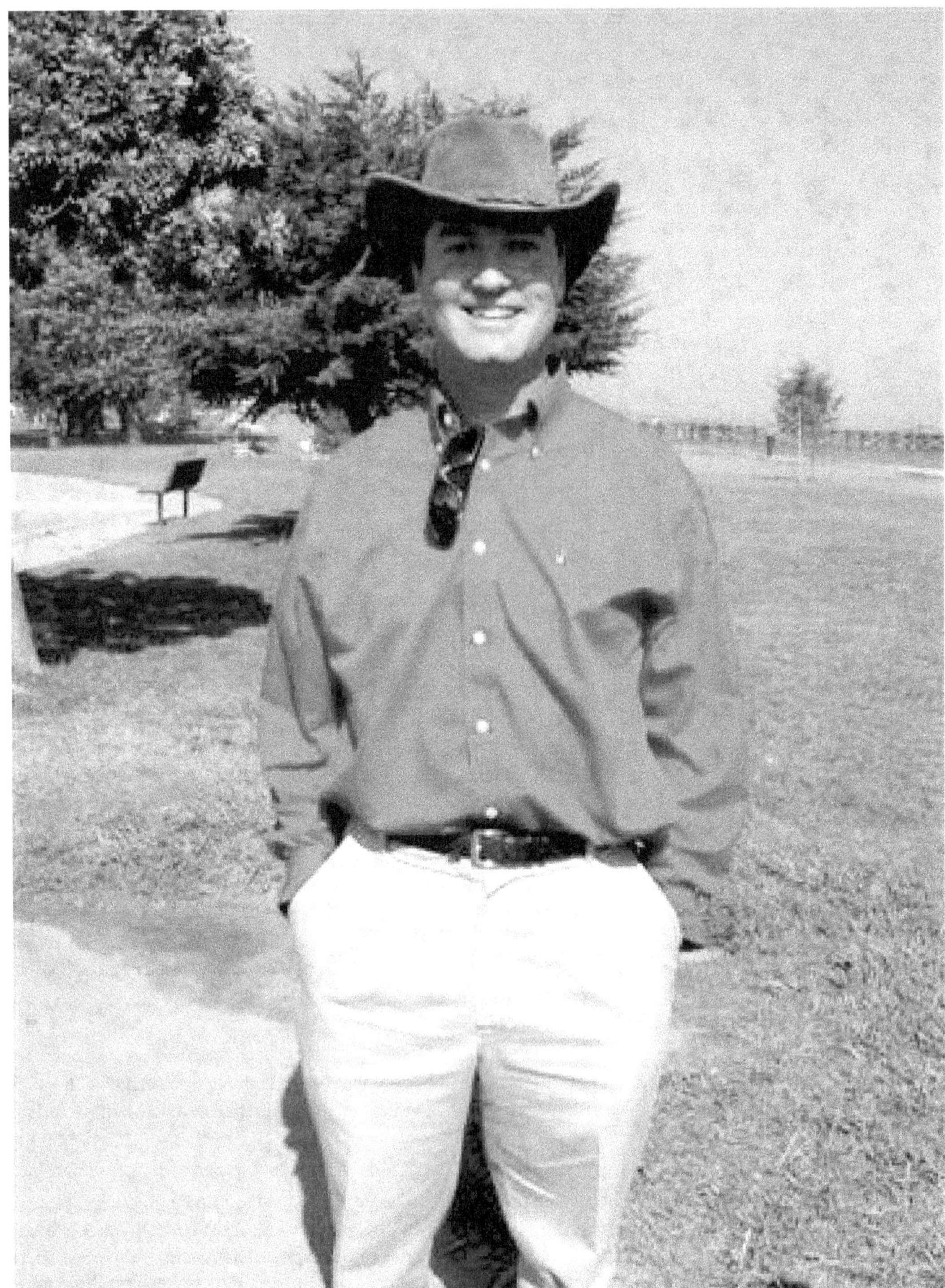

Joseph Polchinski

approximately the same time. The flurry of activity that began at this time is sometimes called the second superstring revolution.[146]

During this period, Tom Banks, Willy Fischler, Stephen Shenker and Leonard Susskind formulated matrix theory, a full holographic description of M-theory using IIA D0 branes.[147] This was the first definition of string theory that was fully non-perturbative and a concrete mathematical realization of the holographic principle. It is an example of a gauge-gravity duality and is now understood to be a special case of the AdS/CFT correspondence. Andrew

Juan Maldacena

Strominger and Cumrun Vafa calculated the entropy of certain configurations of D-branes and found agreement with the semi-classical answer for extreme charged black holes.[148] Petr Hořava and Witten found the eleven-dimensional formulation of the heterotic string theories, showing that orbifolds solve the chirality problem. Witten noted that the effective description of the physics of D-branes at low energies is by a supersymmetric gauge theory, and found geometrical interpretations of mathematical structures in gauge theory that he and Nathan Seiberg had earlier discovered in terms of the location of the branes.

In 1997, Juan Maldacena noted that the low energy excitations of a theory near a black hole consist of objects close to the horizon, which for extreme charged black holes looks like an anti-de Sitter space.[149] He noted that in this limit the gauge theory describes the string excitations near the branes. So he hypothesized that string theory on a near-horizon extreme-charged black-hole geometry, an anti-deSitter space times a sphere with flux, is equally well described by the low-energy limiting gauge theory, the N = 4 supersymmetric Yang–Mills theory. This hypothesis, which is called the AdS/CFT correspondence, was further developed by Steven Gubser, Igor Klebanov and Alexander Polyakov,[150] and by Edward Witten,[151] and it is now well-accepted. It is a concrete realization of the holographic principle, which has far-reaching implications for black holes, locality and information in physics, as well as the nature of the gravitational interaction.[152] Through this relationship, string theory has been shown to be related to gauge theories like quantum chromodynamics and this has led to more quantitative understanding of the behavior of hadrons, bringing string theory back to its roots.[153]

1.8 Criticism

1.8.1 Number of solutions

To construct models of particle physics based on string theory, physicists typically begin by specifying a shape for the extra dimensions of spacetime. Each of these different shapes corresponds to a different possible universe, or "vacuum state", with a different collection of particles and forces. String theory as it is currently understood has an enormous number of vacuum states, typically estimated to be around 10^{500}, and these might be sufficiently diverse to accommodate almost any phenomena that might be observed at low energies.[154]

Many critics of string theory have expressed concerns about the large number of possible universes described by string theory. In his book *Not Even Wrong*, Peter Woit, a lecturer in the mathematics department at Columbia University, has argued that the large number of different physical scenarios renders string theory vacuous as a framework for constructing models of particle physics. According to Woit,

> The possible existence of, say, 10^{500} consistent different vacuum states for superstring theory probably destroys the hope of using the theory to predict anything. If one picks among this large set just those states whose properties agree with present experimental observations, it is likely there still will be such a large number of these that one can get just about whatever value one wants for the results of any new observation.[155]

Some physicists believe this large number of solutions is actually a virtue because it may allow a natural anthropic explanation of the observed values of physical constants, in particular the small value of the cosmological constant.[156] The anthropic principle is the idea that some of the numbers appearing in the laws of physics are not fixed by any fundamental principle but must be compatible with the evolution of intelligent life. In 1987, Steven Weinberg published an article in which he argued that the cosmological constant could not have been too large, or else galaxies and intelligent life would not have been able to develop.[157] Weinberg suggested that there might be a huge number of possible consistent universes, each with a different value of the cosmological constant, and observations indicate a small value of the cosmological constant only because humans happen to live in a universe that has allowed intelligent life, and hence observers, to exist.[158]

String theorist Leonard Susskind has argued that string theory provides a natural anthropic explanation of the small value of the cosmological constant.[159] According to Susskind, the different vacuum states of string theory might be realized as different universes within a larger multiverse. The fact that the observed universe has a small cosmological constant is just a tautological consequence of the fact that a small value is required for life to exist.[160] Many prominent theorists and critics have disagreed with Susskind's conclusions.[161] According to Woit, "in this case [anthropic reasoning] is nothing more than an excuse for failure. Speculative scientific ideas fail not just when they make incorrect predictions, but also when they turn out to be vacuous and incapable of predicting anything."[162]

1.8.2 Background independence

Main article: Background independence

One of the fundamental principles of Einstein's general theory of relativity is the idea that the laws of physics should be background independent. This means that the geometry of spacetime is not specified from the outset but is instead determined dynamically by the theory. In general relativity, the geometry of spacetime and can evolve in time, responding to whatever matter is present.[163]

One of the older criticisms of string theory is that it is not manifestly background independent. In string theory, one must typically specify a fixed reference geometry for spacetime, and all other possible geometries are described as perturbations of this fixed one. In his book *The Trouble With Physics*, physicist Lee Smolin of the Perimeter Institute for Theoretical Physics claims that this is the principal weakness of string theory as a theory of quantum gravity, saying that string theory has failed to incorporate this important insight from general relativity.[164]

Others have disagreed with Smolin's characterization of string theory. In a review of Smolin's book, string theorist Joseph Polchinski writes

> [Smolin] is mistaking an aspect of the mathematical language being used for one of the physics being described. New physical theories are often discovered using a mathematical language that is not the most suitable for them... In string theory it has always been clear that the physics is background-independent even if the language being used is not, and the search for more suitable language continues. Indeed, as Smolin belatedly notes, [AdS/CFT] provides a solution to this problem, one that is unexpected and powerful.[165]

Polchinski notes that an important open problem in quantum gravity is to develop holographic descriptions of gravity which do not require the gravitational field to be asymptotically anti-de Sitter.[166]

1.8.3 Sociological issues

Since the superstring revolutions of the 1980s and 1990s, string theory has become the dominant paradigm of high energy theoretical physics.[167] Some string theorists have expressed the view that there does not exist an equally successful alternative theory addressing the deep questions of fundamental physics. In an interview from 1987, Nobel laureate David Gross made the following controversial comments about the reasons for the popularity of string theory:

> The most important [reason] is that there are no other good ideas around. That's what gets most people into it. When people started to get interested in string theory they didn't know anything about it. In fact, the first reaction of most people is that the theory is extremely ugly and unpleasant, at least that was the case a few years ago when the understanding of string theory was much less developed. It was difficult for people to learn about it and to be turned on. So I think the real reason why people have got attracted by it is because there is no other game in town. All other approaches of constructing grand unified theories, which were more conservative to begin with, and only gradually became more and more radical, have failed, and this game hasn't failed yet.[168]

Several other high profile theorists and commentators have expressed similar views, suggesting that there are no viable alternatives to string theory.[169]

Many critics of string theory have commented on this state of affairs. In his book criticizing string theory, Peter Woit views the status of string theory research as unhealthy and detrimental to the future of fundamental physics. He argues that the extreme popularity of string theory among theoretical physicists is partly a consequence of the financial structure of academia and the fierce competition for scarce resources.[170] In his book *The Road to Reality*, mathematical physicist Roger Penrose expresses similar views, stating "The often frantic competitiveness that this ease of communication engenders leads to 'bandwagon' effects, where researchers fear to be left behind if they do not join in."[171] Penrose also claims that the technical difficulty of modern physics forces young scientists to rely on the preferences of established researchers, rather than forging new paths of their own.[172] Lee Smolin expresses a slightly different position in his critique, claiming that string theory grew out of a tradition of particle physics which discourages speculation about the foundations of physics, while his preferred approach, loop quantum gravity, encourages more radical thinking. According to Smolin,

> String theory is a powerful, well-motivated idea and deserves much of the work that has been devoted to it. If it has so far failed, the principal reason is that its intrinsic flaws are closely tied to its strengths—and, of course, the story is unfinished, since string theory may well turn out to be part of the truth. The real question is not why we have expended so much energy on string theory but why we haven't expended nearly enough on alternative approaches.[173]

Smolin goes on to offer a number of prescriptions for how scientists might encourage a greater diversity of approaches to quantum gravity research.[174]

1.9 References

1.9.1 Notes

[1] For example, physicists are still working to understand the phenomenon of quark confinement, the paradoxes of black holes, and the origin of dark energy.

[2] For example, in the context of the AdS/CFT correspondence, theorists often formulate and study theories of gravity in unphysical numbers of spacetime dimensions.

[3] "Top Cited Articles during 2010 in hep-th". Retrieved 25 July 2013.

[4] More precisely, one cannot apply the methods of perturbative quantum field theory.

[5] Two independent mathematical proofs of mirror symmetry were given by Givental 1996, 1998 and Lian, Liu, Yau 1997, 1999, 2000.

[6] More precisely, a nontrivial group is called *simple* if its only normal subgroups are the trivial group and the group itself. The Jordan–Hölder theorem exhibits finite simple groups as the building blocks for all finite groups.

1.9.2 Citations

[1] Becker, Becker, and Schwarz 2007, p. 1

[2] Becker, Becker, and Schwarz 2007, p. 1

[3] Zwiebach 2009, p. 6

[4] Becker, Becker, and Schwarz 2007, pp. 2–3

[5] Becker, Becker, and Schwarz 2007, pp. 9–12

[6] Becker, Becker, and Schwarz 2007, pp. 14–15

[7] Klebanov and Maldacena 2009

[8] Merali 2011

[9] Sachdev 2013

[10] Becker, Becker, and Schwarz 2007, pp. 3, 15–16

[11] Becker, Becker, and Schwarz 2007, p. 8

[12] Becker, Becker, and Schwarz 13–14

[13] Woit 2006

[14] Zee 2010

[15] Zee 2010

[16] Becker, Becker, and Schwarz 2007, p. 2

[17] Becker, Becker, and Schwarz 2007, p. 6

[18] Zwiebach 2009, p. 12

[19] Becker, Becker, and Schwarz 2007, p. 6

[20] Becker, Becker, and Schwarz 2007, pp. 2–3

[21] Becker, Becker, and Schwarz 2007, p. 4

[22] Zwiebach 2009, p. 324

[23] Wald 1984, p. 4

[24] Zee 2010, Parts V and VI

[25] Zwiebach 2009, p. 9

[26] Zwiebach 2009, p. 8

[27] Yau and Nadis 2010, Ch. 6

[28] Greene 2000, p. 186

[29] Yau and Nadis 2010, Ch. 6

[30] Yau and Nadis 2010, p. ix

[31] Randall and Sundrum 1999

[32] Becker, Becker, and Schwarz 2007

[33] Becker, Becker, and Schwarz 2007

[34] Zwiebach 2009, p. 376

[35] Moore 2005, p. 214

[36] Moore 2005, p. 214

[37] Moore 2005, p. 215

[38] Aspinwall et al. 2009

[39] Kontsevich 1995

[40] Kapustin and Witten 2007

[41] Duff 1998

[42] Duff 1998, p. 64

[43] Nahm 1978

[44] Cremmer, Julia, and Scherk 1978

[45] Duff 1998, p. 65

[46] Duff 1998, p. 65

[47] Duff 1998, p. 65

[48] Duff 1998, p. 65

[49] Duff 1998, p. 65

[50] Sen 1994a

[51] Sen 1994b

[52] Hull and Townsend 1995

[53] Duff 1998, p. 67

[54] Bergshoeff, Sezgin, and Townsend 1987

[55] Duff et al. 1987

[56] Duff 1998, p. 66

[57] Witten 1995

[58] Duff 1998, pp. 67–68

[59] Becker, Becker, and Schwarz 2007, p. 296

[60] Hořava and Witten 1996

[61] Duff 1996, sec. 1

[62] Banks et al. 1997

[63] Banks et al. 1997

[64] Connes 1994

[65] Connes, Douglas, and Schwarz 1998

[66] Nekrasov and Schwarz 1998

[67] Seiberg and Witten 1999

[68] de Haro et al. 2013, p. 2

[69] Yau and Nadis 2010, p. 187–188

[70] Bekenstein 1973

[71] Hawking 1975

[72] Wald 1984, p. 417

[73] Yau and Nadis 2010, p. 189

[74] Strominger and Vafa 1996

[75] Yau and Nadis 2010, pp. 190–192

[76] Maldacena, Strominger, and Witten 1997

[77] Ooguri, Strominger, and Vafa 2004

[78] Yau and Nadis 2010, pp. 192–193

[79] Yau and Nadis 2010, pp. 194–195

[80] Strominger 1998

[81] Guica et al. 2009

[82] Castro, Maloney, and Strominger 2010

[83] Klebanov and Maldacena 2009

[84] Maldacena 1998

[85] Gubser, Klebanov, and Polyakov 1998

[86] Witten 1998

[87] Klebanov and Maldacena 2009, p. 28

[88] Maldacena 2005, p. 60

[89] Maldacena 2005, p. 61

[90] Maldacena 2005, p. 60

[91] Maldacena 2005, p. 61

[92] Maldacena 2005, p. 60

[93] Zwiebach 2009, p. 552

[94] Maldacena 2005, pp. 61–62

[95] de Haro et al. 2013, p. 2

[96] Hawking 1975

[97] Susskind 2008

[98] Zwiebach 2009, p. 554

[99] Maldacena 2005, p. 63

[100] Hawking 2005

[101] Merali 2011

[102] Zwiebach 2009, p. 559

[103] Kovtun, Son, and Starinets 2001

[104] Merali 2011, p. 303

[105] Luzum and Romatschke 2008

[106] Merali 2011, p. 303

[107] Sachdev 2013, p. 51

[108] Woit 2006

[109] Candelas et al. 1985

[110] Yau and Nadis 2010, pp. 147–150

[111] Becker, Becker, and Schwarz 2007, pp. 530–531

[112] Becker, Becker, and Schwarz 2007, p. 531

[113] Becker, Becker, and Schwarz 2007, p. 538

[114] Becker, Becker, and Schwarz 2007, p. 533

[115] Becker, Becker, and Schwarz 2007, pp. 539–543

[116] Deligne et al. 1999, p. 1

[117] Hori et al. 2003, p. xvii

[118] Aspinwall et al. 2009, p. 13

[119] Hori et al. 2003

[120] Aspinwall et al. 2009

[121] Yau and Nadis 2010, p. 167

[122] Yau and Nadis 2010, p. 166

[123] Yau and Nadis 2010, p. 169

[124] Candelas et al. 1991

[125] Yau and Nadis 2010, p. 169

[126] Yau and Nadis 2010, p. 171

[127] Hori et al. 2003, p. xix

[128] Kontsevich 1995

[129] Strominger, Yau, and Zaslow 1996

[130] Dummit and Foote 2004

[131] Dummit and Foote 2004, pp. 102–103

[132] Klarreich 2015

[133] Gannon 2006, p. 2

[134] Gannon 2006, p. 4

[135] Conway and Norton 1979

[136] Gannon 2006, p. 5

[137] Gannon 2006, p. 8

[138] Borcherds 1992

[139] Frenkel, Lepowsky, and Meurman 1988

[140] Gannon 2006, p. 11

[141] Klarreich 2015

[142] Eguchi, Ooguri, and Tachikawa 2010

[143] Cheng, Duncan, and Harvey 2013

[144] Duncan, Griffin, and Ono 2015

[145] Witten 2007

[146] Duff 1998

[147] Banks et al. 1997

[148] Strominger and Vafa 1996

[149] Maldacena 1998

[150] Gubser, Klebanov, and Polyakov 1998

[151] Witten 1998

[152] de Haro et al. 2013, p. 2

[153] Kovtun, Son, and Starinets 2001

[154] Woit 2006, pp. 240–242

[155] Woit 2006, p. 242

[156] Woit 2006, p. 242

[157] Weinberg 1987

[158] Woit 2006, p. 243

[159] Susskind 2005

[160] Woit 2006, pp. 242–243

[161] Woit 2006, p. 240

[162] Woit 2006, p. 249

[163] Smolin 2006, p. 81

[164] Smolin 2006, p. 184

[165] Polchinski 2007

[166] Polchinski 2007

[167] Penrose 2004, p. 1017

[168] Woit 2006, pp. 224–225

[169] Woit 2006, Ch. 16

[170] Woit 2006, p. 239

[171] Penrose 2004, p. 1018

[172] Penrose 2004, pp. 1019–1020

[173] Smolin 2006, p. 349

[174] Smolin 2006, Ch. 20

1.9.3 Bibliography

- Aspinwall, Paul; Bridgeland, Tom; Craw, Alastair; Douglas, Michael; Gross, Mark; Kapustin, Anton; Moore, Gregory; Segal, Graeme; Szendröi, Balázs; Wilson, P.M.H., eds. (2009). *Dirichlet Branes and Mirror Symmetry*. American Mathematical Society. ISBN 978-0-8218-3848-8.

- Banks, Tom; Fischler, Willy; Schenker, Stephen; Susskind, Leonard (1997). "M theory as a matrix model: A conjecture". *Physical Review D* **55** (8): 5112. arXiv:hep-th/9610043. Bibcode:1997PhRvD..55.5112B. doi:10.1103/physrevd.55.5112.

- Becker, Katrin; Becker, Melanie; Schwarz, John (2007). *String theory and M-theory: A modern introduction*. Cambridge University Press. ISBN 978-0-521-86069-7.

- Bekenstein, Jacob (1973). "Black holes and entropy". *Physical Review D* **7** (8): 2333. Bibcode:1973PhRvD...7.2333B. doi:10.1103/PhysRevD.7.2333.

- Bergshoeff, Eric; Sezgin, Ergin; Townsend, Paul (1987). "Supermembranes and eleven-dimensional supergravity". *Physics Letters B* **189** (1): 75–78. Bibcode:1987PhLB..189...75B. doi:10.1016/0370-2693(87)91272-X.

- Borcherds, Richard (1992). "Monstrous moonshine and Lie superalgebras". *Inventiones mathematicae* **109** (1): 405–444. Bibcode:1992InMat.109..405B. doi:10.1007/BF01232032.

- Candelas, Philip; de la Ossa, Xenia; Green, Paul; Parks, Linda (1991). "A pair of Calabi–Yau manifolds as an exactly soluble superconformal field theory". *Nuclear Physics B* **359** (1): 21–74. Bibcode:1991NuPhB.359...21C. doi:10.1016/0550-3213(91)90292-6.

- Candelas, Philip; Horowitz, Gary; Strominger, Andrew; Witten, Edward (1985). "Vacuum configurations for superstrings". *Nuclear Physics B* **258**: 46–74. Bibcode:1985NuPhB.258...46C. doi:10.1016/0550-3213(85)90602-9.

- Castro, Alejandra; Maloney, Alexander; Strominger, Andrew (2010). "Hidden conformal symmetry of the Kerr black hole". *Physical Review D* **82** (2). arXiv:1004.0996. Bibcode:2010PhRvD..82b4008C. doi:10.1103/PhysRevD.82.024(

- Cheng, Miranda; Duncan, John; Harvey, Jeffrey (2013). "Umbral Moonshine". arXiv:1204.2779.

- Connes, Alain (1994). *Noncommutative Geometry*. Academic Press. ISBN 978-0-12-185860-5.

- Connes, Alain; Douglas, Michael; Schwarz, Albert (1998). "Noncommutative geometry and matrix theory". *Journal of High Energy Physics*. 19981 (2): 003. arXiv:hep-th/9711162. Bibcode:1998JHEP...02..003C. doi:10.1088/1126-6708/1998/02/003.

- Conway, John; Norton, Simon (1979). "Monstrous moonshine". *Bull. London. Math. Soc.* **11** (3): 308–339.

- Cremmer, Eugene; Julia, Bernard; Scherk, Joel (1978). "Supergravity theory in eleven dimensions". *Physics Letters B* **76** (4): 409–412. Bibcode:1978PhLB...76..409C. doi:10.1016/0370-2693(78)90894-8.

- de Haro, Sebastian; Dieks, Dennis; 't Hooft, Gerard; Verlinde, Erik (2013). "Forty Years of String Theory Reflecting on the Foundations". *Foundations of Physics* **43** (1): 1–7. Bibcode:2013FoPh...43....1D. doi:10.1007/s10701-012-9691-3.

- Deligne, Pierre; Etingof, Pavel; Freed, Daniel; Jeffery, Lisa; Kazhdan, David; Morgan, John; Morrison, David; Witten, Edward, eds. (1999). *Quantum Fields and Strings: A Course for Mathematicians* **1**. American Mathematical Society. ISBN 978-0821820124.

- Duff, Michael (1996). "M-theory (the theory formerly known as strings)". *International Journal of Modern Physics A* **11** (32): 6523–41. arXiv:hep-th/9608117. Bibcode:1996IJMPA..11.5623D. doi:10.1142/S0217751X96002583.

- Duff, Michael (1998). "The theory formerly known as strings". *Scientific American* **278** (2): 64–9. doi:10.1038/scientificamerican 64.
- Duff, Michael; Howe, Paul; Inami, Takeo; Stelle, Kellogg (1987). "Superstrings in D=10 from supermembranes in D=11". *Nuclear Physics B* **191** (1): 70–74. Bibcode:1987PhLB..191...70D. doi:10.1016/0370-2693(87)91323-2.
- Dummit, David; Foote, Richard (2004). *Abstract Algebra*. Wiley. ISBN 978-0-471-43334-7.
- Duncan, John; Griffin, Michael; Ono, Ken (2015). "Proof of the Umbral Moonshine Conjecture". arXiv:1503.01472.
- Eguchi, Tohru; Ooguri, Hirosi; Tachikawa, Yuji (2011). "Notes on the K3 surface and the Mathieu group M_{24}". *Experimental Mathematics* **20** (1): 91–96. doi:10.1080/10586458.2011.544585.
- Frenkel, Igor; Lepowsky, James; Meurman, Arne (1988). *Vertex Operator Algebras and the Monster*. Pure and Applied Mathematics **134**. Academic Press. ISBN 0-12-267065-5.
- Gannon, Terry. *Moonshine Beyond the Monster: The Bridge Connecting Algebra, Modular Forms, and Physics*. Cambridge University Press.
- Givental, Alexander (1996). "Equivariant Gromov-Witten invariants". *International Mathematics Research Notices* **1996** (13): 613–663. doi:10.1155/S1073792896000414.
- Givental, Alexander (1998). "A mirror theorem for toric complete intersections". *Topological field theory, primitive forms and related topics*: 141–175. doi:10.1007/978-1-4612-0705-4_5. ISBN 978-1-4612-6874-1.
- Gubser, Steven; Klebanov, Igor; Polyakov, Alexander (1998). "Gauge theory correlators from non-critical string theory". *Physics Letters B* **428**: 105–114. arXiv:hep-th/9802109. Bibcode:1998PhLB..428..105G. doi:10.1016/S0370-2693(98)00377-3.
- Guica, Monica; Hartman, Thomas; Song, Wei; Strominger, Andrew (2009). "The Kerr/CFT Correspondence". *Physical Review D* **80** (12). arXiv:0809.4266. Bibcode:2009PhRvD..80l4008G. doi:10.1103/PhysRevD.80.124008.
- Hawking, Stephen (1975). "Particle creation by black holes". *Communications in mathematical physics* **43** (3): 199–220. Bibcode:1975CMaPh..43..199H. doi:10.1007/BF02345020.
- Hawking, Stephen (2005). "Information loss in black holes". *Physical Review D* **72** (8). arXiv:hep-th/0507171. Bibcode:2005PhRvD..72h4013H. doi:10.1103/PhysRevD.72.084013.
- Hořava, Petr; Witten, Edward (1996). "Heterotic and Type I string dynamics from eleven dimensions". *Nuclear Physics B* **460** (3): 506–524. arXiv:hep-th/9510209. Bibcode:1996NuPhB.460..506H. doi:10.1016/0550-3213(95)00621-4.
- Hori, Kentaro; Katz, Sheldon; Klemm, Albrecht; Pandharipande, Rahul; Thomas, Richard; Vafa, Cumrun; Vakil, Ravi; Zaslow, Eric, eds. (2003). *Mirror Symmetry* (PDF). American Mathematical Society. ISBN 0-8218-2955-6.
- Hull, Chris; Townsend, Paul (1995). "Unity of superstring dualities". *Nuclear Physics B* **4381** (1): 109–137. arXiv:hep-th/9410167. Bibcode:1995NuPhB.438..109H. doi:10.1016/0550-3213(94)00559-W.
- Kapustin, Anton; Witten, Edward (2007). "Electric-magnetic duality and the geometric Langlands program". *Communications in Number Theory and Physics* **1** (1): 1–236. arXiv:hep-th/0604151. Bibcode:2007CNTP....1....1K. doi:10.4310/cntp.2007.v1.n1.a1.
- Klarreich, Erica. "Mathematicians chase moonshine's shadow". *Quanta Magazine*. Retrieved March 2015.

- Klebanov, Igor; Maldacena, Juan (2009). "Solving Quantum Field Theories via Curved Spacetimes" (PDF). *Physics Today* **62**: 28. Bibcode:2009PhT....62a..28K. doi:10.1063/1.3074260. Retrieved May 2013.

- Kontsevich, Maxim (1995). "Homological algebra of mirror symmetry". *Proceedings of the International Congress of Mathematicians*: 120–139. arXiv:alg-geom/9411018. Bibcode:1994alg.geom.11018K.

- Kovtun, P. K.; Son, Dam T.; Starinets, A. O. (2001). "Viscosity in strongly interacting quantum field theories from black hole physics". *Physical review letters* **94** (11): 111601. arXiv:hep-th/0405231. Bibcode:2005PhRvL..94k1601K. doi:10.1103/PhysRevLett.94.111601. PMID 15903845.

- Lian, Bong; Liu, Kefeng; Yau, Shing-Tung (1997). "Mirror principle, I". *Asian Journal of Math* **1**: 729–763. arXiv:alg-geom/9712011. Bibcode:1997alg.geom.12011L.

- Lian, Bong; Liu, Kefeng; Yau, Shing-Tung (1999a). "Mirror principle, II". *Asian Journal of Math* **3**: 109–146. arXiv:math/9905006. Bibcode:1999math......5006L.

- Lian, Bong; Liu, Kefeng; Yau, Shing-Tung (1999b). "Mirror principle, III". *Asian Journal of Math* **3**: 771–800. arXiv:math/9912038. Bibcode:1999math.....12038L.

- Lian, Bong; Liu, Kefeng; Yau, Shing-Tung (2000). "Mirror principle, IV". *Surveys in Differential Geometry*: 475–496. arXiv:math/0007104. Bibcode:2000math......7104L.

- Luzum, Matthew; Romatschke, Paul (2008). "Conformal relativistic viscous hydrodynamics: Applications to RHIC results at $\surd sNN$=200 GeV". *Physical Review C* **78** (3). arXiv:0804.4015. doi:10.1103/PhysRevC.78.034915.

- Maldacena, Juan (1998). "The Large *N* limit of superconformal field theories and supergravity". *Advances in Theoretical and Mathematical Physics* **2**: 231–252. arXiv:hep-th/9711200. Bibcode:1998AdTMP...2..231M. doi:10.1063/1.59653.

- Maldacena, Juan (2005). "The Illusion of Gravity" (PDF). *Scientific American* **293** (5): 56–63. Bibcode:2005SciAm.293e..56M. doi:10.1038/scientificamerican1105-56. PMID 16318027. Retrieved July 2013.

- Maldacena, Juan; Strominger, Andrew; Witten, Edward (1997). "Black hole entropy in M-theory". *Journal of High Energy Physics* **1997** (12).

- Merali, Zeeya (2011). "Collaborative physics: string theory finds a bench mate". *Nature* **478** (7369): 302–304. Bibcode:2011Natur.478..302M. doi:10.1038/478302a. PMID 22012369.

- Moore, Gregory (2005). "What is ... a Brane?" (PDF). *Notices of the AMS* **52**: 214. Retrieved June 2013.

- Nahm, Walter (1978). "Supersymmetries and their representations". *Nuclear Physics B* **135** (1): 149–166. Bibcode:1978NuPhB.135..149N. doi:10.1016/0550-3213(78)90218-3.

- Nekrasov, Nikita; Schwarz, Albert (1998). "Instantons on noncommutative $\mathbf{R}^4$ and (2,0) superconformal six dimensional theory". *Communications in Mathematical Physics* **198** (3): 689–703. arXiv:hep-th/9802068. Bibcode:1998CMaPh.198..689N. doi:10.1007/s002200050490.

- Ooguri, Hirosi; Strominger, Andrew; Vafa, Cumrun (2004). "Black hole attractors and the topological string". *Physical Review D* **70** (10).

- Polchinski, Joseph (2007). "All Strung Out?". *American Scientist*. Retrieved April 2015.

- Penrose, Roger (2005). *The Road to Reality: A Complete Guide to the Laws of the Universe*. Knopf. ISBN 0-679-45443-8.

- Randall, Lisa; Sundrum, Raman (1999). "An alternative to compactification". *Physical Review Letters* **83** (23): 4690. arXiv:hep-th/9906064. Bibcode:1999PhRvL..83.4690R. doi:10.1103/PhysRevLett.83.4690.

- Sachdev, Subir (2013). "Strange and stringy". *Scientific American* **308** (44): 44. Bibcode:2012SciAm.308a..44S. doi:10.1038/scientificamerican0113-44.

- Seiberg, Nathan; Witten, Edward (1999). "String Theory and Noncommutative Geometry". *Journal of High Energy Physics* **1999** (9): 032. arXiv:hep-th/9908142. Bibcode:1999JHEP...09..032S. doi:10.1088/1126-6708/1999/09/032.

- Sen, Ashoke (1994a). "Strong-weak coupling duality in four-dimensional string theory". *International Journal of Modern Physics A* **9** (21): 3707–3750. arXiv:hep-th/9402002. Bibcode:1994IJMPA...9.3707S. doi:10.1142/S0217751X94C

- Sen, Ashoke (1994b). "Dyon-monopole bound states, self-dual harmonic forms on the multi-monopole moduli space, and *SL*(2,**Z**) invariance in string theory". *Physics Letters B* **329** (2): 217–221. arXiv:hep-th/9402032. Bibcode:1994PhLB..329..217S. doi:10.1016/0370-2693(94)90763-3.

- Smolin, Lee (2006). *The Trouble with Physics: The Rise of String Theory, the Fall of a Science, and What Comes Next*. New York: Houghton Mifflin Co. ISBN 0-618-55105-0.

- Strominger, Andrew (1998). "Black hole entropy from near-horizon microstates". *Journal of High Energy Physics* **1998** (2). arXiv:hep-th/9712251. Bibcode:1998JHEP...02..009S. doi:10.1088/1126-6708/1998/02/009.

- Strominger, Andrew; Vafa, Cumrun (1996). "Microscopic origin of the Bekenstein–Hawking entropy". *Physics Letters B* **379** (1): 99–104. arXiv:hep-th/9601029. Bibcode:1996PhLB..379...99S. doi:10.1016/0370-2693(96)00345-0.

- Strominger, Andrew; Yau, Shing-Tung; Zaslow, Eric (1996). "Mirror symmetry is T-duality". *Nuclear Physics B* **479** (1): 243–259. arXiv:hep-th/9606040. Bibcode:1996NuPhB.479..243S. doi:10.1016/0550-3213(96)00434-8.

- Susskind, Leonard (2005). *The Cosmic Landscape: String Theory and the Illusion of Intelligent Design*. Back Bay Books. ISBN 978-0316013338.

- Susskind, Leonard (2008). *The Black Hole War: My Battle with Stephen Hawking to Make the World Safe for Quantum Mechanics*. Little, Brown and Company. ISBN 978-0-316-01641-4.

- Wald, Robert (1984). *General Relativity*. University of Chicago Press. ISBN 978-0-226-87033-5.

- Weinberg, Steven (1987). *Anthropic bound on the cosmological constant* **59** (22). Physical Review Letters. p. 2607.

- Witten, Edward (1995). "String theory dynamics in various dimensions". *Nuclear Physics B* **443** (1): 85–126. arXiv:hep-th/9503124. Bibcode:1995NuPhB.443...85W. doi:10.1016/0550-3213(95)00158-O.

- Witten, Edward (1998). "Anti-de Sitter space and holography". *Advances in Theoretical and Mathematical Physics* **2**: 253–291. arXiv:hep-th/9802150. Bibcode:1998AdTMP...2..253W.

- Witten, Edward (2007). "Three-dimensional gravity revisited". arXiv:0706.3359 [hep-th].

- Woit, Peter (2006). *Not Even Wrong: The Failure of String Theory and the Search for Unity in Physical Law*. Basic Books. p. 105. ISBN 0-465-09275-6.

- Yau, Shing-Tung; Nadis, Steve (2010). *The Shape of Inner Space: String Theory and the Geometry of the Universe's Hidden Dimensions*. Basic Books. ISBN 978-0-465-02023-2.

- Zee, Anthony (2010). *Quantum Field Theory in a Nutshell* (2nd ed.). Princeton University Press. ISBN 978-0-691-14034-6.

- Zwiebach, Barton (2009). *A First Course in String Theory*. Cambridge University Press. ISBN 978-0-521-88032-9.

1.10 Further reading

1.10.1 Popularizations

General

- Greene, Brian (2003). *The Elegant Universe: Superstrings, Hidden Dimensions, and the Quest for the Ultimate Theory*. New York: W.W. Norton & Company. ISBN 0-393-05858-1.

- Greene, Brian (2004). *The Fabric of the Cosmos: Space, Time, and the Texture of Reality*. New York: Alfred A. Knopf. ISBN 0-375-41288-3.

Critical

- Penrose, Roger (2005). *The Road to Reality: A Complete Guide to the Laws of the Universe*. Knopf. ISBN 0-679-45443-8.

- Smolin, Lee (2006). *The Trouble with Physics: The Rise of String Theory, the Fall of a Science, and What Comes Next*. New York: Houghton Mifflin Co. ISBN 0-618-55105-0.

- Woit, Peter (2006). *Not Even Wrong: The Failure of String Theory And the Search for Unity in Physical Law*. London: Jonathan Cape &: New York: Basic Books. ISBN 978-0-465-09275-8.

1.10.2 Textbooks

For physicists

- Becker, Katrin; Becker, Melanie; Schwarz, John (2007). *String theory and M-theory: A modern introduction*. Cambridge University Press. ISBN 978-0-521-86069-7.

- Green, Michael; Schwarz, John; Witten, Edward (2012). *Superstring theory. Vol. 1: Introduction*. Cambridge University Press. ISBN 978-1107029118.

- Green, Michael; Schwarz, John; Witten, Edward (2012). *Superstring theory. Vol. 2: Loop amplitudes, anomalies and phenomenology*. Cambridge University Press. ISBN 978-1107029132.

- Polchinski, Joseph (1998). *String Theory Vol. 1: An Introduction to the Bosonic String*. Cambridge University Press. ISBN 0-521-63303-6.

- Polchinski, Joseph (1998). *String Theory Vol. 2: Superstring Theory and Beyond*. Cambridge University Press. ISBN 0-521-63304-4.

- Zwiebach, Barton (2009). *A First Course in String Theory*. Cambridge University Press. ISBN 978-0-521-88032-9.

For mathematicians

- Deligne, Pierre; Etingof, Pavel; Freed, Daniel; Jeffery, Lisa; Kazhdan, David; Morgan, John; Morrison, David; Witten, Edward, eds. (1999). *Quantum Fields and Strings: A Course for Mathematicians, Vol. 2*. American Mathematical Society. ISBN 978-0821819883.

1.11 External links

- *The Elegant Universe*—A three-hour miniseries with Brian Greene by *NOVA* (original PBS Broadcast Dates: October 28, 8–10 p.m. and November 4, 8–9 p.m., 2003). Various images, texts, videos and animations explaining string theory.

- Not Even Wrong—A blog critical of string theory

- The Official String Theory Web Site

- Why String Theory—An introduction to string theory.

Chapter 2

String (physics)

This article is about quantum strings, objects in string theory. For a classical string, such as a guitar string, see vibrating string. For other uses, see string (disambiguation).

In physics, a **string** is a physical object that appears in string theory and related subjects. Unlike elementary particles, which are zero-dimensional or point-like by definition, strings are one-dimensional extended objects. Theories in which the fundamental objects are strings rather than point particles automatically have many properties that are expected to hold in a fundamental theory of physics. Most notably, a theory of strings that evolve and interact according to the rules of quantum mechanics will automatically describe quantum gravity.

In string theory, the strings may be open (forming a segment with two endpoints) or closed (forming a loop like a circle) and may have other special properties. Prior to 1995, there were five known versions of string theory incorporating the idea of supersymmetry, which differed in the type of strings and in other aspects. Today these different string theories are thought to arise as different limiting cases of a single theory called M-theory.

In theories of particle physics based on string theory, the characteristic length scale of strings is typically on the order of the Planck length, the scale at which the effects of quantum gravity are believed to become significant. On much larger length scales, such as the scales visible in physics laboratories, such objects would be indistinguishable from zero-dimensional point particles, and the vibrational state of the string would determine the type of particle. Strings are also sometimes studied in nuclear physics where they are used to model flux tubes.

As it propagates through spacetime, a string sweeps out a two-dimensional surface called its worldsheet. This is analogous to the one-dimensional worldline traced out by a point particle. The physics of a string is described by means of a two-dimensional conformal field theory associated with the worldsheet. The formalism of two dimensional conformal field theory also has many applications outside of string theory, for example in condensed matter physics and parts of pure mathematics.

2.1 Types of strings

2.1.1 Closed and open strings

Strings can be either open or closed. A **closed string** is a string that has no end-points, and therefore is topologically equivalent to a circle. An **open string**, on the other hand, has two end-points and is topologically equivalent to a line interval. Not all string theories contain open strings, but every theory must contain closed strings, as interactions between open strings can always result in closed strings.

The oldest superstring theory containing open strings was type I string theory. However, the developments in string theory in the 1990s have shown that the open strings should always be thought of as ending on a new type of objects called D-branes, and the spectrum of possibilities for open strings has increased greatly.

Open and closed strings are generally associated with characteristic vibrational modes. One of the vibration modes of a closed string can be identified as the graviton. In certain string theories the lowest-energy vibration of an open string is a tachyon and can undergo tachyon condensation. Other vibrational modes of open strings exhibit the properties of photons and gluons.

2.1.2 Orientation

Strings can also possess an **orientation**, which can be thought of as an internal "arrow" which distinguishes the string from one with the opposite orientation. By contrast, an **unoriented string** is one with no such arrow on it.

2.2 See also

- Elementary particle
- Brane
- D-brane

2.3 References

- Schwarz, John (2000). "Introduction to Superstring Theory". Retrieved Dec. 12, 2005.
- "NOVA's strings homepage"

Chapter 3

Compactification (physics)

For the concept of compactification in mathematics, see compactification (mathematics).

In physics, **compactification** means changing a theory with respect to one of its space-time dimensions. Instead of having a theory with this dimension being infinite, one changes the theory so that this dimension has a finite length, and may also be periodic.

Compactification plays an important part in thermal field theory where one compactifies time, in string theory where one compactifies the extra dimensions of the theory, and in two- or one-dimensional solid state physics, where one considers a system which is limited in one of the three usual spatial dimensions.

At the limit where the size of the compact dimension goes to zero, no fields depend on this extra dimension, and the theory is dimensionally reduced.

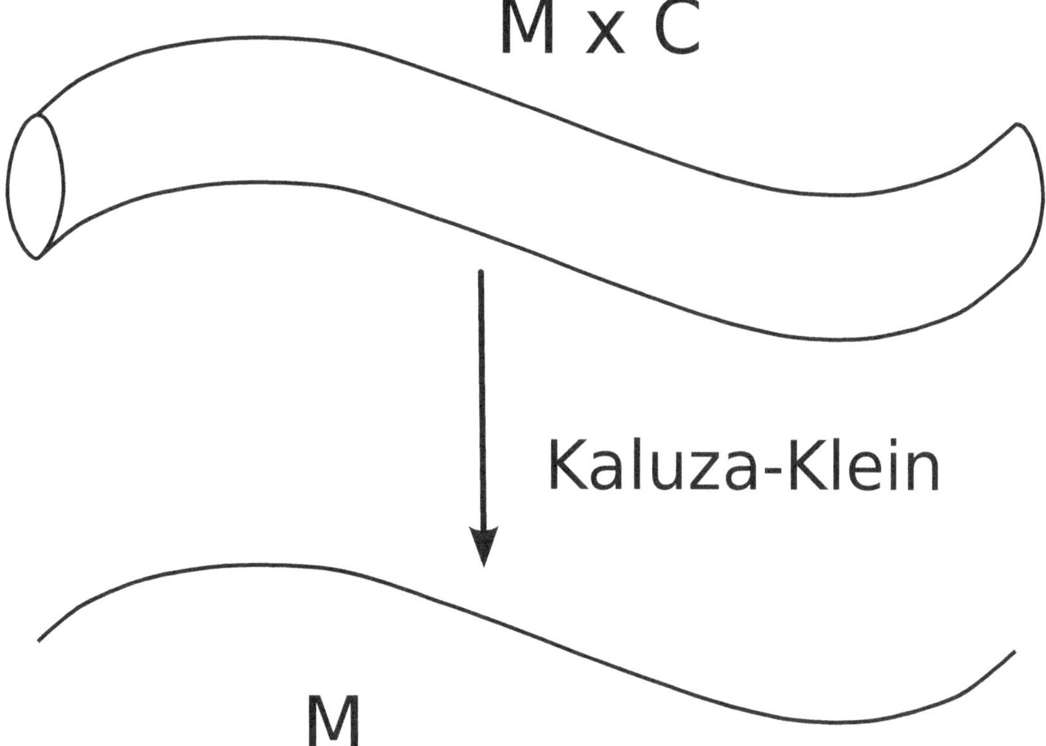

The space $M \times C$ is compactified over the compact C and after Kaluza–Klein decomposition, we have an effective field theory over M.

3.1 Compactification in string theory

In string theory, compactification is a generalization of Kaluza–Klein theory. It tries to conciliate the gap between the conception of our universe based on its four observable dimensions with the ten, eleven, or twenty-six dimensions which theoretical equations lead us to suppose the universe is made with. For this purpose it is assumed the extra dimensions are "wrapped" up on themselves, or "curled" up on Calabi–Yau spaces, or on orbifolds. Models in which the compact directions support fluxes are known as *flux compactifications*. The coupling constant of string theory, which determines the probability of strings to split and reconnect, can be described by a field called dilaton. This in turn can be described as the size of an extra (eleventh) dimension which is compact. In this way, the ten-dimensional type IIA string theory can be described as the compactification of M-theory in eleven dimensions. Furthermore, different versions of string theory are related by different compactifications in a procedure known as T-duality.

The formulation of more precise versions of the meaning of compactification in this context has been promoted by discoveries such as the mysterious duality.

3.2 Flux compactification

A **flux compactification** is a particular way to deal with additional dimensions required by string theory. It assumes that the shape of the internal manifold is a Calabi–Yau manifold or generalized Calabi–Yau manifold which is equipped with non-zero values of fluxes, i.e. differential forms that generalize the concept of an electromagnetic field (see p-form electrodynamics).

The hypothetical concept of the anthropic landscape in string theory follows from a large number of possibilities in which the integers that characterize the fluxes can be chosen without violating rules of string theory. The flux compactifications can be described as F-theory vacua or type IIB string theory vacua with or without D-branes.

3.3 See also

- Dimensional reduction
- Kaluza–Klein theory

3.4 References

- Chapter 16 of Michael Green, John H. Schwarz and Edward Witten (1987) *Superstring theory*. Cambridge University Press. *Vol. 2: Loop amplitudes, anomalies and phenomenology*. ISBN 0-521-35753-5.
- Brian R. Greene, "String Theory on Calabi–Yau Manifolds". arXiv:hep-th/9702155.
- Mariana Graña, "Flux compactifications in string theory: A comprehensive review", *Physics Reports* **423**, 91–158 (2006). arXiv:hep-th/0509003.
- Michael R. Douglas and Shamit Kachru "Flux compactification", *Rev. Mod. Phys.* **79**, 733 (2007). arXiv:hep-th/0610102.
- Ralph Blumenhagen, Boris Körs, Dieter Lüst, Stephan Stieberger, "Four-dimensional string compactifications with D-branes, orientifolds and fluxes", *Physics Reports* **445**, 1–193 (2007). arXiv:hep-th/0610327.

3.5 External links

- Flux compactification on arxiv.org

Chapter 4

Supersymmetry

"SUSY" redirects here. For other uses, see Susy (disambiguation).

For the episode of the American TV series Angel, see Supersymmetry (Angel).

Supersymmetry (**SUSY**), a theory of particle physics, is a proposed type of spacetime symmetry that relates two basic classes of elementary particles: bosons, which have an integer-valued spin, and fermions, which have a half-integer spin.[1] Each particle from one group is associated with a particle from the other, known as its superpartner, the spin of which differs by a half-integer. In a theory with perfectly "unbroken" supersymmetry, each pair of superpartners would share the same mass and internal quantum numbers besides spin. For example, there would be a "selectron" (superpartner electron), a bosonic version of the electron with the same mass as the electron, that would be easy to find in a laboratory. Thus, since no superpartners have been observed, if supersymmetry exists it must be a spontaneously broken symmetry so that superpartners may differ in mass.[2][3] Spontaneously-broken supersymmetry could solve many mysterious problems in particle physics including the hierarchy problem. The simplest realization of spontaneously-broken supersymmetry, the so-called Minimal Supersymmetric Standard Model, is one of the best studied candidates for physics beyond the Standard Model.

There is only indirect evidence and motivation for the existence of supersymmetry. Direct confirmation would entail production of superpartners in collider experiments, such as the Large Hadron Collider (LHC). The first run of the LHC found no evidence for supersymmetry (all results were consistent with the Standard Model), and thus set limits on superpartner masses in supersymmetric theories. Whilst many remain enthusiastic about supersymmetry,[4] this first run at the LHC lead some physicists to explore other ideas.[5] In any case, in 2015 the LHC resumed its search for supersymmetry and other new physics in its second run.

4.1 Motivations

There are numerous phenomenological motivations for supersymmetry close to the electroweak scale, as well as technical motivations for supersymmetry at any scale.

4.1.1 The hierarchy problem

Supersymmetry close to the electroweak scale ameliorates the hierarchy problem that afflicts the Standard Model. In the Standard Model, the electroweak scale receives enormous Planck-scale quantum corrections. The observed hierarchy between the electroweak scale and the Planck scale must be achieved with extraordinary fine tuning. In a supersymmetric theory, on the other hand, Planck-scale quantum corrections cancel between partners and superpartners (owing to a minus sign associated with fermionic loops). The hierarchy between the electroweak scale and the Planck sale is achieved in a natural manner, without miraculous fine-tuning.

4.1.2 Gauge coupling unification

The idea that the gauge symmetry groups unify at high-energy is called Grand unification theory. In the Standard Model, however, the weak, strong and electromagnetic couplings fail to unify at high energy. In a supersymmetry theory, the running of the gauge couplings are modified, and precise high-energy unification of the gauge couplings is achieved. The modified running also provides a natural mechanism for radiative electroweak symmetry breaking.

4.1.3 Dark matter

TeV-scale supersymmetry (augmented with a discrete symmetry) typically provides a candidate dark matter particle at a mass scale consistent with thermal relic abundance calculations.[6][7]

4.1.4 Other technical motivations

Supersymmetry is also motivated by solutions to several theoretical problems, for generally providing many desirable mathematical properties, and for ensuring sensible behavior at high energies. Supersymmetric quantum field theory is often much easier to analyze, as many more problems become exactly solvable. When supersymmetry is imposed as a *local* symmetry, Einstein's theory of general relativity is included automatically, and the result is said to be a theory of supergravity. It is also a necessary feature of the most popular candidate for a theory of everything, superstring theory.

Another theoretically appealing property of supersymmetry is that it offers the only "loophole" to the Coleman–Mandula theorem, which prohibits spacetime and internal symmetries from being combined in any nontrivial way, for quantum field theories like the Standard Model with very general assumptions. The Haag-Lopuszanski-Sohnius theorem demonstrates that supersymmetry is the only way spacetime and internal symmetries can be combined consistently.[8]

4.2 History

A supersymmetry relating mesons and baryons was first proposed, in the context of hadronic physics, by Hironari Miyazawa during 1966. This supersymmetry did not involve spacetime, that is, it concerned internal symmetry, and was broken badly. Miyazawa's work was largely ignored at the time.[9][10][11][12]

J. L. Gervais and B. Sakita (during 1971),[13] Yu. A. Golfand and E. P. Likhtman (also during 1971), and D.V. Volkov and V.P. Akulov (1972),[14] independently rediscovered supersymmetry in the context of quantum field theory, a radically new type of symmetry of spacetime and fundamental fields, which establishes a relationship between elementary particles of different quantum nature, bosons and fermions, and unifies spacetime and internal symmetries of microscopic phenomena. Supersymmetry with a consistent Lie-algebraic graded structure on which the Gervais–Sakita rediscovery was based directly first arose during 1971[15] in the context of an early version of string theory by Pierre Ramond, John H. Schwarz and André Neveu.

Finally, Julius Wess and Bruno Zumino (during 1974)[16] identified the characteristic renormalization features of four-dimensional supersymmetric field theories, which identified them as remarkable QFTs, and they and Abdus Salam and their fellow researchers introduced early particle physics applications. The mathematical structure of supersymmetry (Graded Lie superalgebras) has subsequently been applied successfully to other topics of physics, ranging from nuclear physics,[17][18] critical phenomena,[19] quantum mechanics to statistical physics. It remains a vital part of many proposed theories of physics.

The first realistic supersymmetric version of the Standard Model was proposed during 1977 by Pierre Fayet and is known as the Minimal Supersymmetric Standard Model or MSSM for short. It was proposed to solve, amongst other things, the hierarchy problem.

4.3 Applications

4.3.1 Extension of possible symmetry groups

One reason that physicists explored supersymmetry is because it offers an extension to the more familiar symmetries of quantum field theory. These symmetries are grouped into the Poincaré group and internal symmetries and the Coleman–Mandula theorem showed that under certain assumptions, the symmetries of the S-matrix must be a direct product of the Poincaré group with a compact internal symmetry group or if there is not any mass gap, the conformal group with a compact internal symmetry group. During 1971 Golfand and Likhtman were the first to show that the Poincaré algebra can be extended through introduction of four anticommuting spinor generators (in four dimensions), which later became known as supercharges. During 1975 the Haag-Lopuszanski-Sohnius theorem analyzed all possible superalgebras in the general form, including those with an extended number of the supergenerators and central charges. This extended super-Poincaré algebra paved the way for obtaining a very large and important class of supersymmetric field theories.

The supersymmetry algebra

Main article: Supersymmetry algebra

Traditional symmetries of physics are generated by objects that transform by the tensor representations of the Poincaré group and internal symmetries. Supersymmetries, however, are generated by objects that transform by the spinor representations. According to the spin-statistics theorem, bosonic fields commute while fermionic fields anticommute. Combining the two kinds of fields into a single algebra requires the introduction of a $\mathbf{Z}_2$-grading under which the bosons are the even elements and the fermions are the odd elements. Such an algebra is called a Lie superalgebra.

The simplest supersymmetric extension of the Poincaré algebra is the Super-Poincaré algebra. Expressed in terms of two Weyl spinors, has the following anti-commutation relation:

$$\{Q_\alpha, \bar{Q\beta}\} = 2(\sigma^\mu)_{\alpha\dot{\beta}}P_\mu$$

and all other anti-commutation relations between the *Q*s and commutation relations between the *Q*s and *P*s vanish. In the above expression $P_\mu = -i\partial_\mu$ are the generators of translation and σ^μ are the Pauli matrices.

There are representations of a Lie superalgebra that are analogous to representations of a Lie algebra. Each Lie algebra has an associated Lie group and a Lie superalgebra can sometimes be extended into representations of a Lie supergroup.

4.3.2 The Supersymmetric Standard Model

Main article: Minimal Supersymmetric Standard Model

Incorporating supersymmetry into the Standard Model requires doubling the number of particles since there is no way that any of the particles in the Standard Model can be superpartners of each other. With the addition of new particles, there are many possible new interactions. The simplest possible supersymmetric model consistent with the Standard Model is the Minimal Supersymmetric Standard Model (MSSM) which can include the necessary additional new particles that are able to be superpartners of those in the Standard Model.

One of the main motivations for SUSY comes from the quadratically divergent contributions to the Higgs mass squared. The quantum mechanical interactions of the Higgs boson causes a large renormalization of the Higgs mass and unless there is an accidental cancellation, the natural size of the Higgs mass is the greatest scale possible. This problem is known as the hierarchy problem. Supersymmetry reduces the size of the quantum corrections by having automatic cancellations between fermionic and bosonic Higgs interactions. If supersymmetry is restored at the weak scale, then the Higgs mass is related to supersymmetry breaking which can be induced from small non-perturbative effects explaining the vastly different scales in the weak interactions and gravitational interactions.

In many supersymmetric Standard Models there is a heavy stable particle (such as neutralino) which could serve as a weakly interacting massive particle (WIMP) dark matter candidate. The existence of a supersymmetric dark matter candidate is related closely to R-parity.

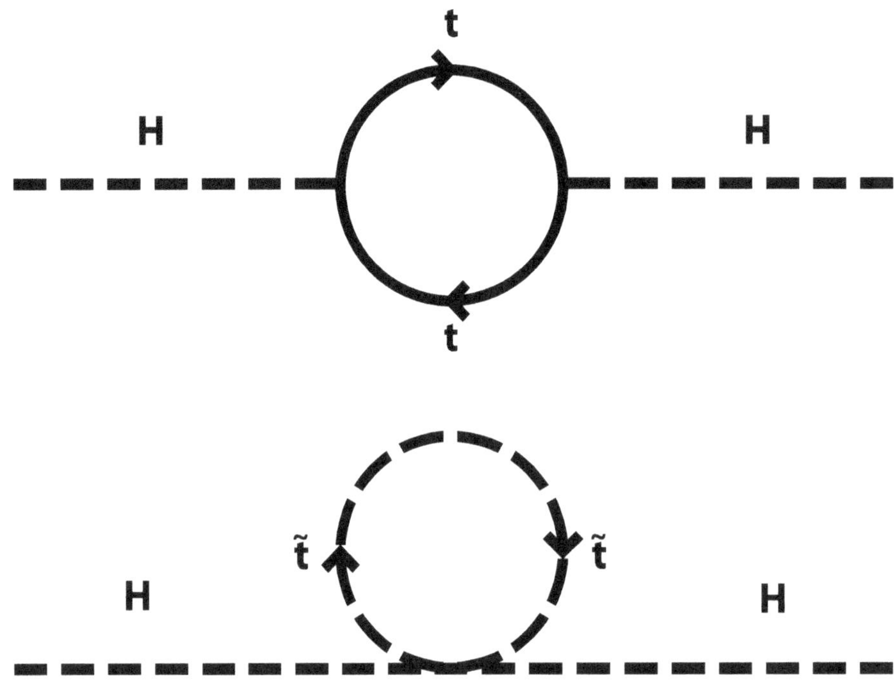

Cancellation of the Higgs boson quadratic mass renormalization between fermionic top quark loop and scalar stop squark tadpole Feynman diagrams in a supersymmetric extension of the Standard Model

The standard paradigm for incorporating supersymmetry into a realistic theory is to have the underlying dynamics of the theory be supersymmetric, but the ground state of the theory does not respect the symmetry and supersymmetry is broken spontaneously. The supersymmetry break can not be done permanently by the particles of the MSSM as they currently appear. This means that there is a new sector of the theory that is responsible for the breaking. The only constraint on this new sector is that it must break supersymmetry permanently and must give superparticles TeV scale masses. There are many models that can do this and most of their details do not matter. In order to parameterize the relevant features of supersymmetry breaking, arbitrary soft SUSY breaking terms are added to the theory which temporarily break SUSY explicitly but could never arise from a complete theory of supersymmetry breaking.

Gauge-coupling unification

Main article: Minimal Supersymmetric Standard Model § Gauge-coupling unification

One piece of evidence for supersymmetry existing is gauge coupling unification. The renormalization group evolution of the three gauge coupling constants of the Standard Model is somewhat sensitive to the present particle content of the theory. These coupling constants do not quite meet together at a common energy scale if we run the renormalization group using the Standard Model.[20] With the addition of minimal SUSY joint convergence of the coupling constants is projected at approximately 10^{16} GeV.[20]

4.3.3 Supersymmetric quantum mechanics

Main article: Supersymmetric quantum mechanics

Supersymmetric quantum mechanics adds the SUSY superalgebra to quantum mechanics as opposed to quantum field theory. Supersymmetric quantum mechanics often becomes relevant when studying the dynamics of supersymmetric solitons, and due to the simplified nature of having fields which are only functions of time (rather than space-time), a great deal of progress has been made in this subject and it is now studied in its own right.

SUSY quantum mechanics involves pairs of Hamiltonians which share a particular mathematical relationship, which are called *partner Hamiltonians*. (The potential energy terms which occur in the Hamiltonians are then known as *partner potentials*.) An introductory theorem shows that for every eigenstate of one Hamiltonian, its partner Hamiltonian has a corresponding eigenstate with the same energy. This fact can be exploited to deduce many properties of the eigenstate spectrum. It is analogous to the original description of SUSY, which referred to bosons and fermions. We can imagine a "bosonic Hamiltonian", whose eigenstates are the various bosons of our theory. The SUSY partner of this Hamiltonian would be "fermionic", and its eigenstates would be the theory's fermions. Each boson would have a fermionic partner of equal energy.

4.3.4 Supersymmetry: Applications to condensed matter physics

SUSY concepts have provided useful extensions to the WKB approximation. Additionally, SUSY has been applied to disorder averaged systems both quantum and non-quantum (through statistical mechanics), the Fokker-Planck equation being an example of a non-quantum theory. The `supersymmetry' in all these systems arises from the fact that one is modelling one particle and as such the`statistics' don't matter. The use of the supersymmetry method provides a mathematical rigorous alternative to the replica trick, but only in non-interacting systems, which attempts to address the so-called `problem of the denominator' under disorder averaging. For more on the applications of supersymmetry in condensed matter physics see the book[21]

4.3.5 Supersymmetry in optics

Integrated optics was recently found[22] to provide a fertile ground on which certain ramifications of SUSY can be explored in readily-accessible laboratory settings. Making use of the analogous mathematical structure of the quantum-mechanical Schrödinger equation and the wave equation governing the evolution of light in one-dimensional settings, one may interpret the refractive index distribution of a structure as a potential landscape in which optical wave packets propagate. In this manner, a new class of functional optical structures with possible applications in phase matching, mode conversion[23] and space-division multiplexing becomes possible. SUSY transformations have been also proposed as a way to address inverse scattering problems in optics and as a one-dimensional transformation optics [24]

4.3.6 Mathematics

SUSY is also sometimes studied mathematically for its intrinsic properties. This is because it describes complex fields satisfying a property known as holomorphy, which allows holomorphic quantities to be exactly computed. This makes supersymmetric models useful "toy models" of more realistic theories. A prime example of this has been the demonstration of S-duality in four-dimensional gauge theories[25] that interchanges particles and monopoles.

The proof of the Atiyah-Singer index theorem is much simplified by the use of supersymmetric quantum mechanics.

4.4 General supersymmetry

Supersymmetry appears in many related contexts of theoretical physics. It is possible to have multiple supersymmetries and also have supersymmetric extra dimensions.

4.4.1 Extended supersymmetry

Main article: Extended supersymmetry

It is possible to have more than one kind of supersymmetry transformation. Theories with more than one supersymmetry transformation are known as extended supersymmetric theories. The more supersymmetry a theory has, the more constrained are the field content and interactions. Typically the number of copies of a supersymmetry is a power of 2, i.e. 1, 2, 4, 8. In four dimensions, a spinor has four degrees of freedom and thus the minimal number of supersymmetry generators is four in four dimensions and having eight copies of supersymmetry means that there are 32 supersymmetry generators.

The maximal number of supersymmetry generators possible is 32. Theories with more than 32 supersymmetry generators automatically have massless fields with spin greater than 2. It is not known how to make massless fields with spin greater than two interact, so the maximal number of supersymmetry generators considered is 32. This corresponds to an $N = 8$ supersymmetry theory. Theories with 32 supersymmetries automatically have a graviton.

For four dimensions there are the following theories, with the corresponding multiplets[26](CPT adds a copy, whenever they are not invariant under such symmetry)

- $N = 1$

Chiral multiplet: $(0,\frac{1}{2})$ Vector multiplet: $(\frac{1}{2},1)$ Gravitino multiplet: $(1,\frac{3}{2})$ Graviton multiplet: $(\frac{3}{2},2)$

- $N = 2$

hypermultiplet: $(-\frac{1}{2},0^2,\frac{1}{2})$ vector multiplet: $(0,\frac{1}{2}^2,1)$ supergravity multiplet: $(1,\frac{3}{2}^2,2)$

- $N = 4$

Vector multiplet: $(-1,-\frac{1}{2}^4,0^6,\frac{1}{2}^4,1)$ Supergravity multiplet: $(0,\frac{1}{2}^4,1^6,\frac{3}{2}^4,2)$

- $N = 8$

Supergravity multiplet: $(-2,-\frac{3}{2}^8,-1^{28},-\frac{1}{2}^{56},0^{70},\frac{1}{2}^{56},1^{28},\frac{3}{2}^8,2)$

4.4.2 Supersymmetry in alternate numbers of dimensions

It is possible to have supersymmetry in dimensions other than four. Because the properties of spinors change drastically between different dimensions, each dimension has its characteristic. In d dimensions, the size of spinors is approximately $2^{d/2}$ or $2^{(d-1)/2}$. Since the maximum number of supersymmetries is 32, the greatest number of dimensions in which a supersymmetric theory can exist is eleven.

4.5 Supersymmetry as a quantum group

Main article: Supersymmetry as a quantum group

Supersymmetry can be reinterpreted in the language of noncommutative geometry and quantum groups. In particular, it involves a mild form of noncommutativity, namely supercommutativity. See the main article for more details.

4.6 Supersymmetry in quantum gravity

Supersymmetry is part of a larger enterprise of theoretical physics to unify everything we know about the universe into a single consistent set of physical principles, known as the quest for a Theory of Everything (TOE). A significant part of this larger enterprise is the quest for a theory of quantum gravity, which would unify the classical theory of general relativity and the Standard Model, which explains the other three basic forces in physics (electromagnetism, the strong interaction, and the weak interaction), and provides a palette of fundamental particles upon which all four forces act. Two of the most active methods of forming a theory of quantum gravity are string theory and loop quantum gravity (LQG), although in theory, supersymmetry could be a component of other theories as well.

For string theory to be consistent, supersymmetry seems to be required at some level (although it may be a strongly broken symmetry). In particle theory, supersymmetry is recognized as a way to stabilize the hierarchy between the unification scale and the electroweak scale (or the Higgs boson mass), and can also provide a natural dark matter candidate. String theory also requires extra spatial dimensions which have to be compactified as in Kaluza–Klein theory.

Loop quantum gravity (LQG) predicts no additional spatial dimensions, nor anything else about particle physics. These theories can be formulated in three spatial dimensions and one dimension of time, although in some LQG theories dimensionality is an emergent property of the theory, rather than a fundamental assumption of the theory. Also, LQG is a theory of quantum gravity which does not require supersymmetry. Lee Smolin, one of the originators of LQG, has proposed that a loop quantum gravity theory incorporating either supersymmetry or extra dimensions, or both, be called "loop quantum gravity II".

If experimental evidence confirms supersymmetry in the form of supersymmetric particles such as the neutralino that is often believed to be the lightest superpartner, some people believe this would be a major boost to string theory. Since supersymmetry is a required component of string theory, any discovered supersymmetry would be consistent with string theory. If the Large Hadron Collider and other major particle physics experiments fail to detect supersymmetric partners or evidence of extra dimensions, many versions of string theory which had predicted certain low mass superpartners to existing particles may need to be significantly revised. The failure of experiments to discover either supersymmetric partners or extra spatial dimensions, as of 2013, has encouraged loop quantum gravity researchers.

4.7 Current status

Supersymmetric models are constrained by a variety of experiments, including measurements of low-energy observables – for example, the anomalous magnetic moment of the muon at Brookhaven; the WMAP dark matter density measurement and direct detection experiments – for example, XENON−100 and LUX; and by particle collider experiments, including B-physics, Higgs phenomenology and direct searches for superpartners (sparticles), at the Large Electron–Positron Collider, Tevatron and the LHC.

Historically, the tightest limits were from direct production at colliders. The first mass limits for squarks and gluinos were made at CERN by the UA1 experiment and the UA2 experiment at the Super Proton Synchrotron. LEP later set very strong limits.[27], which in 2006 were extended by the D0 experiment at the Tevatron.[28][29] From 2003, WMAP's and Planck's dark matter density measurements have strongly constrained supersymmetry models, which, if they explain dark matter, have to be tuned to invoke a particular mechanism to sufficiently reduce the neutralino density.

Prior to the beginning of the LHC, in 2009 fits of available data to CMSSM and NUHM1 indicated that squarks and gluinos were most likely to have masses in the 500 to 800 GeV range, though values as high as 2.5 TeV were allowed with low probabilities. Neutralinos and sleptons were expected to be quite light, with the lightest neutralino and the lightest stau most likely to be found between 100 to 150 GeV.[30]

The first run of the LHC found no evidence for supersymmetry, and, as a result, surpassed existing experimental limits from the Large Electron–Positron Collider and Tevatron and partially excluded the aforementioned expected ranges.[31].

During 2011 and 2012, the LHC discovered a Higgs boson with a mass of about 125 GeV, and with couplings to fermions and bosons which are consistent with the Standard Model. The MSSM predicts that the mass of the lightest Higgs boson should not be much higher than the mass of the Z boson, and, in the absence of fine tuning (with the supersymmetry breaking scale on the order of 1 TeV), should not exceed 130 GeV. Furthermore, for values of the MSSM parameter $\tan\beta \leq 3$, it predicts a Higgs mass below 114 GeV over most of the parameter space.[32] This region of Higgs mass was excluded by LEP by 2000. The LHC result is somewhat problematic for the minimal supersymmetric model, as the value of 125 GeV is relatively large for the model and can only be achieved with large radiative loop corrections from top squarks, which many theorists consider to be "unnatural" (see naturalness and fine tuning).[33] On the other hand, the lightest Higgs boson in the MSSM is Standard Model-like, which is consistent with measurements of the Higgs boson couplings at the LHC.

In spite of the null searches and the heavy Higgs, a recent analysis of the constrained minimal supersymmetric Standard Model, the CMSSM, suggests that the model is still compatible with all present experimental constraints.[34][35] The preferred masses for squarks and gluinos is about 2 TeV. The resulting fine-tuning of the electroweak scale, however, is considered "unnatural" (see little hierarchy problem), and some theorists now favor extended supersymmetry models – for example, the NMSSM.

4.8 See also

- Supersymmetric gauge theory
- Wess–Zumino model
- Minimal Supersymmetric Standard Model
- Supersymmetry as a quantum group
- Quantum group
- Supercharge
- Superfield
- Supergeometry
- Supergravity
- Supergroup
- Superspace

4.9 References

[1] Haber, Howie. "SUPERSYMMETRY, PART I (THEORY)" (PDF). *Reviews, Tables and Plots*. Particle Data Group (PDG). Retrieved 8 July 2015.

[2] Martin, Stephen P. (1997). "A Supersymmetry Primer". arXiv:hep-ph/9709356.

[3] Dine, Michael (2007). *Supersymmetry and String Theory: Beyond the Standard Model*. p. 169.

[4] Ellis, John. "The Physics Landscape after the Higgs Discovery at the LHC". *arXiv*. Invited plenary talk at SILAFAE 2014. Retrieved 8 July 2015.

[5] Wolchover, Natalie (November 20, 2012). "Supersymmetry Fails Test, Forcing Physics to Seek New Ideas". *Quanta Magazine*.

[6] Jonathan Feng: Supersymmetric Dark Matter *(pdf)*, University of California, Irvine, 11 May 2007

[7] Torsten Bringmann: The WIMP "Miracle" *(pdf)* University of Hamburg

[8] R. Haag, J. T. Lopuszanski and M. Sohnius, "All Possible Generators Of Supersymmetries Of The S Matrix", Nucl. Phys. B 88 (1975) 257

[9] H. Miyazawa (1966). "Baryon Number Changing Currents". *Prog. Theor. Phys.* **36** (6): 1266–1276. Bibcode:1966PThPh..36.1266M. doi:10.1143/PTP.36.1266.

[10] H. Miyazawa (1968). "Spinor Currents and Symmetries of Baryons and Mesons". *Phys. Rev.* **170** (5): 1586–1590. Bibcode:1968PhRv..170.1586M. doi:10.1103/PhysRev.170.1586.

[11] Michio Kaku, *Quantum Field Theory*, ISBN 0-19-509158-2, pg 663.

[12] Peter Freund, *Introduction to Supersymmetry*, ISBN 0-521-35675-X, pages 26-27, 138.

[13] Gervais, J. -L.; Sakita, B. (1971). "Field theory interpretation of supergauges in dual models". *Nuclear Physics B* **34** (2): 632. Bibcode:1971NuPhB..34..632G. doi:10.1016/0550-3213(71)90351-8.

[14] D.V. Volkov, V.P. Akulov, Pisma Zh.Eksp.Teor.Fiz. 16 (1972) 621; Phys.Lett. B46 (1973) 109; V.P. Akulov, D.V. Volkov, Teor.Mat.Fiz. 18 (1974) 39

[15] Ramond, P. (1971). "Dual Theory for Free Fermions". *Physical Review D* **3** (10): 2415. Bibcode:1971PhRvD...3.2415R. doi:10.1103/PhysRevD.3.2415.

[16] Wess, J.; Zumino, B. (1974). "Supergauge transformations in four dimensions". *Nuclear Physics B* **70**: 39. Bibcode:1974NuPhB..70...39W. doi:10.1016/0550-3213(74)90355-1.

[17] http://users.physik.fu-berlin.de/~{ }kleinert/kleinert/?p=supersym suggested here

[18] Iachello, F. (1980). "Dynamical Supersymmetries in Nuclei". *Physical Review Letters* **44** (12): 772. Bibcode:1980PhRvL..44..772I. doi:10.1103/PhysRevLett.44.772.

[19] Friedan, D.; Qiu, Z.; Shenker, S. (1984). "Conformal Invariance, Unitarity, and Critical Exponents in Two Dimensions". *Physical Review Letters* **52** (18): 1575. Bibcode:1984PhRvL..52.1575F. doi:10.1103/PhysRevLett.52.1575.

[20] Gordon L. Kane, *The Dawn of Physics Beyond the Standard Model*, Scientific American, June 2003, page 60 and *The frontiers of physics*, special edition, Vol 15, #3, page 8

[21] *Supersymmetry in Disorder and Chaos*, Konstantin Efetov, Cambridge university press, 1997.

[22] Miri, M.-A.; Heinrich, M.; El-Ganainy, R.; Christodoulides, D. N. (2013). "Superymmetric optical structures". *Physical Review Letters* (APS) **110** (23): 233902. arXiv:1304.6646. Bibcode:2013PhRvL.110w3902M. doi:10.1103/PhysRevLett.110.233902. PMID 25167493. Retrieved April 2014.

[23] Heinrich, M.; Miri, M.-A.; Stützer, S.; El-Ganainy, R.; Nolte, S.; Szameit, A.; Christodoulides, D. N. (2014). "Superymmetric mode converters". *Nature Communications* (NPG) **5**: 3698. arXiv:1401.5734. Bibcode:2014NatCo...5E3698H. doi:10.1038/ncomms4698. PMID 24739256. Retrieved April 2014.

[24] Miri, M.-A.; Heinrich, Matthias; Christodoulides, D. N. (2014). "SUSY-inspired one-dimensional transformation optics". *Optica* (OSA) **1** (2): 89. arXiv:1408.0832. doi:10.1364/OPTICA.1.000089. Retrieved August 2014.

[25] Krasnitz, Michael (2002). *Correlation functions in supersymmetric gauge theories from supergravity fluctuafluctuations hHKtions* (PDF). Princeton University Department of Physics: Princeton University Department of Physics. p. 91.

[26] Polchinski,J. *String theory. Vol. 2: Superstring theory and beyond*, Appendix B

[27] LEPSUSYWG, ALEPH, DELPHI, L3 and OPAL experiments, charginos, large m0 LEPSUSYWG/01-03.1

[28] The D0-Collaboration (2009). "Search for associated production of charginos and neutralinos in the trilepton final state using 2.3 fb^{-1} of data". arXiv:0901.0646. Bibcode:2009PhLB..680...34D. doi:10.1016/j.physletb.2009.08.011.

[29] The D0 Collaboration (2006). "Search for squarks and gluinos in events with jets and missing transverse energy using 2.1 fb-1 of pp¯ collision data at s=1.96 TeV". arXiv:0712.3805. Bibcode:2008PhLB..660..449D. doi:10.1016/j.physletb.2008.01.042.

[30] O. Buchmueller et al. (2009). "Likelihood Functions for Supersymmetric Observables in Frequentist Analyses of the CMSSM and NUHM1". *The European Physical Journal C* **64** (3): 391–415. arXiv:0907.5568. Bibcode:2009EPJC...64..391B. doi:10.1140/epjc/s10052-009-1159-z.

[31] Roszkowski, Leszek; Sessolo, Enrico Maria; Williams, Andrew J. (11 August 2014). "What next for the CMSSM and the NUHM: improved prospects for superpartner and dark matter detection". *Journal of High Energy Physics* **2014** (8). doi:10.1007/JHEP08(2014)067.

[32] Marcela Carena and Howard E. Haber; Haber (1970). "Higgs Boson Theory and Phenomenology". *Progress in Particle and Nuclear Physics* **50**: 63. arXiv:hep-ph/0208209v3. Bibcode:2003PrPNP..50...63C. doi:10.1016/S0146-6410(02)00177-1.

[33] Patrick Draper et al. (December 2011). "Implications of a 125 GeV Higgs for the MSSM and Low-Scale SUSY Breaking". *Physical Review D* **85** (9): 095007. arXiv:1112.3068. Bibcode:2012PhRvD..85i5007D. doi:10.1103/PhysRevD.85.095007.

[34] Bechtle, Philip. "How alive is constrained SUSY really?". *arXiv*. Retrieved 8 July 2015.

[35] Jan de Vries, Kees. "SUSY fits with full LHC Run I data". *arXiv*. Retrieved 8 July 2015.

4.10 Further reading

- Supersymmetry and Supergravity page in String Theory Wiki lists more books and reviews.

4.10.1 Theoretical introductions, free and online

- S. Martin (2011). "A Supersymmetry Primer". arXiv:hep-ph/9709356.
- Joseph D. Lykken (1996). "Introduction to Supersymmetry". arXiv:hep-th/9612114.
- Manuel Drees (1996). "An Introduction to Supersymmetry". arXiv:hep-ph/9611409.
- Adel Bilal (2001). "Introduction to Supersymmetry". arXiv:hep-th/0101055.
- An Introduction to Global Supersymmetry by Philip Arygres, 2001

4.10.2 Monographs

- Weak Scale Supersymmetry by Howard Baer and Xerxes Tata, 2006.
- Cooper, F.; Khare, A.; Sukhatme, U. (1995). "Supersymmetry and quantum mechanics". *Physics Reports* **251** (5–6): 267. doi:10.1016/0370-1573(94)00080-M. (arXiv:hep-th/9405029).
- Junker, G. (1996). "Supersymmetric Methods in Quantum and Statistical Physics". doi:10.1007/978-3-642-61194-0. ISBN 978-3-540-61591-0..
- Gordon L. Kane.*Supersymmetry: Unveiling the Ultimate Laws of Nature* Basic Books, New York (2001). ISBN 0-7382-0489-7.
- Gordon L. Kane and Shifman, M., eds. *The Supersymmetric World: The Beginnings of the Theory*, World Scientific, Singapore (2000). ISBN 981-02-4522-X.
- Weinberg, Steven, *The Quantum Theory of Fields, Volume 3: Supersymmetry*, Cambridge University Press, Cambridge, (1999). ISBN 0-521-66000-9.
- Wess, Julius, and Jonathan Bagger, *Supersymmetry and Supergravity*, Princeton University Press, Princeton, (1992). ISBN 0-691-02530-4.
- "Concise Encyclopedia of Supersymmetry". 2003. doi:10.1007/1-4020-4522-0. ISBN 978-1-4020-1338-6.

4.10.3 On experiments

- Bennett GW; Muon (g–2) Collaboration; Bousquet; Brown; Bunce; Carey; Cushman; Danby; Debevec; Deile; Deng; Dhawan; Druzhinin; Duong; Farley; Fedotovich; Gray; Grigoriev; Grosse-Perdekamp; Grossmann; Hare; Hertzog; Huang; Hughes; Iwasaki; Jungmann; Kawall; Khazin; Krienen; Kronkvist et al. (2004). "Measurement of the negative muon anomalous magnetic moment to 0.7 ppm". *Physical Review Letters* **92** (16): 161802. arXiv:hep-ex/0401008. Bibcode:2004PhRvL..92p1802B. doi:10.1103/PhysRevLett.92.161802. PMID 15169217.
- Brookhaven National Laboratory (Jan. 8, 2004). *New g−2 measurement deviates further from Standard Model.* Press Release.
- Fermi National Accelerator Laboratory (Sept 25, 2006). *Fermilab's CDF scientists have discovered the quick-change behavior of the B-sub-s meson.* Press Release.

4.11 External links

- Supersymmetry (physics) at *Encyclopædia Britannica*
- What do current LHC results (mid-August 2011) imply about supersymmetry? Matt Strassler
- ATLAS Experiment Supersymmetry search documents
- CMS Experiment Supersymmetry search documents
- "Particle wobble shakes up supersymmetry", *Cosmos* magazine, September 2006

- LHC results put supersymmetry theory 'on the spot' BBC news 27/8/2011
- SUSY running out of hiding places BBC news 12/11/2012
- Supersymmetry in optics? “Skulls in the Stars” blog 22/08/2013

Chapter 5

Anthropic principle

In astrophysics and cosmology, the **anthropic principle** (from Greek *anthropos*, meaning "human") is the philosophical consideration that observations of the physical Universe must be compatible with the conscious and sapient life that observes it. Some proponents of the anthropic principle reason that it explains why the universe has the age and the fundamental physical constants necessary to accommodate conscious life. As a result, they believe it is unremarkable that the universe's fundamental constants happen to fall within the narrow range thought to be compatible with life.[1][2]

The strong anthropic principle (SAP) as explained by John D. Barrow and Frank Tipler (see variants) states that this is all the case because the universe is compelled, in some sense, to eventually have conscious and sapient life emerge within it. Some critics of the SAP argue in favor of a weak anthropic principle (WAP) similar to the one defined by Brandon Carter, which states that the universe's ostensible fine tuning is the result of selection bias: i.e., only in a universe capable of eventually supporting life will there be living beings capable of observing and reflecting upon any such fine tuning, while a universe less compatible with life will go unbeheld. Most often such arguments draw upon some notion of the multiverse for there to be a statistical population of universes to select from and from which selection bias (our observance of *only* this universe, apparently compatible with life) could occur.

5.1 Definition and basis

The principle was formulated as a response to a series of observations that the laws of nature and parameters of the universe take on values that are consistent with conditions for life as we know it rather than a set of values that would not be consistent with life on Earth. The anthropic principle states that this is a necessity, because if life were impossible, no living entity would be there to observe it, and thus would not be known. That is, it must be possible to observe *some* universe, and hence, the laws and constants of any such universe must accommodate that possibility.

The term *anthropic* in "anthropic principle" has been argued[3] to be a misnomer.[4] While singling out our kind of carbon-based life, none of the finely tuned phenomena require human life or some kind of carbon chauvinism.[5][6] Any form of life or any form of heavy atom, stone, star or galaxy would do; nothing specifically human or anthropic is involved.

The anthropic principle has given rise to some confusion and controversy, partly because the phrase has been applied to several distinct ideas. All versions of the principle have been accused of discouraging the search for a deeper physical understanding of the universe. The anthropic principle is often criticized for lacking falsifiability and therefore critics of the anthropic principle may point out that the anthropic principle is a non-scientific concept, even though the weak anthropic principle, *"conditions that are observed in the universe must allow the observer to exist"*,[7] is "easy" to support in mathematics and philosophy, i.e. it is a tautology or truism. However, building a substantive argument based on a tautological foundation is problematic. Stronger variants of the anthropic principle are not tautologies and thus make claims considered controversial by some and that are contingent upon empirical verification.[8][9]

5.2 Anthropic coincidences

Main article: Fine-tuned Universe

In 1961, Robert Dicke noted that the age of the universe, as seen by living observers, cannot be random.[10] Instead, biological factors constrain the universe to be more or less in a "golden age," neither too young nor too old.[11] If the universe were one tenth as old as its present age, there would not have been sufficient time to build up appreciable levels of metallicity (levels of elements besides hydrogen and helium) especially carbon, by nucleosynthesis. Small rocky planets did not yet exist. If the universe were 10 times older than it actually is, most stars would be too old to remain on the main sequence and would have turned into white dwarfs, aside from the dimmest red dwarfs, and stable planetary systems would have already come to an end. Thus, Dicke explained the coincidence between large dimensionless numbers constructed from the constants of physics and the age of the universe, a coincidence which had inspired Dirac's varying-G theory.

Dicke later reasoned that the density of matter in the universe must be almost exactly the critical density needed to prevent the Big Crunch (the "Dicke coincidences" argument). The most recent measurements may suggest that the observed density of baryonic matter, and some theoretical predictions of the amount of dark matter account for about 30% of this critical density, with the rest contributed by a cosmological constant. Steven Weinberg[12] gave an anthropic explanation for this fact: he noted that the cosmological constant has a remarkably low value, some 120 orders of magnitude smaller than the value particle physics predicts (this has been described as the "worst prediction in physics").[13] However, if the cosmological constant were only one order of magnitude larger than its observed value, the universe would suffer catastrophic inflation, which would preclude the formation of stars, and hence life.

The observed values of the dimensionless physical constants (such as the fine-structure constant) governing the four fundamental interactions are balanced as if fine-tuned to permit the formation of commonly found matter and subsequently the emergence of life.[14] A slight increase in the strong interaction would bind the dineutron and the diproton, and nuclear fusion would have converted all hydrogen in the early universe to helium. Water, as well as sufficiently long-lived stable stars, both essential for the emergence of life as we know it, would not exist. More generally, small changes in the relative strengths of the four fundamental interactions can greatly affect the universe's age, structure, and capacity for life.

5.3 Origin

The phrase "anthropic principle" first appeared in Brandon Carter's contribution to a 1973 Kraków symposium honouring Copernicus's 500th birthday. Carter, a theoretical astrophysicist, articulated the Anthropic Principle in reaction to the Copernican Principle, which states that humans do not occupy a privileged position in the Universe. As Carter said: "Although our situation is not necessarily *central*, it is inevitably privileged to some extent."[15] Specifically, Carter disagreed with using the Copernican principle to justify the Perfect Cosmological Principle, which states that all large regions *and times* in the universe must be statistically identical. The latter principle underlay the steady-state theory, which had recently been falsified by the 1965 discovery of the cosmic microwave background radiation. This discovery was unequivocal evidence that the universe has changed radically over time (for example, via the Big Bang).

Carter defined two forms of the anthropic principle, a "weak" one which referred only to anthropic selection of privileged spacetime locations in the universe, and a more controversial "strong" form which addressed the values of the fundamental constants of physics.

Roger Penrose explained the weak form as follows:

> The argument can be used to explain why the conditions happen to be just right for the existence of (intelligent) life on the Earth at the present time. For if they were not just right, then we should not have found ourselves to be here now, but somewhere else, at some other appropriate time. This principle was used very effectively by Brandon Carter and Robert Dicke to resolve an issue that had puzzled physicists for a good many years. The issue concerned various striking numerical relations that are observed to hold between the physical constants (the gravitational constant, the mass of the proton, the age of the universe, etc.). A puzzling aspect of this was that some of the relations hold only at the present epoch in the Earth's history, so we appear, coincidentally, to be living at a very special time (give or take a few million years!). This was later explained, by Carter and Dicke, by the fact that this epoch coincided

> with the lifetime of what are called main-sequence stars, such as the Sun. At any other epoch, so the argument ran, there would be no intelligent life around in order to measure the physical constants in question — so the coincidence had to hold, simply because there would be intelligent life around only at the particular time that the coincidence did hold!
> —*The Emperor's New Mind*, Chapter 10

One reason this is plausible is that there are many other places and times in which we can imagine finding ourselves. But when applying the strong principle, we only have one universe, with one set of fundamental parameters, so what exactly is the point being made? Carter offers two possibilities: First, we can use our own existence to make "predictions" about the parameters. But second, "as a last resort", we can convert these predictions into *explanations* by assuming that there *is* more than one universe, in fact a large and possibly infinite collection of universes, something that is now called the multiverse ("world ensemble" was Carter's term), in which the parameters (and perhaps the laws of physics) vary across universes. The strong principle then becomes an example of a selection effect, exactly analogous to the weak principle. Postulating a multiverse is certainly a radical step, but taking it could provide at least a partial answer to a question which had seemed to be out of the reach of normal science: "why do the fundamental laws of physics take the particular form we observe and not another?"

Since Carter's 1973 paper, the term "anthropic principle" has been extended to cover a number of ideas which differ in important ways from those he espoused. Particular confusion was caused in 1986 by the book *The Anthropic Cosmological Principle* by John D. Barrow and Frank Tipler,[16] published that year which distinguished between "weak" and "strong" anthropic principle in a way very different from Carter's, as discussed in the next section.

Carter was not the first to invoke some form of the anthropic principle. In fact, the evolutionary biologist Alfred Russel Wallace anticipated the anthropic principle as long ago as 1904: "Such a vast and complex universe as that which we know exists around us, may have been absolutely required ... in order to produce a world that should be precisely adapted in every detail for the orderly development of life culminating in man."[17] In 1957, Robert Dicke wrote: "The age of the Universe 'now' is not random but conditioned by biological factors ... [changes in the values of the fundamental constants of physics] would preclude the existence of man to consider the problem."[18]

5.4 Variants

Weak anthropic principle (WAP) (Carter): "we must be prepared to take account of the fact that our location in the universe is *necessarily* privileged to the extent of being compatible with our existence as observers." Note that for Carter, "location" refers to our location in time as well as space.

Strong anthropic principle (SAP) (Carter): "the universe (and hence the fundamental parameters on which it depends) must be such as to admit the creation of observers within it at some stage. To paraphrase Descartes, *cogito ergo mundus talis est*."
The Latin tag ("I think, therefore the world is such [as it is]") makes it clear that "must" indicates a deduction from the fact of our existence; the statement is thus a truism.

In their 1986 book, *The Anthropic Cosmological Principle*, John Barrow and Frank Tipler depart from Carter and define the WAP and SAP as follows:[19][20]

Weak anthropic principle (WAP) (Barrow and Tipler): "The observed values of all physical and cosmological quantities are not equally probable but they take on values restricted by the requirement that there exist sites where carbon-based life can evolve and by the requirements that the universe be old enough for it to have already done so."[21]
Unlike Carter they restrict the principle to carbon-based life, rather than just "observers." A more important difference is that they apply the WAP to the fundamental physical constants, such as the fine structure constant, the number of spacetime dimensions, and the cosmological constant — topics that fall under Carter's SAP.

Strong anthropic principle (SAP) (Barrow and Tipler): "The Universe must have those properties which allow life to develop within it at some stage in its history."[22]
This looks very similar to Carter's SAP, but unlike the case with Carter's SAP, the "must" is an imperative, as shown by the following three possible elaborations of the SAP, each proposed by Barrow and Tipler:[23]

- "There exists one possible Universe 'designed' with the goal of generating and sustaining 'observers'."

This can be seen as simply the classic design argument restated in the garb of contemporary cosmology. It implies that the purpose of the universe is to give rise to intelligent life, with the laws of nature and their fundamental physical constants set to ensure that life as we know it will emerge and evolve.

- "Observers are necessary to bring the Universe into being."

  Barrow and Tipler believe that this is a valid conclusion from quantum mechanics, as John Archibald Wheeler has suggested, especially via his idea that information is the fundamental reality, see It from bit, and his **Participatory Anthropic Principle (PAP)** which is an interpretation of quantum mechanics associated with the ideas of John von Neumann and Eugene Wigner.

- "An ensemble of other different universes is necessary for the existence of our Universe."

  By contrast, Carter merely says that an ensemble of universes is necessary for the SAP to count as an explanation.

Modified anthropic principle (MAP) (Schmidhuber): The 'problem' of existence is only relevant to a species capable of formulating the question. Prior to *Homo sapiens* intellectual evolution to the point where the nature of the observed universe - and humans' place within same - spawned deep inquiry into its origins, the 'problem' simply did not exist.[24]

The philosophers John Leslie[25] and Nick Bostrom[26] reject the Barrow and Tipler SAP as a fundamental misreading of Carter. For Bostrom, Carter's anthropic principle just warns us to make allowance for **anthropic bias**, that is, the bias created by anthropic selection effects (which Bostrom calls "observation" selection effects) — the necessity for observers to exist in order to get a result. He writes:

> Many 'anthropic principles' are simply confused. Some, especially those drawing inspiration from Brandon Carter's seminal papers, are sound, but... they are too weak to do any real scientific work. In particular, I argue that existing methodology does not permit any observational consequences to be derived from contemporary cosmological theories, though these theories quite plainly can be and are being tested empirically by astronomers. What is needed to bridge this methodological gap is a more adequate formulation of how observation selection effects are to be taken into account.
> —*Anthropic Bias*, Introduction[27]

Strong self-sampling assumption (SSSA) (Bostrom): "Each observer-moment should reason as if it were randomly selected from the class of all observer-moments in its reference class."
Analysing an observer's experience into a sequence of "observer-moments" helps avoid certain paradoxes; but the main ambiguity is the selection of the appropriate "reference class": for Carter's WAP this might correspond to all real or potential observer-moments in our universe; for the SAP, to all in the multiverse. Bostrom's mathematical development shows that choosing either too broad or too narrow a reference class leads to counter-intuitive results, but he is not able to prescribe an ideal choice.

According to Jürgen Schmidhuber, the anthropic principle essentially just says that the conditional probability of finding yourself in a universe compatible with your existence is always 1. It does not allow for any additional nontrivial predictions such as "gravity won't change tomorrow." To gain more predictive power, additional assumptions on the prior distribution of alternative universes are necessary.[24][28]

Playwright and novelist Michael Frayn describes a form of the Strong Anthropic Principle in his 2006 book *The Human Touch*, which explores what he characterises as "the central oddity of the Universe":

> It's this simple paradox. The Universe is very old and very large. Humankind, by comparison, is only a tiny disturbance in one small corner of it - and a very recent one. Yet the Universe is only very large and very old because we are here to say it is... And yet, of course, we all know perfectly well that it is what it is whether we are here or not.[29]

5.5 Character of anthropic reasoning

Carter chose to focus on a tautological aspect of his ideas, which has resulted in much confusion. In fact, anthropic reasoning interests scientists because of something that is only implicit in the above formal definitions, namely that we should give serious consideration to there being other universes with different values of the "fundamental parameters" — that is, the dimensionless physical constants and initial conditions for the Big Bang. Carter and others have argued that life as we know it would not be possible in most such universes. In other words, the universe we are in is fine tuned to permit life. Collins & Hawking (1973) characterized Carter's then-unpublished big idea as the postulate that "there is not one universe but a whole infinite ensemble of universes with all possible initial conditions".[30] If this is granted, the anthropic principle provides a plausible explanation for the fine tuning of our universe: the "typical" universe is not fine-tuned, but given enough universes, a small fraction thereof will be capable of supporting intelligent life. Ours must be one of these, and so the observed fine tuning should be no cause for wonder.

Although philosophers have discussed related concepts for centuries, in the early 1970s the only genuine physical theory yielding a multiverse of sorts was the many-worlds interpretation of quantum mechanics. This would allow variation in initial conditions, but not in the truly fundamental constants. Since that time a number of mechanisms for producing a multiverse have been suggested: see the review by Max Tegmark.[31] An important development in the 1980s was the combination of inflation theory with the hypothesis that some parameters are determined by symmetry breaking in the early universe, which allows parameters previously thought of as "fundamental constants" to vary over very large distances, thus eroding the distinction between Carter's weak and strong principles. At the beginning of the 21st century, the string landscape emerged as a mechanism for varying essentially all the constants, including the number of spatial dimensions.[32]

The anthropic idea that fundamental parameters are selected from a multitude of different possibilities (each actual in some universe or other) contrasts with the traditional hope of physicists for a theory of everything having no free parameters: as Einstein said, "What really interests me is whether God had any choice in the creation of the world." In 2002, proponents of the leading candidate for a "theory of everything", string theory, proclaimed "the end of the anthropic principle"[33] since there would be no free parameters to select. Ironically, string theory now seems to offer no hope of predicting fundamental parameters, and now some who advocate it invoke the anthropic principle as well (see below).

The modern form of a design argument is put forth by Intelligent design. Proponents of intelligent design often cite the fine-tuning observations that (in part) preceded the formulation of the anthropic principle by Carter as a proof of an intelligent designer. Opponents of intelligent design are not limited to those who hypothesize that other universes exist; they may also argue, anti-anthropically, that the universe is less fine-tuned than often claimed, or that accepting fine tuning as a brute fact is less astonishing than the idea of an intelligent creator. Furthermore, even accepting fine tuning, Sober (2005)[34] and Ikeda and Jefferys,[35][36] argue that the Anthropic Principle as conventionally stated actually undermines intelligent design; see fine-tuned universe.

Paul Davies's book *The Goldilocks Enigma* (2006) reviews the current state of the fine tuning debate in detail, and concludes by enumerating the following responses to that debate:

1. The absurd universe: Our universe just happens to be the way it is.

2. The unique universe: There is a deep underlying unity in physics which necessitates the Universe being the way it is. Some Theory of Everything will explain why the various features of the Universe must have exactly the values that we see.

3. The multiverse: Multiple universes exist, having all possible combinations of characteristics, and we inevitably find ourselves within a universe that allows us to exist.

4. Intelligent Design: A creator designed the Universe with the purpose of supporting complexity and the emergence of intelligence.

5. The life principle: There is an underlying principle that constrains the Universe to evolve towards life and mind.

6. The self-explaining universe: A closed explanatory or causal loop: "perhaps only universes with a capacity for consciousness can exist." This is Wheeler's Participatory Anthropic Principle (PAP).

7. The fake universe: We live inside a virtual reality simulation.

Omitted here is Lee Smolin's model of cosmological natural selection, also known as "fecund universes," which proposes that universes have "offspring" which are more plentiful if they resemble our universe. Also see Gardner (2005).[37]

Clearly each of these hypotheses resolve some aspects of the puzzle, while leaving others unanswered. Followers of Carter would admit only option 3 as an anthropic explanation, whereas 3 through 6 are covered by different versions of Barrow and Tipler's SAP (which would also include 7 if it is considered a variant of 4, as in Tipler 1994).

The anthropic principle, at least as Carter conceived it, can be applied on scales much smaller than the whole universe. For example, Carter (1983)[38] inverted the usual line of reasoning and pointed out that when interpreting the evolutionary record, one must take into account cosmological and astrophysical considerations. With this in mind, Carter concluded that given the best estimates of the age of the universe, the evolutionary chain culminating in *Homo sapiens* probably admits only one or two low probability links. Antonio Feoli and Salvatore Rampone dispute this conclusion, arguing instead that the estimated size of our universe and the number of planets in it allows for a higher bound, so that there is no need to invoke intelligent design to explain evolution.[39]

5.6 Observational evidence

No possible observational evidence bears on Carter's WAP, as it is merely advice to the scientist and asserts nothing debatable. The obvious test of Barrow's SAP, which says that the universe is "required" to support life, is to find evidence of life in universes other than ours. Any other universe is, by most definitions, unobservable (otherwise it would be included in *our* portion of *this* universe). Thus, in principle Barrow's SAP cannot be falsified by observing a universe in which an observer cannot exist.

Philosopher John Leslie[40] states that the Carter SAP (with multiverse) predicts the following:

- Physical theory will evolve so as to strengthen the hypothesis that early phase transitions occur probabilistically rather than deterministically, in which case there will be no deep physical reason for the values of fundamental constants;
- Various theories for generating multiple universes will prove robust;
- Evidence that the universe is fine tuned will continue to accumulate;
- No life with a non-carbon chemistry will be discovered;
- Mathematical studies of galaxy formation will confirm that it is sensitive to the rate of expansion of the universe.

Hogan[41] has emphasised that it would be very strange if all fundamental constants were strictly determined, since this would leave us with no ready explanation for apparent fine tuning. In fact we might have to resort to something akin to Barrow and Tipler's SAP: there would be no option for such a universe *not* to support life.

Probabilistic predictions of parameter values can be made given:

1. a particular multiverse with a "measure", i.e. a well defined "density of universes" (so, for parameter X, one can calculate the prior probability $P(X_0)\,dX$ that X is in the range $X_0 < X < X_0 + dX$), and
2. an estimate of the number of observers in each universe, $N(X)$ (e.g., this might be taken as proportional to the number of stars in the universe).

The probability of observing value X is then proportional to $N(X)\,P(X)$. (A more sophisticated analysis is that of Nick Bostrom.)[42] A generic feature of an analysis of this nature is that the expected values of the fundamental physical constants should not be "over-tuned," i.e. if there is some perfectly tuned predicted value (e.g. zero), the observed value need be no closer to that predicted value than what is required to make life possible. The small but finite value of the cosmological constant can be regarded as a successful prediction in this sense.

One thing that would *not* count as evidence for the Anthropic Principle is evidence that the Earth or the solar system occupied a privileged position in the universe, in violation of the Copernican principle (for possible counterevidence to this principle, see Copernican principle), unless there was some reason to think that that position was a necessary condition for our existence as observers.

5.7 Applications of the principle

5.7.1 The nucleosynthesis of carbon-12

Fred Hoyle may have invoked anthropic reasoning to predict an astrophysical phenomenon. He is said to have reasoned from the prevalence on Earth of life forms whose chemistry was based on carbon-12 atoms, that there must be an undiscovered resonance in the carbon-12 nucleus facilitating its synthesis in stellar interiors via the triple-alpha process. He then calculated the energy of this undiscovered resonance to be 7.6 million electron-volts.[43][44] Willie Fowler's research group soon found this resonance, and its measured energy was close to Hoyle's prediction.

However, a recently released paper argues that Hoyle did not use anthropic reasoning to make this prediction.[45]

5.7.2 Cosmic inflation

Main article: Cosmic inflation

Don Page criticized the entire theory of cosmic inflation as follows.[46] He emphasized that initial conditions which made possible a thermodynamic arrow of time in a universe with a Big Bang origin, must include the assumption that at the initial singularity, the entropy of the universe was low and therefore extremely improbable. Paul Davies rebutted this criticism by invoking an inflationary version of the anthropic principle.[47] While Davies accepted the premise that the initial state of the visible universe (which filled a microscopic amount of space before inflating) had to possess a very low entropy value — due to random quantum fluctuations — to account for the observed thermodynamic arrow of time, he deemed this fact an advantage for the theory. That the tiny patch of space from which our observable universe grew had to be extremely orderly, to allow the post-inflation universe to have an arrow of time, makes it unnecessary to adopt any “ad hoc” hypotheses about the initial entropy state, hypotheses other Big Bang theories require.

5.7.3 String theory

Main article: String theory landscape

String theory predicts a large number of possible universes, called the “backgrounds” or “vacua.” The set of these vacua is often called the "multiverse" or "anthropic landscape" or “string landscape.” Leonard Susskind has argued that the existence of a large number of vacua puts anthropic reasoning on firm ground: only universes whose properties are such as to allow observers to exist are observed, while a possibly much larger set of universes lacking such properties go unnoticed.

Steven Weinberg[48] believes the Anthropic Principle may be appropriated by cosmologists committed to nontheism, and refers to that Principle as a “turning point” in modern science because applying it to the string landscape "...may explain how the constants of nature that we observe can take values suitable for life without being fine-tuned by a benevolent creator.” Others, most notably David Gross but also Lubos Motl, Peter Woit, and Lee Smolin, argue that this is not predictive. Max Tegmark,[49] Mario Livio, and Martin Rees[50] argue that only some aspects of a physical theory need be observable and/or testable for the theory to be accepted, and that many well-accepted theories are far from completely testable at present.

Jürgen Schmidhuber (2000–2002) points out that Ray Solomonoff's theory of universal inductive inference and its extensions already provide a framework for maximizing our confidence in any theory, given a limited sequence of physical observations, and some prior distribution on the set of possible explanations of the universe.

5.7.4 Spacetime

Main article: Spacetime

In 1920, Paul Ehrenfest showed that if there is only one time dimension and greater than three spatial dimensions, the orbit of a planet about its Sun cannot remain stable. The same is true of a star’s orbit around the center of its galaxy.[51] Ehrenfest also showed that if there are an even number of spatial dimensions, then the different parts of

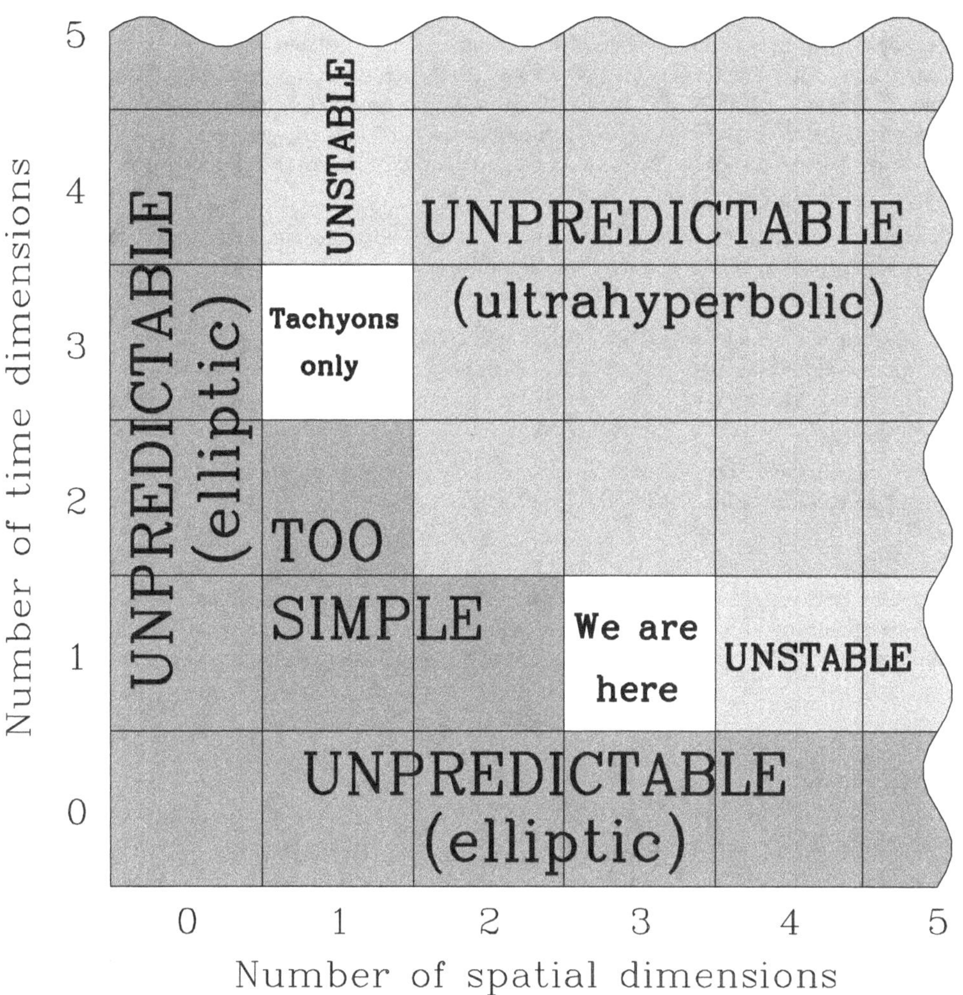

Properties of n+m-*dimensional spacetimes*

a wave impulse will travel at different speeds. If there are $5 + 2k$ spatial dimensions, where k is a whole number, then wave impulses become distorted. In 1922, Hermann Weyl showed that Maxwell's theory of electromagnetism works only with three dimensions of space and one of time.[52] Finally, Tangherlini showed in 1963 that when there are more than three spatial dimensions, electron orbitals around nuclei cannot be stable; electrons would either fall into the nucleus or disperse.[53]

Max Tegmark expands on the preceding argument in the following anthropic manner.[54] If T differs from 1, the behavior of physical systems could not be predicted reliably from knowledge of the relevant partial differential equations. In such a universe, intelligent life capable of manipulating technology could not emerge. Moreover, if $T > 1$, Tegmark maintains that protons and electrons would be unstable and could decay into particles having greater mass than themselves. (This is not a problem if the particles have a sufficiently low temperature).

5.8 *The Anthropic Cosmological Principle*

A thorough extant study of the anthropic principle is the book *The Anthropic Cosmological Principle* by John D. Barrow, a cosmologist, and Frank J. Tipler, a cosmologist and mathematical physicist. This book sets out in detail the many known anthropic coincidences and constraints, including many found by its authors. While the book is primarily a work of theoretical astrophysics, it also touches on quantum physics, chemistry, and earth science. An entire chapter argues that *Homo sapiens* is, with high probability, the only intelligent species in the Milky Way.

The book begins with an extensive review of many topics in the history of ideas the authors deem relevant to the anthropic principle, because the authors believe that principle has important antecedents in the notions of teleology and intelligent design. They discuss the writings of Fichte, Hegel, Bergson, and Alfred North Whitehead, and the Omega Point cosmology of Teilhard de Chardin. Barrow and Tipler carefully distinguish teleological reasoning from *eutaxiological* reasoning; the former asserts that order must have a consequent purpose; the latter asserts more modestly that order must have a planned cause. They attribute this important but nearly always overlooked distinction to an obscure 1883 book by L. E. Hicks.[55]

Seeing little sense in a principle requiring intelligent life to emerge while remaining indifferent to the possibility of its eventual extinction, Barrow and Tipler propose the final anthropic principle (FAP): Intelligent information-processing must come into existence in the universe, and, once it comes into existence, it will never die out.[56]

Barrow and Tipler submit that the FAP is both a valid physical statement and "closely connected with moral values." FAP places strong constraints on the structure of the universe, constraints developed further in Tipler's *The Physics of Immortality*.[57] One such constraint is that the universe must end in a big crunch, which seems unlikely in view of the tentative conclusions drawn since 1998 about dark energy, based on observations of very distant supernovas.

In his review[58] of Barrow and Tipler, Martin Gardner ridiculed the FAP by quoting the last two sentences of their book as defining a Completely Ridiculous Anthropic Principle (CRAP):

> At the instant the Omega Point is reached, life will have gained control of *all* matter and forces not only in a single universe, but in all universes whose existence is logically possible; life will have spread into *all* spatial regions in all universes which could logically exist, and will have stored an infinite amount of information, including *all* bits of knowledge which it is logically possible to know. And this is the end.[59]

5.9 Criticisms

Carter has frequently regretted his own choice of the word "anthropic," because it conveys the misleading impression that the principle involves humans specifically, rather than intelligent observers in general.[60] Others[61] have criticised the word "principle" as being too grandiose to describe straightforward applications of selection effects.

A common criticism of Carter's SAP is that it is an easy deus ex machina which discourages searches for physical explanations. To quote Penrose again: "it tends to be invoked by theorists whenever they do not have a good enough theory to explain the observed facts."[62]

Carter's SAP and Barrow and Tipler's WAP have been dismissed as truisms or trivial tautologies, that is, statements true solely by virtue of their logical form (the conclusion is identical to the premise) and not because a substantive claim is made and supported by observation of reality. As such, they are criticized as an elaborate way of saying "if things were different, they would be different," which is a valid statement, but does not make a claim of some factual alternative over another.

Critics of the Barrow and Tipler SAP claim that it is neither testable nor falsifiable, and thus is not a scientific statement but rather a philosophical one. The same criticism has been leveled against the hypothesis of a multiverse, although some argue that it does make falsifiable predictions. A modified version of this criticism is that we understand so little about the emergence of life, especially intelligent life, that it is effectively impossible to calculate the number of observers in each universe. Also, the prior distribution of universes as a function of the fundamental constants is easily modified to get any desired result.[63]

Many criticisms focus on versions of the strong anthropic principle, such as Barrow and Tipler's *anthropic cosmological principle*, which are teleological notions that tend to describe the existence of life as a *necessary prerequisite* for the observable constants of physics. Similarly, Stephen Jay Gould,[64][65] Michael Shermer,[66] and others claim that the stronger versions of the anthropic principle seem to reverse known causes and effects. Gould compared the claim that the universe is fine-tuned for the benefit of our kind of life to saying that sausages were made long and narrow so that they could fit into modern hotdog buns, or saying that ships had been invented to house barnacles. These critics cite the vast physical, fossil, genetic, and other biological evidence consistent with life having been fine-tuned through natural selection to adapt to the physical and geophysical environment in which life exists. Life appears to have adapted to the universe, and not vice versa.

Some applications of the anthropic principle have been criticized as an argument by lack of imagination, for tacitly assuming that carbon compounds and water are the only possible chemistry of life (sometimes called "carbon chauvinism", see also alternative biochemistry).[67] The range of fundamental physical constants consistent with the evolution of carbon-based life may also be wider than those who advocate a fine tuned universe have argued.[68] For instance, Harnik et al.[69] propose a weakless universe in which the weak nuclear force is eliminated. They show that this has no significant effect on the other fundamental interactions, provided some adjustments are made in how those interactions work. However, if some of the fine-tuned details of our universe were violated, that would rule out complex structures of any kind — stars, planets, galaxies, etc.

Lee Smolin has offered a theory designed to improve on the lack of imagination that anthropic principles have been accused of. He puts forth his fecund universes theory, which assumes universes have “offspring” through the creation of black holes whose offspring universes have values of physical constants that depend on those of the mother universe.[70]

Some versions of the anthropic principle are only interesting if the range of physical constants that allow certain kinds of life are unlikely in a landscape of possible universes. But Lee Smolin assumes that conditions for carbon based life are similar to conditions for black hole creation, which would change the a priori distribution of universes such that universes containing life would be likely. In Smolin vs. Susskind: The Anthropic Principle[71] the string theorist Leonard Susskind disagrees about some assumptions in Lee Smolin's theory, while Smolin defends his theory.

The philosophers of cosmology John Earman,[72] Ernan McMullin,[73] and Jesús Mosterín contend that “in its weak version, the anthropic principle is a mere tautology, which does not allow us to explain anything or to predict anything that we did not already know. In its strong version, it is a gratuitous speculation”.[74] A further criticism by Mosterín concerns the flawed “anthropic” inference from the assumption of an infinity of worlds to the existence of one like ours:

> The suggestion that an infinity of objects characterized by certain numbers or properties implies the existence among them of objects with any combination of those numbers or characteristics [...] is mistaken. An infinity does not imply at all that any arrangement is present or repeated. [...] The assumption that all possible worlds are realized in an infinite universe is equivalent to the assertion that any infinite set of numbers contains all numbers (or at least all Gödel numbers of the [defining] sequences), which is obviously false.

5.10 See also

- Big Bounce
- Boltzmann brain
- Doomsday argument
- Goldilocks planet
- Goldilocks principle
- The Great Filter
- Infinite monkey theorem
- Inverse gambler’s fallacy
- Mediocrity principle
- Metaphysical naturalism
- Neocatastrophism
- Puddle thinking
- Quark mass and congeniality to life
- Rare Earth hypothesis
- Triple-alpha process
- Teleology

5.11 Footnotes

[1] Anthropic Principle

[2] James Schombert, Department of Physics at University of Oregon

[3] Mosterín J., (2005), *Antropic Explanations in Cosmology*, in Hajek, Valdés & Westerstahl (eds.), *Proceedings of the 12th International Congress of Logic, Methodology and Philosophy of Science*; http://philsci-archive.pitt.edu/1658/"

[4] "anthropic" means "of or pertaining to mankind or humans"

[5] The Anthropic Principle, Victor J. Stenger

[6] Anthropic Bias, Nick Bostrom, p.6

[7] Merriam-Webster Online Dictionary

[8] The Strong Anthropic Principle and the Final Anthropic Principle

[9] On Knowing, Sagan from Pale Blue Dot

[10] Dicke, R. H. (1961). "Dirac's Cosmology and Mach's Principle". *Nature* **192** (4801): 440–441. Bibcode:1961Natur.192..440D. doi:10.1038/192440a0.

[11] Davies, P. (2006). *The Goldilocks Enigma*. Allen Lane. ISBN 0-7139-9883-0.

[12] Weinberg, S. (1987). "Anthropic bound on the cosmological constant". *Physical Review Letters* **59** (22): 2607–2610. Bibcode:1987PhRvL..59.2607W. doi:10.1103/PhysRevLett.59.2607. PMID 10035596.

[13] New Scientist Space Blog: Physicists debate the nature of space-time - New Scientist

[14] How Many Fundamental Constants Are There? John Baez, mathematical physicist. U. C. Riverside, April 22, 2011

[15] Carter, B. (1974). "Large Number Coincidences and the Anthropic Principle in Cosmology". *IAU Symposium 63: Confrontation of Cosmological Theories with Observational Data*. Dordrecht: Reidel. pp. 291–298.; republished in General Relativity and Gravitation (Nov. 2011), Vol. 43, Iss. 11, p. 3225-3233, with an introduction by George Ellis (available on Arxiv

[16] Barrow, John D.; Tipler, Frank J. (1988). *The Anthropic Cosmological Principle*. Oxford University Press. ISBN 978-0-19-282147-8. LCCN 87028148.

[17] Wallace, A. R. (1904). *Man's place in the universe: a study of the results of scientific research in relation to the unity or plurality of worlds, 4th ed*. London: George Bell & Sons. pp. 256–7.

[18] Dicke, R. H. (1957). "Gravitation without a Principle of Equivalence". *Reviews of Modern Physics* **29** (3): 363–376. Bibcode:1957RvMP...29..363D. doi:10.1103/RevModPhys.29.363.

[19] Barrow, John D. (1997). "Anthropic Definitions". *Quarterly Journal of the Royal Astronomical Society* **24**: 146–53. Bibcode:1983QJRAS..24..146B.

[20] Barrow & Tipler's definitions are quoted verbatim at *Genesis of Eden Diversity Encyclopedia*.

[21] Barrow and Tipler 1986: 16.

[22] Barrow and Tipler 1986: 21.

[23] Barrow and Tipler 1986: 22.

[24] Jürgen Schmidhuber, 2000, "Algorithmic theories of everything."

[25] Leslie, J. (1986). "Probabilistic Phase Transitions and the Anthropic Principle". *Origin and Early History of the Universe: LIEGE 26*. Knudsen. pp. 439–444.

[26] Bostrom, N. (2002). *Anthropic Bias: Observation Selection Effects in Science and Philosophy*. Routledge. ISBN 0-415-93858-9. 5 chapters available online.

[27] Bostrom, N. (2002), op. cit.

[28] Jürgen Schmidhuber, 2002, "The Speed Prior: A New Simplicity Measure Yielding Near-Optimal Computable Predictions." *Proc. 15th Annual Conference on Computational Learning Theory* (COLT 2002), Sydney, Australia, Lecture Notes in Artificial Intelligence. Springer: 216-28.

[29] Michael Frayn, *The Human Touch*. Faber & Faber ISBN 0-571-23217-5

[30] Collins C. B., Hawking, S. W. (1973). "Why is the universe isotropic?". *Astrophysical Journal* **180**: 317–334. Bibcode:1973ApJ...180..317C. doi:10.1086/151965.

[31] Tegmark, M. (1998). "Is 'the theory of everything' merely the ultimate ensemble theory?". *Annals of Physics* **270**: 1–51. arXiv:gr-qc/9704009. Bibcode:1998AnPhy.270....1T. doi:10.1006/aphy.1998.5855.

[32] Strictly speaking, the number of non-compact dimensions, see String theory.

[33] Kane, Gordon L., Perry, Malcolm J., and Zytkow, Anna N. (2002). "The Beginning of the End of the Anthropic Principle". *New Astronomy* **7**: 45–53. arXiv:astro-ph/0001197. Bibcode:2002NewA....7...45K. doi:10.1016/S1384-1076(01)00088-4.

[34] Sober, Elliott, 2005, "The Design Argument" in Mann, W. E., ed., *The Blackwell Guide to the Philosophy of Religion*. Blackwell Publishers.

[35] Ikeda, M. and Jefferys, W., "The Anthropic Principle Does Not Support Supernaturalism," in *The Improbability of God*, Michael Martin and Ricki Monnier, Editors, pp. 150-166. Amherst, N.Y.: Prometheus Press. ISBN 1-59102-381-5

[36] Ikeda, M. and Jefferys, W. (2006). Unpublished FAQ "The Anthropic Principle Does Not Support Supernaturalism."

[37] Gardner, James N., 2005, "The Physical Constants as Biosignature: An anthropic retrodiction of the Selfish Biocosm Hypothesis," *International Journal of Astrobiology*.

[38] Carter, B.; McCrea, W. H. (1983). "The anthropic principle and its implications for biological evolution". *Philosophical Transactions of the Royal Society* **A310** (1512): 347–363. Bibcode:1983RSPTA.310..347C. doi:10.1098/rsta.1983.0096.

[39] Feoli, A. and Rampone, S.; Rampone (1999). "Is the Strong Anthropic Principle too weak?". *Nuovo Cim.* **B114**: 281–289. arXiv:gr-qc/9812093. Bibcode:1999NCimB.114..281F.

[40] Leslie, J. (1986) op. cit.

[41] Hogan, Craig (2000). "Why is the universe just so?". *Reviews of Modern Physics* **72** (4): 1149–1161. arXiv:astro-ph/9909295. Bibcode:2000RvMP...72.1149H. doi:10.1103/RevModPhys.72.1149.

[42] Bostrom (2002), op. cit.

[43] University of Birmingham Life, Bent Chains and the Anthropic Principle

[44] *Rev. Mod. Phys.* 29 (1957) 547

[45] Kragh, Helge (2010) When is a prediction anthropic? Fred Hoyle and the 7.65 MeV carbon resonance. http://philsci-archive.pitt.edu/5332/

[46] Page, D.N. (1983). "Inflation does not explain time asymmetry". *Nature* **304** (5921): 39. Bibcode:1983Natur.304...39P. doi:10.1038/304039a0.

[47] Davies, P.C.W. (1984). "Inflation to the universe and time asymmetry". *Nature* **312** (5994): 524. Bibcode:1984Natur.312..524D. doi:10.1038/312524a0.

[48] Weinberg, S. (2007). "Living in the multiverse". In B. Carr (ed). *Universe or multiverse?*. Cambridge University Press. ISBN 0-521-84841-5. preprint

[49] Tegmark (1998) op. cit.

[50] Livio, M. and Rees, M. J. (2003). "Anthropic reasoning". *Science* **309** (5737): 1022–3. Bibcode:2005Sci...309.1022L. doi:10.1126/science.1111446. PMID 16099967.

[51] Ehrenfest, Paul (1920). "How do the fundamental laws of physics make manifest that Space has 3 dimensions?". *Annalen der Physik* **61** (5): 440. Bibcode:1920AnP...366..440E. doi:10.1002/andp.19203660503.. Also see Ehrenfest, P. (1917) "In what way does it become manifest in the fundamental laws of physics that space has three dimensions?" *Proceedings of the Amsterdam Academy*20: 200.

[52] Weyl, H. (1922) *Space, time, and matter*. Dover reprint: 284.

[53] Tangherlini, F. R. (1963). "Atoms in Higher Dimensions". *Nuovo Cimento* **14** (27): 636.

[54] Tegmark, Max (April 1997). "On the dimensionality of spacetime" (PDF). *Classical and Quantum Gravity* **14** (4): L69–L75. arXiv:gr-qc/9702052. Bibcode:1997CQGra..14L..69T. doi:10.1088/0264-9381/14/4/002. Retrieved 2006-12-16.

[55] Hicks, L. E. (1883). *A Critique of Design Arguments.* New York: Scribner's.

[56] Barrow and Tipler 1986: 23

[57] Tipler, F. J. (1994). *The Physics of Immortality.* DoubleDay. ISBN 0-385-46798-2.

[58] Gardner, M., "WAP, SAP, PAP, and FAP," *The New York Review of Books 23*, No. 8 (May 8, 1986): 22-25.

[59] Barrow and Tipler 1986: 677

[60] e.g. Carter (2004) op. cit.

[61] e.g. message from Martin Rees presented at the Kavli-CERCA conference (see video in External links)

[62] Penrose, R. (1989). *The Emperor's New Mind.* Oxford University Press. ISBN 0-19-851973-7. Chapter 10.

[63] Starkman, G. D., Trotta, R. (2006). "Why Anthropic Reasoning Cannot Predict Λ". *Physical Review Letters* **97** (20): 201301. arXiv:astro-ph/0607227. Bibcode:2006PhRvL..97t1301S. doi:10.1103/PhysRevLett.97.201301. PMID 17155671. See also this news story.

[64] Gould, Stephen Jay (1998). "Clear Thinking in the Sciences". *Lectures at Harvard University.*

[65] Gould, Stephen Jay (2002). *Why People Believe Weird Things: Pseudoscience, Superstition, and Other Confusions of Our Time.* ISBN 0-7167-3090-1.

[66] Shermer, Michael (2007). *Why Darwin Matters.* ISBN 0-8050-8121-6.

[67] e.g. Carr, B. J., Rees, M. J. (1979). "The anthropic principle and the structure of the physical world". *Nature* **278** (5705): 605–612. Bibcode:1979Natur.278..605C. doi:10.1038/278605a0.

[68] Stenger, Victor J (2000). *Timeless Reality: Symmetry, Simplicity, and Multiple Universes.* Prometheus Books. ISBN 1-57392-859-3.

[69] Harnik, R., Kribs, G., Perez, G. (2006). "A Universe without Weak interactions". *Physical Review* **D74** (3): 035006. arXiv:hep-ph/0604027. Bibcode:2006PhRvD..74c5006H. doi:10.1103/PhysRevD.74.035006.

[70] Lee Smolin (2001). Tyson, Neil deGrasse and Soter, Steve, ed. *Cosmic Horizons: Astronomy at the Cutting Edge.* The New Press. pp. 148–152. ISBN 978-1-56584-602-9.

[71] Smolin vs. Susskind: The Anthropic Principle

[72] Earman John (1987). "The SAP also rises: A critical examination of the anthropic principle". *American Philosophical Quarterly* **24**: 307–317.

[73] McMullin, Ernan. (1994). "Fine-tuning the Universe?" In M. Shale & G. Shields (ed.), *Science, Technology, and Religious Ideas*, Lanham: University Press of America.

[74] Mosterín, Jesús. (2005). Op. cit.

5.12 References

- Barrow, John D.; Tipler, Frank J. (1988). *The Anthropic Cosmological Principle.* Oxford University Press. ISBN 978-0-19-282147-8. LCCN 87028148.

- Cirkovic, M. M. (2002). "On the First Anthropic Argument in Astrobiology". *Earth, Moon, and Planets* **91** (4): 243–254. doi:10.1023/A:1026266630823.

- Cirkovic, M. M. (2004). "The Anthropic Principle and the Duration of the Cosmological Past". *Astronomical and Astrophysical Transactions* **23** (6): 567–597. arXiv:astro-ph/0505005. Bibcode:2004A&AT...23..567C. doi:10.1080/10556790412331335327.

- Conway Morris, Simon (2003). *Life's Solution: Inevitable Humans in a Lonely Universe.* Cambridge University Press.

- Craig, William Lane (1987). "Critical review of *The Anthropic Cosmological Principle*". *International Philosophical Quarterly* **27**: 437–47. doi:10.5840/ipq198727433.

- Hawking, Stephen W. (1988). *A Brief History of Time.* New York: Bantam Books. p. 174. ISBN 0-553-34614-8.
- Stenger, Victor J. (1999), "Anthropic design," *The Skeptical Inquirer 23* (August 31, 1999): 40-43
- Mosterín, Jesús (2005). "Anthropic Explanations in Cosmology." In P. Háyek, L. Valdés and D. Westerstahl (ed.), *Logic, Methodology and Philosophy of Science, Proceedings of the 12th International Congress of the LMPS.* London: King's College Publications, pp. 441–473. ISBN 1-904987-21-4.
- Taylor, Stuart Ross (1998). *Destiny or Chance: Our Solar System and Its Place in the Cosmos.* Cambridge University Press. ISBN 0-521-78521-9.
- Tegmark, Max (1997). "On the dimensionality of spacetime". *Classical and Quantum Gravity* **14** (4): L69–L75. arXiv:gr-qc/9702052. Bibcode:1997CQGra..14L..69T. doi:10.1088/0264-9381/14/4/002. A simple anthropic argument for why there are 3 spatial and 1 temporal dimensions.
- Tipler, F. J. (2003). "Intelligent Life in Cosmology". *International Journal of Astrobiology* **2** (2): 141–48. arXiv:0704.0058. Bibcode:2003IJAsB...2..141T. doi:10.1017/S1473550403001526.
- Walker, M. A., and Cirkovic, M. M. (2006). "Anthropic Reasoning, Naturalism and the Contemporary Design Argument". *International Studies in the Philosophy of Science* **20** (3): 285–307. doi:10.1080/02698590600960945. Shows that some of the common criticisms of AP based on its relationship with numerology or the theological Design Argument are wrong.
- Ward, P. D., and Brownlee, D. (2000). *Rare Earth: Why Complex Life is Uncommon in the Universe.* Springer Verlag. ISBN 0-387-98701-0.
- Vilenkin, Alex (2006). *Many Worlds in One: The Search for Other Universes.* Hill and Wang. ISBN 978-0-8090-9523-0.

5.13 External links

- Nick Bostrom: web site devoted to the Anthropic Principle.
- Gijsbers, Victor. (2000). Theistic Anthropic Principle Refuted Positive Atheism Magazine.
- Chown, Marcus, Anything Goes, *New Scientist*, 6 June 1998. On Max Tegmark's work.
- Stephen Hawking, Steven Weinberg, Alexander Vilenkin, David Gross and Lawrence Krauss: Debate on Anthropic Reasoning Kavli-CERCA Conference Video Archive.
- Sober, Elliott R. 2009, "Absence of Evidence and Evidence of Absence -- Evidential Transitivity in Connection with Fossils, Fishing, Fine-Tuning, and Firing Squads." Philosophical Studies, 2009, 143: 63-90.
- "Anthropic Coincidence"—the anthropic controversy as a segue to Lee Smolin's theory of cosmological natural selection.
- Leonard Susskind and Lee Smolin debate the Anthropic Principle.
- debate among scientists on arxiv.org.
- Evolutionary Probability and Fine Tuning
- Benevolent Design and the Anthropic Principle at MathPages
- Critical review of "The Privileged Planet"
- The Anthropic Principle - a review.
- Berger, Daniel, 2002, "An impertinent resumé of the Anthropic Cosmological Principle." A critique of Barrow & Tipler.
- Jürgen Schmidhuber: Papers on algorithmic theories of everything and the Anthropic Principle's lack of predictive power.
- Paul Davies: Cosmic Jackpot Interview about the Anthropic Principle (starts at 40 min), 15 May 2007.

Chapter 6

Standard Model

This article is about the Standard Model of particle physics. For other uses, see Standard model (disambiguation).
This article is a non-mathematical general overview of the Standard Model. For a mathematical description, see the article Standard Model (mathematical formulation).
For the Standard Model of Big Bang cosmology, Lambda-CDM model.

The **Standard Model** of particle physics is a theory concerning the electromagnetic, weak, and strong nuclear

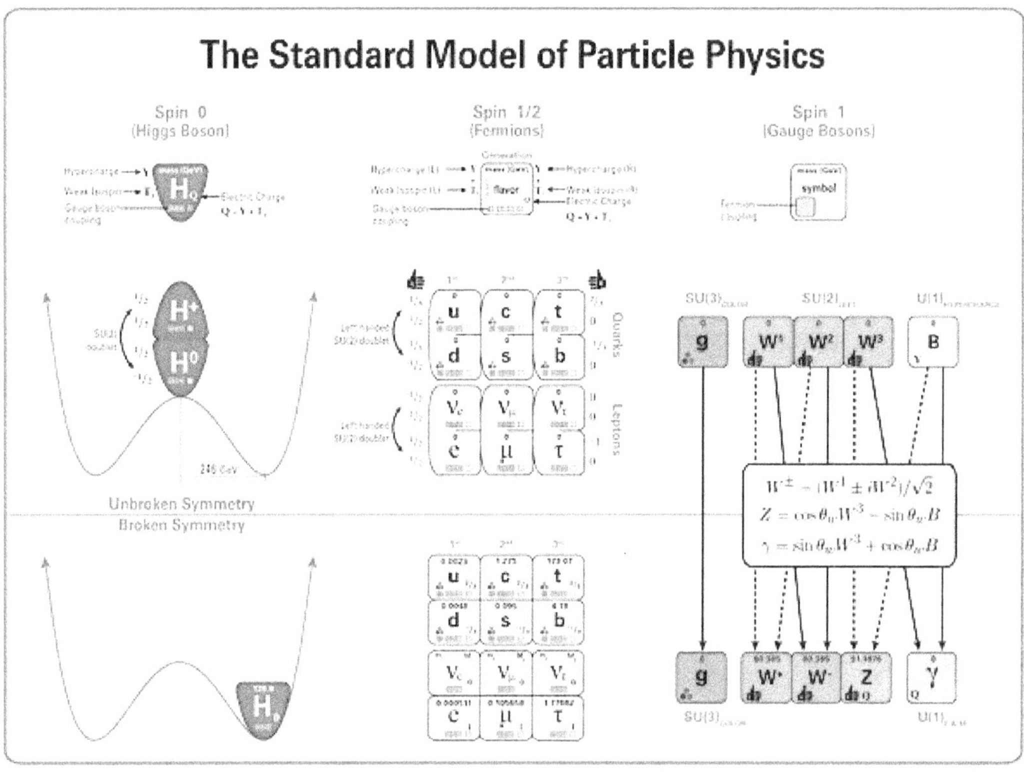

Standard Model of Particle Physics. The diagram shows the elementary particles of the Standard Model (the Higgs boson, the three generations of quarks and leptons, and the gauge bosons), including their names, masses, spins, charges, chiralities, and interactions with the strong, weak and electromagnetic forces. It also depicts the crucial role of the Higgs boson in electroweak symmetry breaking, and shows how the properties of the various particles differ in the (high-energy) symmetric phase (top) and the (low-energy) broken-symmetry phase (bottom).

interactions, as well as classifying all the subatomic particles known. It was developed throughout the latter half of the 20th century, as a collaborative effort of scientists around the world.[1] The current formulation was finalized in the mid-1970s upon experimental confirmation of the existence of quarks. Since then, discoveries of the top quark (1995), the tau neutrino (2000), and more recently the Higgs boson (2013), have given further credence to the

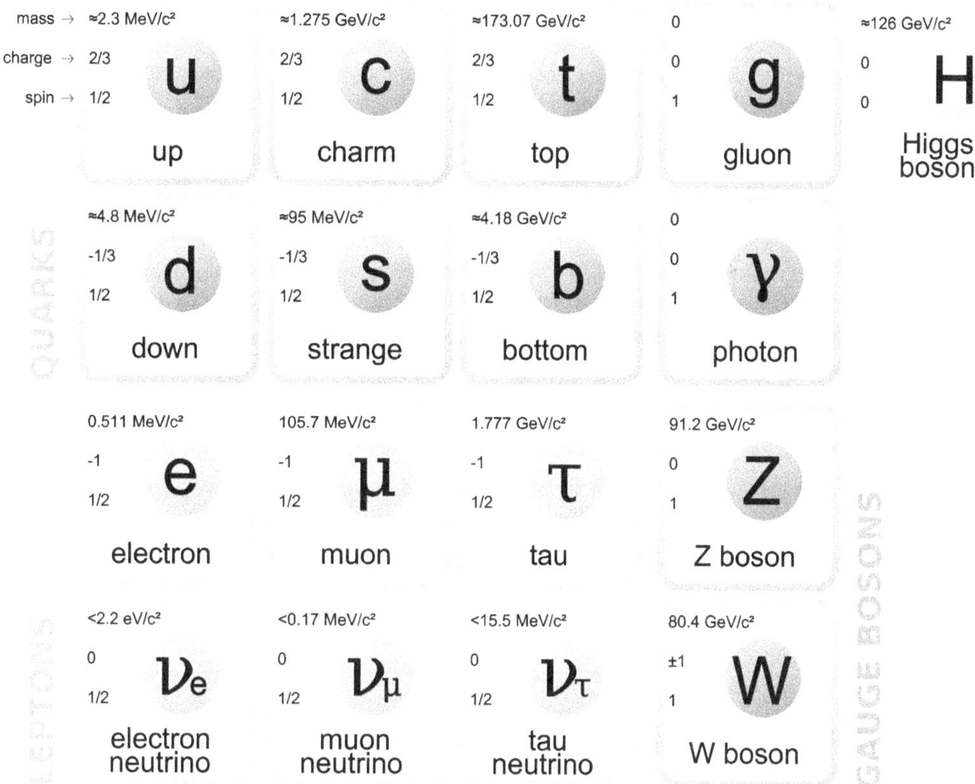

The Standard Model of elementary particles (more schematic depiction), with the three generations of matter, gauge bosons in the fourth column, and the Higgs boson in the fifth.

Standard Model. Because of its success in explaining a wide variety of experimental results, the Standard Model is sometimes regarded as a "theory of almost everything".

Although the Standard Model is believed to be theoretically self-consistent[2] and has demonstrated huge and continued successes in providing experimental predictions, it does leave some phenomena unexplained and it falls short of being a complete theory of fundamental interactions. It does not incorporate the full theory of gravitation[3] as described by general relativity, or account for the accelerating expansion of the universe (as possibly described by dark energy). The model does not contain any viable dark matter particle that possesses all of the required properties deduced from observational cosmology. It also does not incorporate neutrino oscillations (and their non-zero masses).

The development of the Standard Model was driven by theoretical and experimental particle physicists alike. For theorists, the Standard Model is a paradigm of a quantum field theory, which exhibits a wide range of physics including spontaneous symmetry breaking, anomalies, non-perturbative behavior, etc. It is used as a basis for building more exotic models that incorporate hypothetical particles, extra dimensions, and elaborate symmetries (such as supersymmetry) in an attempt to explain experimental results at variance with the Standard Model, such as the existence of dark matter and neutrino oscillations.

6.1 Historical background

The first step towards the Standard Model was Sheldon Glashow's discovery in 1961 of a way to combine the electromagnetic and weak interactions.[4] In 1967 Steven Weinberg[5] and Abdus Salam[6] incorporated the Higgs mechanism[7][8][9] into Glashow's electroweak theory, giving it its modern form.

The Higgs mechanism is believed to give rise to the masses of all the elementary particles in the Standard Model. This includes the masses of the W and Z bosons, and the masses of the fermions, i.e. the quarks and leptons.

After the neutral weak currents caused by Z boson exchange were discovered at CERN in 1973,[10][11][12][13] the electroweak theory became widely accepted and Glashow, Salam, and Weinberg shared the 1979 Nobel Prize in

Physics for discovering it. The W and Z bosons were discovered experimentally in 1981, and their masses were found to be as the Standard Model predicted.

The theory of the strong interaction, to which many contributed, acquired its modern form around 1973–74, when experiments confirmed that the hadrons were composed of fractionally charged quarks.

6.2 Overview

At present, matter and energy are best understood in terms of the kinematics and interactions of elementary particles. To date, physics has reduced the laws governing the behavior and interaction of all known forms of matter and energy to a small set of fundamental laws and theories. A major goal of physics is to find the "common ground" that would unite all of these theories into one integrated theory of everything, of which all the other known laws would be special cases, and from which the behavior of all matter and energy could be derived (at least in principle).[14]

6.3 Particle content

The Standard Model includes members of several classes of elementary particles (fermions, gauge bosons, and the Higgs boson), which in turn can be distinguished by other characteristics, such as color charge.

6.3.1 Fermions

The Standard Model includes 12 elementary particles of spin-½ known as fermions. According to the spin-statistics theorem, fermions respect the Pauli exclusion principle. Each fermion has a corresponding antiparticle.

The fermions of the Standard Model are classified according to how they interact (or equivalently, by what charges they carry). There are six quarks (up, down, charm, strange, top, bottom), and six leptons (electron, electron neutrino, muon, muon neutrino, tau, tau neutrino). Pairs from each classification are grouped together to form a generation, with corresponding particles exhibiting similar physical behavior (see table).

The defining property of the quarks is that they carry color charge, and hence, interact via the strong interaction. A phenomenon called color confinement results in quarks being very strongly bound to one another, forming color-neutral composite particles (hadrons) containing either a quark and an antiquark (mesons) or three quarks (baryons). The familiar proton and the neutron are the two baryons having the smallest mass. Quarks also carry electric charge and weak isospin. Hence they interact with other fermions both electromagnetically and via the weak interaction.

The remaining six fermions do not carry colour charge and are called leptons. The three neutrinos do not carry electric charge either, so their motion is directly influenced only by the weak nuclear force, which makes them notoriously difficult to detect. However, by virtue of carrying an electric charge, the electron, muon, and tau all interact electromagnetically.

Each member of a generation has greater mass than the corresponding particles of lower generations. The first generation charged particles do not decay; hence all ordinary (baryonic) matter is made of such particles. Specifically, all atoms consist of electrons orbiting around atomic nuclei, ultimately constituted of up and down quarks. Second and third generations charged particles, on the other hand, decay with very short half lives, and are observed only in very high-energy environments. Neutrinos of all generations also do not decay, and pervade the universe, but rarely interact with baryonic matter.

6.3.2 Gauge bosons

In the Standard Model, gauge bosons are defined as force carriers that mediate the strong, weak, and electromagnetic fundamental interactions.

Interactions in physics are the ways that particles influence other particles. At a macroscopic level, electromagnetism allows particles to interact with one another via electric and magnetic fields, and gravitation allows particles with mass to attract one another in accordance with Einstein's theory of general relativity. The Standard Model explains such forces as resulting from matter particles exchanging other particles, generally referred to as *force mediating particles*. When a force-mediating particle is exchanged, at a macroscopic level the effect is equivalent to a force influencing

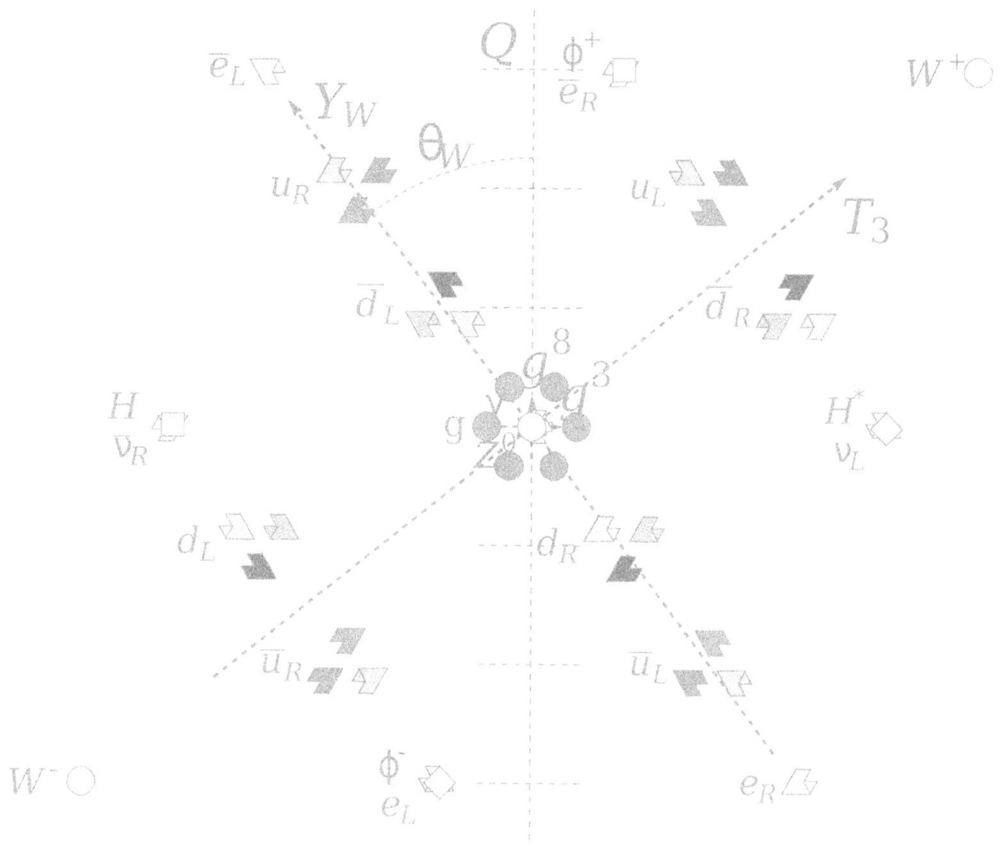

The pattern of weak isospin, T_3, weak hypercharge, YW, and color charge of all known elementary particles, rotated by the weak mixing angle to show electric charge, Q, roughly along the vertical. The neutral Higgs field (gray square) breaks the electroweak symmetry and interacts with other particles to give them mass.

both of them, and the particle is therefore said to have *mediated* (i.e., been the agent of) that force. The Feynman diagram calculations, which are a graphical representation of the perturbation theory approximation, invoke "force mediating particles", and when applied to analyze high-energy scattering experiments are in reasonable agreement with the data. However, perturbation theory (and with it the concept of a "force-mediating particle") fails in other situations. These include low-energy quantum chromodynamics, bound states, and solitons.

The gauge bosons of the Standard Model all have spin (as do matter particles). The value of the spin is 1, making them bosons. As a result, they do not follow the Pauli exclusion principle that constrains fermions: thus bosons (e.g. photons) do not have a theoretical limit on their spatial density (number per volume). The different types of gauge bosons are described below.

- Photons mediate the electromagnetic force between electrically charged particles. The photon is massless and is well-described by the theory of quantum electrodynamics.

- The W+, W−, and Z gauge bosons mediate the weak interactions between particles of different flavors (all quarks and leptons). They are massive, with the Z being more massive than the W±. The weak interactions involving the W± exclusively act on *left-handed* particles and *right-handed* antiparticles. Furthermore, the W±

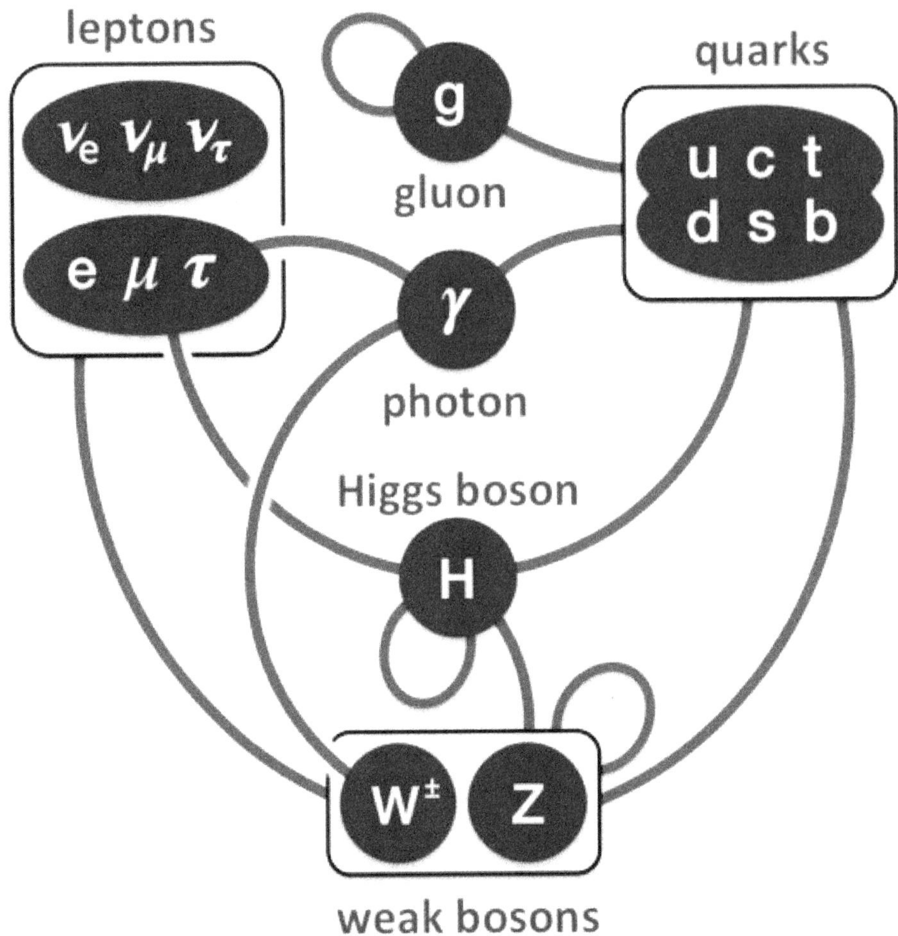

Summary of interactions between particles described by the Standard Model.

carries an electric charge of +1 and −1 and couples to the electromagnetic interaction. The electrically neutral Z boson interacts with both left-handed particles and antiparticles. These three gauge bosons along with the photons are grouped together, as collectively mediating the electroweak interaction.

- The eight gluons mediate the strong interactions between color charged particles (the quarks). Gluons are massless. The eightfold multiplicity of gluons is labeled by a combination of color and anticolor charge (e.g. red–antigreen).[nb 1] Because the gluons have an effective color charge, they can also interact among themselves. The gluons and their interactions are described by the theory of quantum chromodynamics.

The interactions between all the particles described by the Standard Model are summarized by the diagrams on the right of this section.

6.3.3 Higgs boson

Main article: Higgs boson

The Higgs particle is a massive scalar elementary particle theorized by Robert Brout, François Englert, Peter Higgs, Gerald Guralnik, C. R. Hagen, and Tom Kibble in 1964 (see 1964 PRL symmetry breaking papers) and is a key building block in the Standard Model.[7][8][9][15] It has no intrinsic spin, and for that reason is classified as a boson (like the gauge bosons, which have integer spin).

Standard Model Interactions
(Forces Mediated by Gauge Bosons)

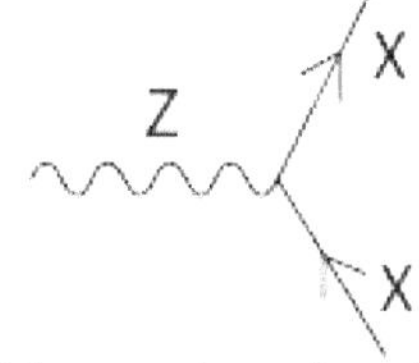

X is any fermion in the Standard Model.

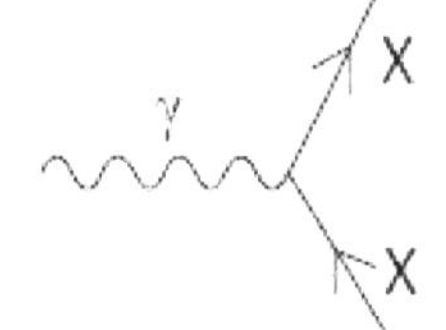

X is electrically charged.

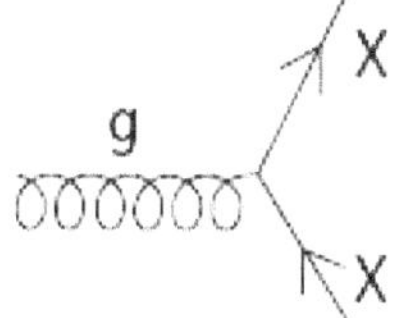

X is any quark.

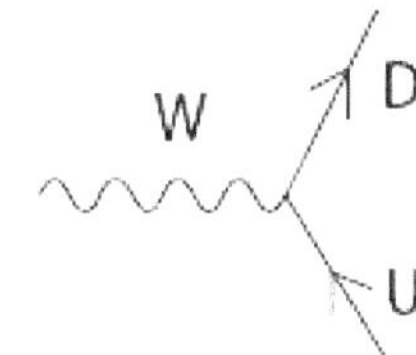

U is a up-type quark;
D is a down-type quark.

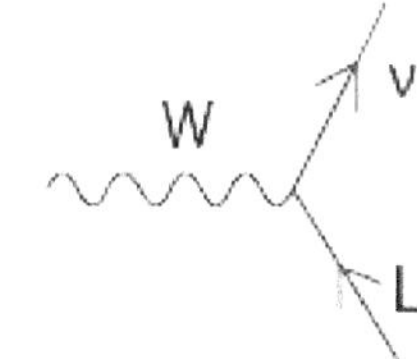

L is a lepton and v is the corresponding neutrino.

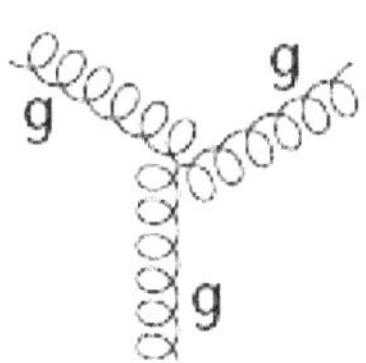

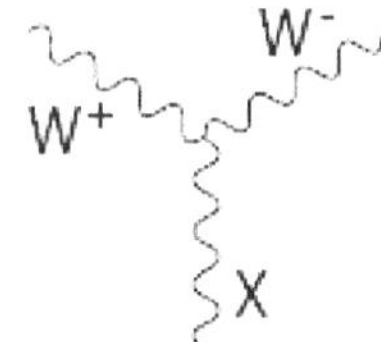

X is a photon or Z-boson.

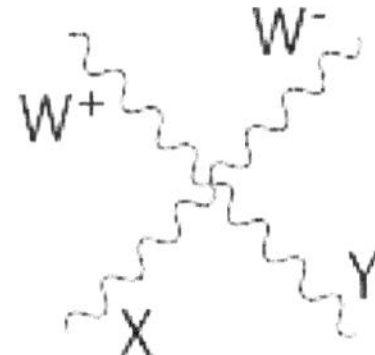

X and Y are any two electroweak bosons such that charge is conserved.

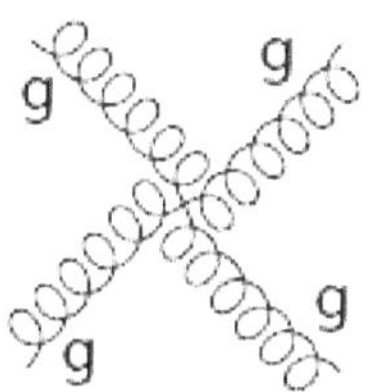

The above interactions form the basis of the standard model. Feynman diagrams in the standard model are built from these vertices. Modifications involving Higgs boson interactions and neutrino oscillations are omitted. The charge of the W bosons is dictated by the fermions they interact with; the conjugate of each listed vertex (i.e. reversing the direction of arrows) is also allowed.

The Higgs boson plays a unique role in the Standard Model, by explaining why the other elementary particles, except the photon and gluon, are massive. In particular, the Higgs boson explains why the photon has no mass, while the W and Z bosons are very heavy. Elementary particle masses, and the differences between electromagnetism (mediated by the photon) and the weak force (mediated by the W and Z bosons), are critical to many aspects of the structure of microscopic (and hence macroscopic) matter. In electroweak theory, the Higgs boson generates the masses of the leptons (electron, muon, and tau) and quarks. As the Higgs boson is massive, it must interact with itself.

Because the Higgs boson is a very massive particle and also decays almost immediately when created, only a very high-energy particle accelerator can observe and record it. Experiments to confirm and determine the nature of the Higgs boson using the Large Hadron Collider (LHC) at CERN began in early 2010, and were performed at Fermilab's Tevatron until its closure in late 2011. Mathematical consistency of the Standard Model requires that any mechanism capable of generating the masses of elementary particles become visible at energies above 1.4 TeV;[16] therefore, the LHC (designed to collide two 7 to 8 TeV proton beams) was built to answer the question of whether the Higgs boson actually exists.[17]

On 4 July 2012, the two main experiments at the LHC (ATLAS and CMS) both reported independently that they found a new particle with a mass of about 125 GeV/c^2 (about 133 proton masses, on the order of 10^{-25} kg), which is "consistent with the Higgs boson." Although it has several properties similar to the predicted "simplest" Higgs,[18] they acknowledged that further work would be needed to conclude that it is indeed the Higgs boson, and exactly which version of the Standard Model Higgs is best supported if confirmed.[19][20][21][22][23]

On 14 March 2013 the Higgs Boson was tentatively confirmed to exist.[24]

6.3.4 Total particle count

Counting particles by a rule that distinguishes between particles and their corresponding antiparticles, and among the many color states of quarks and gluons, gives a total of 61 elementary particles.[25]

6.4 Theoretical aspects

Main article: Standard Model (mathematical formulation)

6.4.1 Construction of the Standard Model Lagrangian

Technically, quantum field theory provides the mathematical framework for the Standard Model, in which a Lagrangian controls the dynamics and kinematics of the theory. Each kind of particle is described in terms of a dynamical field that pervades space-time. The construction of the Standard Model proceeds following the modern method of constructing most field theories: by first postulating a set of symmetries of the system, and then by writing down the most general renormalizable Lagrangian from its particle (field) content that observes these symmetries.

The global Poincaré symmetry is postulated for all relativistic quantum field theories. It consists of the familiar translational symmetry, rotational symmetry and the inertial reference frame invariance central to the theory of special relativity. The local SU(3)×SU(2)×U(1) gauge symmetry is an internal symmetry that essentially defines the Standard Model. Roughly, the three factors of the gauge symmetry give rise to the three fundamental interactions. The fields fall into different representations of the various symmetry groups of the Standard Model (see table). Upon writing the most general Lagrangian, one finds that the dynamics depend on 19 parameters, whose numerical values are established by experiment. The parameters are summarized in the table above (note: with the Higgs mass is at 125 GeV, the Higgs self-coupling strength $\lambda \sim 1/8$).

Quantum chromodynamics sector

Main article: Quantum chromodynamics

The quantum chromodynamics (QCD) sector defines the interactions between quarks and gluons, with SU(3) symmetry, generated by T^a. Since leptons do not interact with gluons, they are not affected by this sector. The Dirac Lagrangian of the quarks coupled to the gluon fields is given by

$$\mathcal{L}_{QCD} = i\overline{U}(\partial_\mu - ig_s G^a_\mu T^a)\gamma^\mu U + i\overline{D}(\partial_\mu - ig_s G^a_\mu T^a)\gamma^\mu D.$$

G^a_μ is the SU(3) gauge field containing the gluons, γ^μ are the Dirac matrices, D and U are the Dirac spinors associated with up- and down-type quarks, and g_s is the strong coupling constant.

Electroweak sector

Main article: Electroweak interaction

The electroweak sector is a Yang–Mills gauge theory with the simple symmetry group U(1)×SU(2)L,

$$\mathcal{L}_{\rm EW} = \sum_\psi \bar\psi\gamma^\mu \left(i\partial_\mu - g'\frac{1}{2}Y_{\rm W}B_\mu - g\frac{1}{2}\vec\tau_{\rm L}\vec W_\mu\right)\psi$$

where $B\mu$ is the U(1) gauge field; YW is the weak hypercharge—the generator of the U(1) group; $\vec{W}_\mu$ is the three-component SU(2) gauge field; $\vec{\tau}_{\rm L}$ are the Pauli matrices—infinitesimal generators of the SU(2) group. The subscript L indicates that they only act on left fermions; g' and g are coupling constants.

Higgs sector

Main article: Higgs mechanism

In the Standard Model, the Higgs field is a complex scalar of the group SU(2)L:

$$\varphi = \frac{1}{\sqrt{2}} \begin{pmatrix} \varphi^+ \\ \varphi^0 \end{pmatrix},$$

where the indices + and 0 indicate the electric charge (Q) of the components. The weak isospin (YW) of both components is 1.

Before symmetry breaking, the Higgs Lagrangian is:

$$\mathcal{L}_{\rm H} = \varphi^\dagger \left(\partial^\mu - \frac{i}{2}\left(g'Y_{\rm W}B^\mu + g\vec\tau\vec W^\mu\right)\right)\left(\partial_\mu + \frac{i}{2}\left(g'Y_{\rm W}B_\mu + g\vec\tau\vec W_\mu\right)\right)\varphi \ - \ \frac{\lambda^2}{4}\left(\varphi^\dagger\varphi - v^2\right)^2,$$

which can also be written as:

$$\mathcal{L}_{\rm H} = \left|\left(\partial_\mu + \frac{i}{2}\left(g'Y_{\rm W}B_\mu + g\vec\tau\vec W_\mu\right)\right)\varphi\right|^2 \ - \ \frac{\lambda^2}{4}\left(\varphi^\dagger\varphi - v^2\right)^2.$$

6.5 Fundamental forces

Main article: Fundamental interaction

The Standard Model classified all four fundamental forces in nature. In the Standard Model, a force is described as an exchange of bosons between the objects affected, such as a photon for the electromagnetic force and a gluon for the strong interaction. Those particles are called force carriers.[26]

6.6 Tests and predictions

The Standard Model (SM) predicted the existence of the W and Z bosons, gluon, and the top and charm quarks before these particles were observed. Their predicted properties were experimentally confirmed with good precision. To give an idea of the success of the SM, the following table compares the measured masses of the W and Z bosons with the masses predicted by the SM:

The SM also makes several predictions about the decay of Z bosons, which have been experimentally confirmed by the Large Electron-Positron Collider at CERN.

In May 2012 BaBar Collaboration reported that their recently analyzed data may suggest possible flaws in the Standard Model of particle physics.[28][29] These data show that a particular type of particle decay called "B to D-star-tau-nu" happens more often than the Standard Model says it should. In this type of decay, a particle called the B-bar meson decays into a D meson, an antineutrino and a tau-lepton. While the level of certainty of the excess (3.4 sigma) is not enough to claim a break from the Standard Model, the results are a potential sign of something amiss and are likely to impact existing theories, including those attempting to deduce the properties of Higgs bosons.[30]

On December 13, 2012, physicists reported the constancy, over space and time, of a basic physical constant of nature that supports the *standard model of physics*. The scientists, studying methanol molecules in a distant galaxy, found the change ($\Delta\mu/\mu$) in the proton-to-electron mass ratio μ to be equal to "$(0.0 \pm 1.0) \times 10^{-7}$ at redshift $z = 0.89$" and consistent with "a null result".[31][32]

6.7 Challenges

See also: Physics beyond the Standard Model

Self-consistency of the Standard Model (currently formulated as a non-abelian gauge theory quantized through path-integrals) has not been mathematically proven. While regularized versions useful for approximate computations (for example lattice gauge theory) exist, it is not known whether they converge (in the sense of S-matrix elements) in the limit that the regulator is removed. A key question related to the consistency is the Yang–Mills existence and mass gap problem.

Experiments indicate that neutrinos have mass, which the classic Standard Model did not allow.[33] To accommodate this finding, the classic Standard Model can be modified to include neutrino mass.

If one insists on using only Standard Model particles, this can be achieved by adding a non-renormalizable interaction of leptons with the Higgs boson.[34] On a fundamental level, such an interaction emerges in the seesaw mechanism where heavy right-handed neutrinos are added to the theory. This is natural in the left-right symmetric extension of the Standard Model[35][36] and in certain grand unified theories.[37] As long as new physics appears below or around 10^{14} GeV, the neutrino masses can be of the right order of magnitude.

Theoretical and experimental research has attempted to extend the Standard Model into a Unified field theory or a Theory of everything, a complete theory explaining all physical phenomena including constants. Inadequacies of the Standard Model that motivate such research include:

- It does not attempt to explain gravitation, although a theoretical particle known as a graviton would help explain it, and unlike for the strong and electroweak interactions of the Standard Model, there is no known way of describing general relativity, the canonical theory of gravitation, consistently in terms of quantum field theory. The reason for this is, among other things, that quantum field theories of gravity generally break down before reaching the Planck scale. As a consequence, we have no reliable theory for the very early universe;

- Some consider it to be *ad hoc* and inelegant, requiring 19 numerical constants whose values are unrelated and arbitrary. Although the Standard Model, as it now stands, can explain why neutrinos have masses, the specifics of neutrino mass are still unclear. It is believed that explaining neutrino mass will require an additional 7 or 8 constants, which are also arbitrary parameters;

- The Higgs mechanism gives rise to the hierarchy problem if some new physics (coupled to the Higgs) is present at high energy scales. In these cases in order for the weak scale to be much smaller than the Planck scale, severe fine tuning of the parameters is required; there are, however, other scenarios that include quantum gravity in which such fine tuning can be avoided.[38]There are also issues of Quantum triviality, which suggests that it may not be possible to create a consistent quantum field theory involving elementary scalar particles.

- It should be modified so as to be consistent with the emerging "Standard Model of cosmology." In particular, the Standard Model cannot explain the observed amount of cold dark matter (CDM) and gives contributions to dark energy which are many orders of magnitude too large. It is also difficult to accommodate the observed predominance of matter over antimatter (matter/antimatter asymmetry). The isotropy and homogeneity of the visible universe over large distances seems to require a mechanism like cosmic inflation, which would also constitute an extension of the Standard Model.

- The existence of ultra-high-energy cosmic rays are difficult to explain under the Standard Model.

Currently, no proposed Theory of Everything has been widely accepted or verified.

6.8 See also

- Fundamental interaction:
    - Quantum electrodynamics
    - Strong interaction: Color charge, Quantum chromodynamics, Quark model
    - Weak interaction: Electroweak theory, Fermi theory of beta decay, Weak hypercharge, Weak isospin
- Gauge theory: Nontechnical introduction to gauge theory
- Generation
- Higgs mechanism: Higgs boson, Higgsless model
- J. C. Ward
- J. J. Sakurai Prize for Theoretical Particle Physics
- Lagrangian
- Open questions: BTeV experiment, CP violation, Neutrino masses, Quark matter, Quantum triviality
- Penguin diagram
- Quantum field theory
- Standard Model: Mathematical formulation of, Physics beyond the Standard Model

6.9 Notes and references

[1] Technically, there are nine such color–anticolor combinations. However, there is one color-symmetric combination that can be constructed out of a linear superposition of the nine combinations, reducing the count to eight.

6.10 References

[1] R. Oerter (2006). *The Theory of Almost Everything: The Standard Model, the Unsung Triumph of Modern Physics* (Kindle ed.). Penguin Group. p. 2. ISBN 0-13-236678-9.

[2] In fact, there are mathematical issues regarding quantum field theories still under debate (see e.g. Landau pole), but the predictions extracted from the Standard Model by current methods applicable to current experiments are all self-consistent. For a further discussion see e.g. Chapter 25 of R. Mann (2010). *An Introduction to Particle Physics and the Standard Model*. CRC Press. ISBN 978-1-4200-8298-2.

[3] Sean Carroll, Ph.D., Cal Tech, 2007, The Teaching Company, *Dark Matter, Dark Energy: The Dark Side of the Universe*, Guidebook Part 2 page 59, Accessed Oct. 7, 2013, "...Standard Model of Particle Physics: The modern theory of elementary particles and their interactions ... It does not, strictly speaking, include gravity, although it's often convenient to include gravitons among the known particles of nature..."

[4] S.L. Glashow (1961). "Partial-symmetries of weak interactions". *Nuclear Physics* **22** (4): 579–588. Bibcode:1961NucPh..22..579G. doi:10.1016/0029-5582(61)90469-2.

[5] S. Weinberg (1967). "A Model of Leptons". *Physical Review Letters* **19** (21): 1264–1266. Bibcode:1967PhRvL..19.1264W. doi:10.1103/PhysRevLett.19.1264.

[6] A. Salam (1968). N. Svartholm, ed. *Elementary Particle Physics: Relativistic Groups and Analyticity*. Eighth Nobel Symposium. Stockholm: Almquvist and Wiksell. p. 367.

[7] F. Englert, R. Brout (1964). "Broken Symmetry and the Mass of Gauge Vector Mesons". *Physical Review Letters* **13** (9): 321–323. Bibcode:1964PhRvL..13..321E. doi:10.1103/PhysRevLett.13.321.

[8] P.W. Higgs (1964). "Broken Symmetries and the Masses of Gauge Bosons". *Physical Review Letters* **13** (16): 508–509. Bibcode:1964PhRvL..13..508H. doi:10.1103/PhysRevLett.13.508.

[9] G.S. Guralnik, C.R. Hagen, T.W.B. Kibble (1964). "Global Conservation Laws and Massless Particles". *Physical Review Letters* **13** (20): 585–587. Bibcode:1964PhRvL..13..585G. doi:10.1103/PhysRevLett.13.585.

[10] F.J. Hasert et al. (1973). "Search for elastic muon-neutrino electron scattering". *Physics Letters B* **46** (1): 121. Bibcode:1973PhLB...46..121H. doi:10.1016/0370-2693(73)90494-2.

[11] F.J. Hasert et al. (1973). "Observation of neutrino-like interactions without muon or electron in the Gargamelle neutrino experiment". *Physics Letters B* **46** (1): 138. Bibcode:1973PhLB...46..138H. doi:10.1016/0370-2693(73)90499-1.

[12] F.J. Hasert et al. (1974). "Observation of neutrino-like interactions without muon or electron in the Gargamelle neutrino experiment". *Nuclear Physics B* **73** (1): 1. Bibcode:1974NuPhB..73....1H. doi:10.1016/0550-3213(74)90038-8.

[13] D. Haidt (4 October 2004). "The discovery of the weak neutral currents". *CERN Courier*. Retrieved 8 May 2008.

[14] "Details can be worked out if the situation is simple enough for us to make an approximation, which is almost never, but often we can understand more or less what is happening." from *The Feynman Lectures on Physics*, Vol 1. pp. 2–7

[15] G.S. Guralnik (2009). "The History of the Guralnik, Hagen and Kibble development of the Theory of Spontaneous Symmetry Breaking and Gauge Particles". *International Journal of Modern Physics A* **24** (14): 2601–2627. arXiv:0907.3466. Bibcode:2009IJMPA..24.2601G. doi:10.1142/S0217751X09045431.

[16] B.W. Lee, C. Quigg, H.B. Thacker (1977). "Weak interactions at very high energies: The role of the Higgs-boson mass". *Physical Review D* **16** (5): 1519–1531. Bibcode:1977PhRvD..16.1519L. doi:10.1103/PhysRevD.16.1519.

[17] "Huge $10 billion collider resumes hunt for 'God particle'". CNN. 11 November 2009. Retrieved 2010-05-04.

[18] M. Strassler (10 July 2012). "Higgs Discovery: Is it a Higgs?". Retrieved 2013-08-06.

[19] "CERN experiments observe particle consistent with long-sought Higgs boson". CERN. 4 July 2012. Retrieved 2012-07-04.

[20] "Observation of a New Particle with a Mass of 125 GeV". CERN. 4 July 2012. Retrieved 2012-07-05.

[21] "ATLAS Experiment". ATLAS. 1 January 2006. Retrieved 2012-07-05.

[22] "Confirmed: CERN discovers new particle likely to be the Higgs boson". *YouTube*. Russia Today. 4 July 2012. Retrieved 2013-08-06.

[23] D. Overbye (4 July 2012). "A New Particle Could Be Physics' Holy Grail". *New York Times*. Retrieved 2012-07-04.

[24] "New results indicate that new particle is a Higgs boson". CERN. 14 March 2013. Retrieved 2013-08-06.

[25] S. Braibant, G. Giacomelli, M. Spurio (2009). *Particles and Fundamental Interactions: An Introduction to Particle Physics*. Springer. pp. 313–314. ISBN 978-94-007-2463-1.

[26] http://home.web.cern.ch/about/physics/standard-model Official CERN website

[27] http://www.pha.jhu.edu/~{ }dfehling/particle.gif

[28] "BABAR Data in Tension with the Standard Model". SLAC. 31 May 2012. Retrieved 2013-08-06.

[29] BaBar Collaboration (2012). "Evidence for an excess of $B \to D^{(*)} \tau^- \nu\tau$ decays". *Physical Review Letters* **109** (10): 101802. arXiv:1205.5442. Bibcode:2012PhRvL.109j1802L. doi:10.1103/PhysRevLett.109.101802.

[30] "BaBar data hint at cracks in the Standard Model". *e! Science News*. 18 June 2012. Retrieved 2013-08-06.

[31] J. Bagdonaite et al. (2012). "A Stringent Limit on a Drifting Proton-to-Electron Mass Ratio from Alcohol in the Early Universe". *Science* **339** (6115): 46. Bibcode:2013Sci...339...46B. doi:10.1126/science.1224898.

[32] C. Moskowitz (13 December 2012). "Phew! Universe's Constant Has Stayed Constant". Space.com. Retrieved 2012-12-14.

[33] "Particle chameleon caught in the act of changing". CERN. 31 May 2010. Retrieved 2012-07-05.

[34] S. Weinberg (1979). "Baryon and Lepton Nonconserving Processes". *Physical Review Letters* **43** (21): 1566. Bibcode:1979PhRvL..43.1566W. doi:10.1103/PhysRevLett.43.1566.

[35] P. Minkowski (1977). "$\mu \to e \gamma$ at a Rate of One Out of 10^9 Muon Decays?". *Physics Letters B* **67** (4): 421. Bibcode:1977PhLB...67..421M. doi:10.1016/0370-2693(77)90435-X.

[36] R. N. Mohapatra, G. Senjanovic (1980). "Neutrino Mass and Spontaneous Parity Nonconservation". *Physical Review Letters* **44** (14): 912–915. Bibcode:1980PhRvL..44..912M. doi:10.1103/PhysRevLett.44.912.

[37] M. Gell-Mann, P. Ramond and R. Slansky (1979). F. van Nieuwenhuizen and D. Z. Freedman, ed. *Supergravity*. North Holland. pp. 315–321. ISBN 0-444-85438-X.

[38] Salvio, Strumia (2014-03-17). "Agravity". *JHEP 1406 (2014) 080*. arXiv:1403.4226. Bibcode:2014JHEP...06..080S. doi:10.1007/JHEP06(2014)080.

6.11 Further reading

- R. Oerter (2006). *The Theory of Almost Everything: The Standard Model, the Unsung Triumph of Modern Physics*. Plume.
- B.A. Schumm (2004). *Deep Down Things: The Breathtaking Beauty of Particle Physics*. Johns Hopkins University Press. ISBN 0-8018-7971-X.

Introductory textbooks

- I. Aitchison, A. Hey (2003). *Gauge Theories in Particle Physics: A Practical Introduction*. Institute of Physics. ISBN 978-0-585-44550-2.
- W. Greiner, B. Müller (2000). *Gauge Theory of Weak Interactions*. Springer. ISBN 3-540-67672-4.
- G.D. Coughlan, J.E. Dodd, B.M. Gripaios (2006). *The Ideas of Particle Physics: An Introduction for Scientists*. Cambridge University Press.
- D.J. Griffiths (1987). *Introduction to Elementary Particles*. John Wiley & Sons. ISBN 0-471-60386-4.
- G.L. Kane (1987). *Modern Elementary Particle Physics*. Perseus Books. ISBN 0-201-11749-5.

Advanced textbooks

- T.P. Cheng, L.F. Li (2006). *Gauge theory of elementary particle physics*. Oxford University Press. ISBN 0-19-851961-3. Highlights the gauge theory aspects of the Standard Model.
- J.F. Donoghue, E. Golowich, B.R. Holstein (1994). *Dynamics of the Standard Model*. Cambridge University Press. ISBN 978-0-521-47652-2. Highlights dynamical and phenomenological aspects of the Standard Model.
- L. O'Raifeartaigh (1988). *Group structure of gauge theories*. Cambridge University Press. ISBN 0-521-34785-8.
- Nagashima Y. Elementary Particle Physics: Foundations of the Standard Model, Volume 2. (Wiley 2013) 920 рапуы
- Schwartz, M.D. Quantum Field Theory and the Standard Model (Cambridge University Press 2013) 952 pages
- Langacker P. The standard model and beyond. (CRC Press, 2010) 670 pages Highlights group-theoretical aspects of the Standard Model.

Journal articles

- E.S. Abers, B.W. Lee (1973). "Gauge theories". *Physics Reports* **9**: 1–141. Bibcode:1973PhR.....9....1A. doi:10.1016/0370-1573(73)90027-6.
- M. Baak et al. (2012). "The Electroweak Fit of the Standard Model after the Discovery of a New Boson at the LHC". *The European Physical Journal C* **72** (11). arXiv:1209.2716. Bibcode:2012EPJC...72.2205B. doi:10.1140/epjc/s10052-012-2205-9.
- Y. Hayato et al. (1999). "Search for Proton Decay through $p \to \nu K^+$ in a Large Water Cherenkov Detector". *Physical Review Letters* **83** (8): 1529. arXiv:hep-ex/9904020. Bibcode:1999PhRvL..83.1529H. doi:10.1103/PhysRevLett.83.1529.
- S.F. Novaes (2000). "Standard Model: An Introduction". arXiv:hep-ph/0001283 [hep-ph].

- D.P. Roy (1999). “Basic Constituents of Matter and their Interactions — A Progress Report”. arXiv:hep-ph/9912523 [hep-ph].
- F. Wilczek (2004). “The Universe Is A Strange Place”. *Nuclear Physics B - Proceedings Supplements* **134**: 3. arXiv:astro-ph/0401347. Bibcode:2004NuPhS.134....3W. doi:10.1016/j.nuclphysbps.2004.08.001.

6.12 External links

- "The Standard Model explained in Detail by CERN’s John Ellis" omega tau podcast.
- "LHC sees hint of lightweight Higgs boson" "New Scientist".
- "Standard Model may be found incomplete," *New Scientist.*
- "Observation of the Top Quark" at Fermilab.
- "The Standard Model Lagrangian." After electroweak symmetry breaking, with no explicit Higgs boson.
- "Standard Model Lagrangian" with explicit Higgs terms. PDF, PostScript, and LaTeX versions.
- "The particle adventure." Web tutorial.
- Nobes, Matthew (2002) “Introduction to the Standard Model of Particle Physics” on Kuro5hin: Part 1, Part 2, Part 3a, Part 3b.
- "The Standard Model" The Standard Model on the CERN web site explains how the basic building blocks of matter interact, governed by four fundamental forces.

Chapter 7

Quantum field theory

"Relativistic quantum field theory" redirects here. For other uses, see Relativity.

In theoretical physics, **quantum field theory** (**QFT**) is a theoretical framework for constructing quantum mechanical models of subatomic particles in particle physics and quasiparticles in condensed matter physics. A QFT treats particles as excited states of an underlying physical field, so these are called field quanta.

In QFT, quantum mechanical interactions between particles are described by interaction terms between the corresponding underlying fields.

7.1 Definition

Quantum electrodynamics (QED) has one electron field and one photon field; quantum chromodynamics (QCD) has one field for each type of quark; and, in condensed matter, there is an atomic displacement field that gives rise to phonon particles. Edward Witten describes QFT as "by far" the most difficult theory in modern physics.[1]

7.1.1 Dynamics

See also: Relativistic dynamics

Ordinary quantum mechanical systems have a fixed number of particles, with each particle having a finite number of degrees of freedom. In contrast, the excited states of a QFT can represent any number of particles. This makes quantum field theories especially useful for describing systems where the particle count/number may change over time, a crucial feature of relativistic dynamics.

7.1.2 States

QFT interaction terms are similar in spirit to those between charges with electric and magnetic fields in Maxwell's equations. However, unlike the classical fields of Maxwell's theory, fields in QFT generally exist in quantum superpositions of states and are subject to the laws of quantum mechanics.

Because the fields are continuous quantities over space, there exist excited states with arbitrarily large numbers of particles in them, providing QFT systems with an effectively infinite number of degrees of freedom. Infinite degrees of freedom can easily lead to divergences of calculated quantities (i.e., the quantities become infinite). Techniques such as renormalization of QFT parameters or discretization of spacetime, as in lattice QCD, are often used to avoid such infinities so as to yield physically meaningful results.

7.1.3 Fields and radiation

The gravitational field and the electromagnetic field are the only two fundamental fields in nature that have infinite range and a corresponding classical low-energy limit, which greatly diminishes and hides their "particle-like" excitations. Albert Einstein in 1905, attributed "particle-like" and discrete exchanges of momenta and energy, characteristic of "field quanta", to the electromagnetic field. Originally, his principal motivation was to explain the thermodynamics of radiation. Although the photoelectric effect and Compton scattering strongly suggest the existence of the photon, it might alternately be explained by a mere quantization of emission; more definitive evidence of the quantum nature of radiation is now taken up into modern quantum optics as in the antibunching effect.[2]

7.2 Theories

There is currently no complete quantum theory of the remaining fundamental force, gravity. Many of the proposed theories to describe gravity as a QFT postulate the existence of a graviton particle that mediates the gravitational force. Presumably, the as yet unknown correct quantum field-theoretic treatment of the gravitational field will behave like Einstein's general theory of relativity in the low-energy limit. Quantum field theory of the fundamental forces itself has been postulated to be the low-energy effective field theory limit of a more fundamental theory such as superstring theory.

Most theories in standard particle physics are formulated as **relativistic quantum field theories**, such as QED, QCD, and the Standard Model. QED, the quantum field-theoretic description of the electromagnetic field, approximately reproduces Maxwell's theory of electrodynamics in the low-energy limit, with small non-linear corrections to the Maxwell equations required due to virtual electron–positron pairs.

In the perturbative approach to quantum field theory, the full field interaction terms are approximated as a perturbative expansion in the number of particles involved. Each term in the expansion can be thought of as forces between particles being mediated by other particles. In QED, the electromagnetic force between two electrons is caused by an exchange of photons. Similarly, intermediate vector bosons mediate the weak force and gluons mediate the strong force in QCD. The notion of a force-mediating particle comes from perturbation theory, and does not make sense in the context of non-perturbative approaches to QFT, such as with bound states.

7.3 History

Main article: History of quantum field theory

7.3.1 Foundations

The early development of the field involved Dirac, Fock, Pauli, Heisenberg and Bogolyubov. This phase of development culminated with the construction of the theory of quantum electrodynamics in the 1950s.

7.3.2 Gauge theory

Gauge theory was formulated and quantized, leading to the **unification of forces** embodied in the standard model of particle physics. This effort started in the 1950s with the work of Yang and Mills, was carried on by Martinus Veltman and a host of others during the 1960s and completed by the 1970s through the work of Gerard 't Hooft, Frank Wilczek, David Gross and David Politzer.

7.3.3 Grand synthesis

Parallel developments in the understanding of phase transitions in condensed matter physics led to the study of the renormalization group. This in turn led to the **grand synthesis** of theoretical physics, which unified theories of particle and condensed matter physics through quantum field theory. This involved the work of Michael Fisher and

Leo Kadanoff in the 1970s, which led to the seminal reformulation of quantum field theory by Kenneth G. Wilson in 1975.

7.4 Principles

7.4.1 Classical and quantum fields

Main article: Classical field theory

A classical field is a function defined over some region of space and time.[3] Two physical phenomena which are described by classical fields are Newtonian gravitation, described by Newtonian gravitational field **g**(**x**, t), and classical electromagnetism, described by the electric and magnetic fields **E**(**x**, t) and **B**(**x**, t). Because such fields can in principle take on distinct values at each point in space, they are said to have infinite degrees of freedom.[3]

Classical field theory does not, however, account for the quantum-mechanical aspects of such physical phenomena. For instance, it is known from quantum mechanics that certain aspects of electromagnetism involve discrete particles—photons—rather than continuous fields. The business of *quantum* field theory is to write down a field that is, like a classical field, a function defined over space and time, but which also accommodates the observations of quantum mechanics. This is a *quantum field.*

It is not immediately clear *how* to write down such a quantum field, since quantum mechanics has a structure very unlike a field theory. In its most general formulation, quantum mechanics is a theory of abstract operators (observables) acting on an abstract state space (Hilbert space), where the observables represent physically observable quantities and the state space represents the possible states of the system under study.[4] For instance, the fundamental observables associated with the motion of a single quantum mechanical particle are the position and momentum operators $\hat{x}$ and $\hat{p}$. Field theory, in contrast, treats x as a way to index the field rather than as an operator.[5]

There are two common ways of developing a quantum field: the path integral formalism and canonical quantization.[6] The latter of these is pursued in this article.

Lagrangian formalism

Quantum field theory frequently makes use of the Lagrangian formalism from classical field theory. This formalism is analogous to the Lagrangian formalism used in classical mechanics to solve for the motion of a particle under the influence of a field. In classical field theory, one writes down a Lagrangian density, $\mathcal{L}$, involving a field, $\varphi(\mathbf{x},t)$, and possibly its first derivatives ($\partial\varphi/\partial t$ and $\nabla\varphi$), and then applies a field-theoretic form of the Euler–Lagrange equation. Writing coordinates $(t, \mathbf{x}) = (x^0, x^1, x^2, x^3) = x^\mu$, this form of the Euler–Lagrange equation is[3]

$$\frac{\partial}{\partial x^\mu}\left[\frac{\partial\mathcal{L}}{\partial(\partial\phi/\partial x^\mu)}\right]-\frac{\partial\mathcal{L}}{\partial\phi}=0,$$

where a sum over μ is performed according to the rules of Einstein notation.

By solving this equation, one arrives at the "equations of motion" of the field.[3] For example, if one begins with the Lagrangian density

$$\mathcal{L}(\phi,\nabla\phi)=-\rho(t,\mathbf{x})\,\phi(t,\mathbf{x})-\frac{1}{8\pi G}|\nabla\phi|^2,$$

and then applies the Euler–Lagrange equation, one obtains the equation of motion

$$4\pi G\rho(t,\mathbf{x})=\nabla^2\phi.$$

This equation is Newton's law of universal gravitation, expressed in differential form in terms of the gravitational potential $\varphi(t, \mathbf{x})$ and the mass density $\rho(t, \mathbf{x})$. Despite the nomenclature, the "field" under study is the gravitational potential, φ, rather than the gravitational field, **g**. Similarly, when classical field theory is used to study electromagnetism, the "field" of interest is the electromagnetic four-potential (V/c, **A**), rather than the electric and magnetic fields **E** and **B**.

Quantum field theory uses this same Lagrangian procedure to determine the equations of motion for quantum fields. These equations of motion are then supplemented by commutation relations derived from the canonical quantization procedure described below, thereby incorporating quantum mechanical effects into the behavior of the field.

7.4.2 Single- and many-particle quantum mechanics

Main articles: Quantum mechanics and First quantization

In quantum mechanics, a particle (such as an electron or proton) is described by a complex wavefunction, $\psi(x, t)$, whose time-evolution is governed by the Schrödinger equation:

$$-\frac{\hbar^2}{2m}\frac{\partial^2}{\partial x^2}\psi(x,t) + V(x)\psi(x,t) = i\hbar\frac{\partial}{\partial t}\psi(x,t).$$

Here m is the particle's mass and $V(x)$ is the applied potential. Physical information about the behavior of the particle is extracted from the wavefunction by constructing expected values for various quantities; for example, the expected value of the particle's position is given by integrating $\psi^*(x)\, x\, \psi(x)$ over all space, and the expected value of the particle's momentum is found by integrating $-i\hbar\psi^*(x)\mathrm{d}\psi/\mathrm{d}x$. The quantity $\psi^*(x)\psi(x)$ is itself in the Copenhagen interpretation of quantum mechanics interpreted as a probability density function. This treatment of quantum mechanics, where a particle's wavefunction evolves against a classical background potential $V(x)$, is sometimes called *first quantization*.

This description of quantum mechanics can be extended to describe the behavior of multiple particles, so long as the number and the type of particles remain fixed. The particles are described by a wavefunction $\psi(x_1, x_2, \ldots, xN, t)$, which is governed by an extended version of the Schrödinger equation.

Often one is interested in the case where N particles are all of the same type (for example, the 18 electrons orbiting a neutral argon nucleus). As described in the article on identical particles, this implies that the state of the entire system must be either symmetric (bosons) or antisymmetric (fermions) when the coordinates of its constituent particles are exchanged. This is achieved by using a Slater determinant as the wavefunction of a fermionic system (and a Slater permanent for a bosonic system), which is equivalent to an element of the symmetric or antisymmetric subspace of a tensor product.

For example, the general quantum state of a system of N bosons is written as

$$|\phi_1 \cdots \phi_N\rangle = \sqrt{\frac{\prod_j N_j!}{N!}} \sum_{p\in S_N} |\phi_{p(1)}\rangle \otimes \cdots \otimes |\phi_{p(N)}\rangle,$$

where $|\phi_i\rangle$ are the single-particle states, Nj is the number of particles occupying state j, and the sum is taken over all possible permutations p acting on N elements. In general, this is a sum of $N!$ (N factorial) distinct terms. $\sqrt{\frac{\prod_j N_j!}{N!}}$ is a normalizing factor.

There are several shortcomings to the above description of quantum mechanics, which are addressed by quantum field theory. First, it is unclear how to extend quantum mechanics to include the effects of special relativity.[7] Attempted replacements for the Schrödinger equation, such as the Klein–Gordon equation or the Dirac equation, have many unsatisfactory qualities; for instance, they possess energy eigenvalues that extend to $-\infty$, so that there seems to be no easy definition of a ground state. It turns out that such inconsistencies arise from relativistic wavefunctions not having a well-defined probabilistic interpretation in position space, as probability conservation is not a relativistically covariant concept. The second shortcoming, related to the first, is that in quantum mechanics there is no mechanism to describe particle creation and annihilation;[8] this is crucial for describing phenomena such as pair production, which result from the conversion between mass and energy according to the relativistic relation $E = mc^2$.

7.4.3 Second quantization

Main article: Second quantization

In this section, we will describe a method for constructing a quantum field theory called **second quantization**. This basically involves choosing a way to index the quantum mechanical degrees of freedom in the space of multiple identical-particle states. It is based on the Hamiltonian formulation of quantum mechanics.

Several other approaches exist, such as the Feynman path integral,[9] which uses a Lagrangian formulation. For an overview of some of these approaches, see the article on quantization.

Bosons

For simplicity, we will first discuss second quantization for bosons, which form perfectly symmetric quantum states. Let us denote the mutually orthogonal single-particle states which are possible in the system by $|\phi_1\rangle, |\phi_2\rangle, |\phi_3\rangle$, and so on. For example, the 3-particle state with one particle in state $|\phi_1\rangle$ and two in state $|\phi_2\rangle$ is

$$\frac{1}{\sqrt{3}}\left[|\phi_1\rangle|\phi_2\rangle|\phi_2\rangle + |\phi_2\rangle|\phi_1\rangle|\phi_2\rangle + |\phi_2\rangle|\phi_2\rangle|\phi_1\rangle\right].$$

The first step in second quantization is to express such quantum states in terms of **occupation numbers**, by listing the number of particles occupying each of the single-particle states $|\phi_1\rangle, |\phi_2\rangle$, etc. This is simply another way of labelling the states. For instance, the above 3-particle state is denoted as

$$|1, 2, 0, 0, 0, \dots\rangle.$$

An N-particle state belongs to a space of states describing systems of N particles. The next step is to combine the individual N-particle state spaces into an extended state space, known as Fock space, which can describe systems of any number of particles. This is composed of the state space of a system with no particles (the so-called vacuum state, written as $|0\rangle$), plus the state space of a 1-particle system, plus the state space of a 2-particle system, and so forth. States describing a definite number of particles are known as Fock states: a general element of Fock space will be a linear combination of Fock states. There is a one-to-one correspondence between the occupation number representation and valid boson states in the Fock space.

At this point, the quantum mechanical system has become a quantum field in the sense we described above. The field's elementary degrees of freedom are the occupation numbers, and each occupation number is indexed by a number j indicating which of the single-particle states $|\phi_1\rangle, |\phi_2\rangle, \dots, |\phi_j\rangle, \dots$ it refers to:

$$|N_1, N_2, N_3, \dots, N_j, \dots\rangle.$$

The properties of this quantum field can be explored by defining creation and annihilation operators, which add and subtract particles. They are analogous to ladder operators in the quantum harmonic oscillator problem, which added and subtracted energy quanta. However, these operators literally create and annihilate particles of a given quantum state. The bosonic annihilation operator a_2 and creation operator $a_2^\dagger$ are easily defined in the occupation number representation as having the following effects:

$$a_2|N_1, N_2, N_3, \dots\rangle = \sqrt{N_2} \mid N_1, (N_2 - 1), N_3, \dots\rangle,$$

$$a_2^\dagger|N_1, N_2, N_3, \dots\rangle = \sqrt{N_2 + 1} \mid N_1, (N_2 + 1), N_3, \dots\rangle.$$

It can be shown that these are operators in the usual quantum mechanical sense, i.e. linear operators acting on the Fock space. Furthermore, they are indeed Hermitian conjugates, which justifies the way we have written them. They can be shown to obey the commutation relation

$$[a_i, a_j] = 0 \quad , \quad \left[a_i^\dagger, a_j^\dagger\right] = 0 \quad , \quad \left[a_i, a_j^\dagger\right] = \delta_{ij},$$

where δ stands for the Kronecker delta. These are precisely the relations obeyed by the ladder operators for an infinite set of independent quantum harmonic oscillators, one for each single-particle state. Adding or removing bosons from each state is therefore analogous to exciting or de-exciting a quantum of energy in a harmonic oscillator.

Applying an annihilation operator a_k followed by its corresponding creation operator $a_k^\dagger$ returns the number N_k of particles in the kth single-particle eigenstate:

$$a_k^\dagger\, a_k|\ldots, N_k, \ldots\rangle = N_k|\ldots, N_k, \ldots\rangle.$$

The combination of operators $a_k^\dagger a_k$ is known as the number operator for the kth eigenstate.

The Hamiltonian operator of the quantum field (which, through the Schrödinger equation, determines its dynamics) can be written in terms of creation and annihilation operators. For instance, for a field of free (non-interacting) bosons, the total energy of the field is found by summing the energies of the bosons in each energy eigenstate. If the kth single-particle energy eigenstate has energy E_k and there are N_k bosons in this state, then the total energy of these bosons is $E_k N_k$. The energy in the *entire* field is then a sum over k :

$$E_{\text{tot}} = \sum_k E_k N_k$$

This can be turned into the Hamiltonian operator of the field by replacing N_k with the corresponding number operator, $a_k^\dagger a_k$. This yields

$$H = \sum_k E_k\, a_k^\dagger\, a_k.$$

Fermions

It turns out that a different definition of creation and annihilation must be used for describing fermions. According to the Pauli exclusion principle, fermions cannot share quantum states, so their occupation numbers Ni can only take on the value 0 or 1. The fermionic annihilation operators c and creation operators $c^\dagger$ are defined by their actions on a Fock state thus

$$c_j|N_1, N_2, \ldots, N_j = 0, \ldots\rangle = 0$$

$$c_j|N_1, N_2, \ldots, N_j = 1, \ldots\rangle = (-1)^{(N_1+\cdots+N_{j-1})}|N_1, N_2, \ldots, N_j = 0, \ldots\rangle$$

$$c_j^\dagger|N_1, N_2, \ldots, N_j = 0, \ldots\rangle = (-1)^{(N_1+\cdots+N_{j-1})}|N_1, N_2, \ldots, N_j = 1, \ldots\rangle$$

$$c_j^\dagger|N_1, N_2, \ldots, N_j = 1, \ldots\rangle = 0.$$

These obey an anticommutation relation:

$$\{c_i, c_j\} = 0 \quad , \quad \left\{c_i^\dagger, c_j^\dagger\right\} = 0 \quad , \quad \left\{c_i, c_j^\dagger\right\} = \delta_{ij}.$$

One may notice from this that applying a fermionic creation operator twice gives zero, so it is impossible for the particles to share single-particle states, in accordance with the exclusion principle.

Field operators

We have previously mentioned that there can be more than one way of indexing the degrees of freedom in a quantum field. Second quantization indexes the field by enumerating the single-particle quantum states. However, as we have discussed, it is more natural to think about a "field", such as the electromagnetic field, as a set of degrees of freedom indexed by position.

To this end, we can define *field operators* that create or destroy a particle at a particular point in space. In particle physics, these operators turn out to be more convenient to work with, because they make it easier to formulate theories that satisfy the demands of relativity.

Single-particle states are usually enumerated in terms of their momenta (as in the particle in a box problem.) We can construct field operators by applying the Fourier transform to the creation and annihilation operators for these states. For example, the bosonic field annihilation operator $\phi(\mathbf{r})$ is

$$\phi(\mathbf{r}) \stackrel{\text{def}}{=} \sum_j e^{i\mathbf{k}_j \cdot \mathbf{r}} a_j.$$

The bosonic field operators obey the commutation relation

$$[\phi(\mathbf{r}), \phi(\mathbf{r}')] = 0 \quad , \quad [\phi^\dagger(\mathbf{r}), \phi^\dagger(\mathbf{r}')] = 0 \quad , \quad [\phi(\mathbf{r}), \phi^\dagger(\mathbf{r}')] = \delta^3(\mathbf{r} - \mathbf{r}')$$

where $\delta(x)$ stands for the Dirac delta function. As before, the fermionic relations are the same, with the commutators replaced by anticommutators.

The field operator is not the same thing as a single-particle wavefunction. The former is an operator acting on the Fock space, and the latter is a quantum-mechanical amplitude for finding a particle in some position. However, they are closely related, and are indeed commonly denoted with the same symbol. If we have a Hamiltonian with a space representation, say

$$H = -\frac{\hbar^2}{2m} \sum_i \nabla_i^2 + \sum_{i<j} U(|\mathbf{r}_i - \mathbf{r}_j|)$$

where the indices *i* and *j* run over all particles, then the field theory Hamiltonian (in the non-relativistic limit and for negligible self-interactions) is

$$H = -\frac{\hbar^2}{2m} \int d^3r \; \phi^\dagger(\mathbf{r}) \nabla^2 \phi(\mathbf{r}) + \frac{1}{2} \int d^3r \int d^3r' \; \phi^\dagger(\mathbf{r}) \phi^\dagger(\mathbf{r}') U(|\mathbf{r} - \mathbf{r}'|) \phi(\mathbf{r}') \phi(\mathbf{r}).$$

This looks remarkably like an expression for the expectation value of the energy, with ϕ playing the role of the wavefunction. This relationship between the field operators and wavefunctions makes it very easy to formulate field theories starting from space-projected Hamiltonians.

7.4.4 Dynamics

Once the Hamiltonian operator is obtained as part of the canonical quantization process, the time dependence of the state is described with the Schrödinger equation, just as with other quantum theories. Alternatively, the Heisenberg picture can be used where the time dependence is in the operators rather than in the states.

7.4.5 Implications

Unification of fields and particles

The "second quantization" procedure that we have outlined in the previous section takes a set of single-particle quantum states as a starting point. Sometimes, it is impossible to define such single-particle states, and one must proceed directly to quantum field theory. For example, a quantum theory of the electromagnetic field *must* be a quantum field theory, because it is impossible (for various reasons) to define a wavefunction for a single photon.[10] In such situations, the quantum field theory can be constructed by examining the mechanical properties of the classical field and guessing the corresponding quantum theory. For free (non-interacting) quantum fields, the quantum field theories obtained in this way have the same properties as those obtained using second quantization, such as well-defined creation and annihilation operators obeying commutation or anticommutation relations.

Quantum field theory thus provides a unified framework for describing "field-like" objects (such as the electromagnetic field, whose excitations are photons) and "particle-like" objects (such as electrons, which are treated as excitations of an underlying electron field), so long as one can treat interactions as "perturbations" of free fields. There are still unsolved problems relating to the more general case of interacting fields that may or may not be adequately described by perturbation theory. For more on this topic, see Haag's theorem.

Physical meaning of particle indistinguishability

The second quantization procedure relies crucially on the particles being identical. We would not have been able to construct a quantum field theory from a distinguishable many-particle system, because there would have been no way of separating and indexing the degrees of freedom.

Many physicists prefer to take the converse interpretation, which is that *quantum field theory explains what identical particles are.* In ordinary quantum mechanics, there is not much theoretical motivation for using symmetric (bosonic) or antisymmetric (fermionic) states, and the need for such states is simply regarded as an empirical fact. From the point of view of quantum field theory, particles are identical if and only if they are excitations of the same underlying quantum field. Thus, the question "why are all electrons identical?" arises from mistakenly regarding individual electrons as fundamental objects, when in fact it is only the electron field that is fundamental.

Particle conservation and non-conservation

During second quantization, we started with a Hamiltonian and state space describing a fixed number of particles (N), and ended with a Hamiltonian and state space for an arbitrary number of particles. Of course, in many common situations N is an important and perfectly well-defined quantity, e.g. if we are describing a gas of atoms sealed in a box. From the point of view of quantum field theory, such situations are described by quantum states that are eigenstates of the number operator $\hat{N}$, which measures the total number of particles present. As with any quantum mechanical observable, $\hat{N}$ is conserved if it commutes with the Hamiltonian. In that case, the quantum state is trapped in the N-particle subspace of the total Fock space, and the situation could equally well be described by ordinary N-particle quantum mechanics. (Strictly speaking, this is only true in the noninteracting case or in the low energy density limit of renormalized quantum field theories)

For example, we can see that the free-boson Hamiltonian described above conserves particle number. Whenever the Hamiltonian operates on a state, each particle destroyed by an annihilation operator a_k is immediately put back by the creation operator $a_k^\dagger$.

On the other hand, it is possible, and indeed common, to encounter quantum states that are *not* eigenstates of $\hat{N}$, which do not have well-defined particle numbers. Such states are difficult or impossible to handle using ordinary quantum mechanics, but they can be easily described in quantum field theory as quantum superpositions of states having different values of N. For example, suppose we have a bosonic field whose particles can be created or destroyed by interactions with a fermionic field. The Hamiltonian of the combined system would be given by the Hamiltonians of the free boson and free fermion fields, plus a "potential energy" term such as

$$H_I = \sum_{k,q} V_q (a_q + a_{-q}^\dagger) c_{k+q}^\dagger c_k,$$

where $a_k^\dagger$ and a_k denotes the bosonic creation and annihilation operators, $c_k^\dagger$ and c_k denotes the fermionic creation and annihilation operators, and V_q is a parameter that describes the strength of the interaction. This "interaction term" describes processes in which a fermion in state k either absorbs or emits a boson, thereby being kicked into a different eigenstate $k+q$. (In fact, this type of Hamiltonian is used to describe interaction between conduction electrons and phonons in metals. The interaction between electrons and photons is treated in a similar way, but is a little more complicated because the role of spin must be taken into account.) One thing to notice here is that even if we start out with a fixed number of bosons, we will typically end up with a superposition of states with different numbers of bosons at later times. The number of fermions, however, is conserved in this case.

In condensed matter physics, states with ill-defined particle numbers are particularly important for describing the various superfluids. Many of the defining characteristics of a superfluid arise from the notion that its quantum state is a superposition of states with different particle numbers. In addition, the concept of a coherent state (used to model the laser and the BCS ground state) refers to a state with an ill-defined particle number but a well-defined phase.

7.4.6 Axiomatic approaches

The preceding description of quantum field theory follows the spirit in which most physicists approach the subject. However, it is not mathematically rigorous. Over the past several decades, there have been many attempts to put quantum field theory on a firm mathematical footing by formulating a set of axioms for it. These attempts fall into two broad classes.

The first class of axioms, first proposed during the 1950s, include the Wightman, Osterwalder–Schrader, and Haag–Kastler systems. They attempted to formalize the physicists' notion of an "operator-valued field" within the context of functional analysis, and enjoyed limited success. It was possible to prove that any quantum field theory satisfying these axioms satisfied certain general theorems, such as the spin-statistics theorem and the CPT theorem. Unfortunately, it proved extraordinarily difficult to show that any realistic field theory, including the Standard Model, satisfied these axioms. Most of the theories that could be treated with these analytic axioms were physically trivial, being restricted to low-dimensions and lacking interesting dynamics. The construction of theories satisfying one of these sets of axioms falls in the field of constructive quantum field theory. Important work was done in this area in the 1970s by Segal, Glimm, Jaffe and others.

During the 1980s, a second set of axioms based on geometric ideas was proposed. This line of investigation, which restricts its attention to a particular class of quantum field theories known as topological quantum field theories, is associated most closely with Michael Atiyah and Graeme Segal, and was notably expanded upon by Edward Witten, Richard Borcherds, and Maxim Kontsevich. However, most of the physically relevant quantum field theories, such as the Standard Model, are not topological quantum field theories; the quantum field theory of the fractional quantum Hall effect is a notable exception. The main impact of axiomatic topological quantum field theory has been on mathematics, with important applications in representation theory, algebraic topology, and differential geometry.

Finding the proper axioms for quantum field theory is still an open and difficult problem in mathematics. One of the Millennium Prize Problems—proving the existence of a mass gap in Yang–Mills theory—is linked to this issue.

7.5 Associated phenomena

In the previous part of the article, we described the most general features of quantum field theories. Some of the quantum field theories studied in various fields of theoretical physics involve additional special ideas, such as renormalizability, gauge symmetry, and supersymmetry. These are described in the following sections.

7.5.1 Renormalization

Main article: Renormalization

Early in the history of quantum field theory, it was found that many seemingly innocuous calculations, such as the perturbative shift in the energy of an electron due to the presence of the electromagnetic field, give infinite results. The reason is that the perturbation theory for the shift in an energy involves a sum over all other energy levels, and there are infinitely many levels at short distances that each give a finite contribution which results in a divergent series.

Many of these problems are related to failures in classical electrodynamics that were identified but unsolved in the 19th century, and they basically stem from the fact that many of the supposedly "intrinsic" properties of an electron are tied to the electromagnetic field that it carries around with it. The energy carried by a single electron—its self energy—is not simply the bare value, but also includes the energy contained in its electromagnetic field, its attendant cloud of photons. The energy in a field of a spherical source diverges in both classical and quantum mechanics, but as discovered by Weisskopf with help from Furry, in quantum mechanics the divergence is much milder, going only as the logarithm of the radius of the sphere.

The solution to the problem, presciently suggested by Stueckelberg, independently by Bethe after the crucial experiment by Lamb, implemented at one loop by Schwinger, and systematically extended to all loops by Feynman and Dyson, with converging work by Tomonaga in isolated postwar Japan, comes from recognizing that all the infinities in the interactions of photons and electrons can be isolated into redefining a finite number of quantities in the equations by replacing them with the observed values: specifically the electron's mass and charge: this is called renormalization. The technique of renormalization recognizes that the problem is essentially purely mathematical, that extremely short distances are at fault. In order to define a theory on a continuum, first place a cutoff on the fields, by postulating that quanta cannot have energies above some extremely high value. This has the effect of replacing continuous space by a structure where very short wavelengths do not exist, as on a lattice. Lattices break rotational symmetry, and one of the crucial contributions made by Feynman, Pauli and Villars, and modernized by 't Hooft and Veltman, is a symmetry-preserving cutoff for perturbation theory (this process is called regularization). There is no known symmetrical cutoff outside of perturbation theory, so for rigorous or numerical work people often use an actual lattice.

On a lattice, every quantity is finite but depends on the spacing. When taking the limit of zero spacing, we make sure that the physically observable quantities like the observed electron mass stay fixed, which means that the constants in the Lagrangian defining the theory depend on the spacing. Hopefully, by allowing the constants to vary with the lattice spacing, all the results at long distances become insensitive to the lattice, defining a continuum limit.

The renormalization procedure only works for a certain class of quantum field theories, called **renormalizable quantum field theories**. A theory is **perturbatively renormalizable** when the constants in the Lagrangian only diverge at worst as logarithms of the lattice spacing for very short spacings. The continuum limit is then well defined in perturbation theory, and even if it is not fully well defined non-perturbatively, the problems only show up at distance scales that are exponentially small in the inverse coupling for weak couplings. The Standard Model of particle physics is perturbatively renormalizable, and so are its component theories (quantum electrodynamics/electroweak theory and quantum chromodynamics). Of the three components, quantum electrodynamics is believed to not have a continuum limit, while the asymptotically free SU(2) and SU(3) weak hypercharge and strong color interactions are nonperturbatively well defined.

The renormalization group describes how renormalizable theories emerge as the long distance low-energy effective field theory for any given high-energy theory. Because of this, renormalizable theories are insensitive to the precise nature of the underlying high-energy short-distance phenomena. This is a blessing because it allows physicists to formulate low energy theories without knowing the details of high energy phenomenon. It is also a curse, because once a renormalizable theory like the standard model is found to work, it gives very few clues to higher energy processes. The only way high energy processes can be seen in the standard model is when they allow otherwise forbidden events, or if they predict quantitative relations between the coupling constants.

7.5.2 Haag's theorem

See also: Haag's theorem

From a mathematically rigorous perspective, there exists no interaction picture in a Lorentz-covariant quantum field theory. This implies that the perturbative approach of Feynman diagrams in QFT is not strictly justified, despite producing vastly precise predictions validated by experiment. This is called Haag's theorem, but most particle physicists relying on QFT largely shrug it off.

7.5.3 Gauge freedom

A gauge theory is a theory that admits a symmetry with a local parameter. For example, in every quantum theory the global phase of the wave function is arbitrary and does not represent something physical. Consequently, the theory is invariant under a global change of phases (adding a constant to the phase of all wave functions, everywhere); this is a global symmetry. In quantum electrodynamics, the theory is also invariant under a *local* change of phase, that is – one may shift the phase of all wave functions so that the shift may be different at every point in space-time. This is a *local* symmetry. However, in order for a well-defined derivative operator to exist, one must introduce a new field, the gauge field, which also transforms in order for the local change of variables (the phase in our example) not to affect the derivative. In quantum electrodynamics this gauge field is the electromagnetic field. The change of local gauge of variables is termed gauge transformation. It is worth noting that by Noether's theorem, for every such symmetry there exists an associated conserved current. The aforementioned symmetry of the wavefunction under global phase changes implies the conservation of electric charge.

In quantum field theory the excitations of fields represent particles. The particle associated with excitations of the gauge field is the gauge boson, which is the photon in the case of quantum electrodynamics.

The degrees of freedom in quantum field theory are local fluctuations of the fields. The existence of a gauge symmetry reduces the number of degrees of freedom, simply because some fluctuations of the fields can be transformed to zero by gauge transformations, so they are equivalent to having no fluctuations at all, and they therefore have no physical meaning. Such fluctuations are usually called "non-physical degrees of freedom" or *gauge artifacts*; usually some of them have a negative norm, making them inadequate for a consistent theory. Therefore, if a classical field theory has a gauge symmetry, then its quantized version (i.e. the corresponding quantum field theory) will have this symmetry as well. In other words, a gauge symmetry cannot have a quantum anomaly. If a gauge symmetry is anomalous (i.e. not kept in the quantum theory) then the theory is non-consistent: for example, in quantum electrodynamics, had there been a gauge anomaly, this would require the appearance of photons with longitudinal polarization and polarization in the time direction, the latter having a negative norm, rendering the theory inconsistent; another possibility would be for these photons to appear only in intermediate processes but not in the final products of any interaction, making the theory non-unitary and again inconsistent (see optical theorem).

In general, the gauge transformations of a theory consist of several different transformations, which may not be commutative. These transformations are together described by a mathematical object known as a gauge group. Infinitesimal gauge transformations are the gauge group generators. Therefore the number of gauge bosons is the group dimension (i.e. number of generators forming a basis).

All the fundamental interactions in nature are described by gauge theories. These are:

- Quantum chromodynamics, whose gauge group is **SU**(3). The gauge bosons are eight gluons.
- The electroweak theory, whose gauge group is **U**(1) × **SU**(2), (a direct product of **U**(1) and **SU**(2)).
- Gravity, whose classical theory is general relativity, admits the equivalence principle, which is a form of gauge symmetry. However, it is explicitly non-renormalizable.

7.5.4 Multivalued gauge transformations

The gauge transformations which leave the theory invariant involve, by definition, only single-valued gauge functions $\Lambda(x_i)$ which satisfy the Schwarz integrability criterion

$$\partial_{x_i x_j}\Lambda = \partial_{x_j x_i}\Lambda.$$

An interesting extension of gauge transformations arises if the gauge functions $\Lambda(x_i)$ are allowed to be multivalued functions which violate the integrability criterion. These are capable of changing the physical field strengths and are therefore not proper symmetry transformations. Nevertheless, the transformed field equations describe correctly the physical laws in the presence of the newly generated field strengths. See the textbook by H. Kleinert cited below for the applications to phenomena in physics.

7.5.5 Supersymmetry

Main article: Supersymmetry

Supersymmetry assumes that every fundamental fermion has a superpartner that is a boson and vice versa. It was introduced in order to solve the so-called Hierarchy Problem, that is, to explain why particles not protected by any symmetry (like the Higgs boson) do not receive radiative corrections to its mass driving it to the larger scales (GUT, Planck...). It was soon realized that supersymmetry has other interesting properties: its gauged version is an extension of general relativity (Supergravity), and it is a key ingredient for the consistency of string theory.

The way supersymmetry protects the hierarchies is the following: since for every particle there is a superpartner with the same mass, any loop in a radiative correction is cancelled by the loop corresponding to its superpartner, rendering the theory UV finite.

Since no superpartners have yet been observed, if supersymmetry exists it must be broken (through a so-called soft term, which breaks supersymmetry without ruining its helpful features). The simplest models of this breaking require that the energy of the superpartners not be too high; in these cases, supersymmetry is expected to be observed by experiments at the Large Hadron Collider. The Higgs particle has been detected at the LHC, and no such superparticles have been discovered.

7.6 See also

- Abraham–Lorentz force
- Basic concepts of quantum mechanics
- Common integrals in quantum field theory
- Constructive quantum field theory
- Einstein–Maxwell–Dirac equations
- Feynman path integral
- Form factor (quantum field theory)
- Fundamental equation of unified field theory
- Green–Kubo relations
- Green's function (many-body theory)
- Invariance mechanics
- List of quantum field theories
- Pauli exclusion principle
- Photon polarization
- Pseudoscalar Field
- Quantum field theory in curved spacetime
- Quantum flavordynamics
- Quantum geometrodynamics
- Quantum hydrodynamics
- Quantum magnetodynamics
- Quantum triviality
- Relation between Schrödinger's equation and the path integral formulation of quantum mechanics
- Relationship between string theory and quantum field theory
- Schwinger–Dyson equation
- Static forces and virtual-particle exchange
- Symmetry in quantum mechanics
- Theoretical and experimental justification for the Schrödinger equation
- Ward–Takahashi identity
- Wheeler–Feynman absorber theory
- Wigner's classification
- Wigner's theorem

7.7 Notes

7.8 References

[1] "Beautiful Minds, Vol. 20: Ed Witten". la Repubblica. 2010. Retrieved 22 June 2012. See here.

[2] J. J. Thorn et al. (2004). Observing the quantum behavior of light in an undergraduate laboratory. . J. J. Thorn, M. S. Neel, V. W. Donato, G. S. Bergreen, R. E. Davies, and M. Beck. American Association of Physics Teachers, 2004.DOI: 10.1119/1.1737397.

[3] David Tong, *Lectures on Quantum Field Theory*, chapter 1.

[4] Srednicki, Mark. *Quantum Field Theory* (1st ed.). p. 19.

[5] Srednicki, Mark. *Quantum Field Theory* (1st ed.). pp. 25–6.

[6] Zee, Anthony. *Quantum Field Theory in a Nutshell* (2nd ed.). p. 61.

[7] David Tong, *Lectures on Quantum Field Theory*, Introduction.

[8] Zee, Anthony. *Quantum Field Theory in a Nutshell* (2nd ed.). p. 3.

[9] Abraham Pais, *Inward Bound: Of Matter and Forces in the Physical World* ISBN 0-19-851997-4. Pais recounts how his astonishment at the rapidity with which Feynman could calculate using his method. Feynman's method is now part of the standard methods for physicists.

[10] Newton, T.D.; Wigner, E.P. (1949). "Localized states for elementary systems". *Reviews of Modern Physics* **21** (3): 400–406. Bibcode:1949RvMP...21..400N. doi:10.1103/RevModPhys.21.400.

7.9 Further reading

General readers

- Feynman, R.P. (2001) [1964]. *The Character of Physical Law.* MIT Press. ISBN 0-262-56003-8.
- Feynman, R.P. (2006) [1985]. *QED: The Strange Theory of Light and Matter.* Princeton University Press. ISBN 0-691-12575-9.
- Gribbin, J. (1998). *Q is for Quantum: Particle Physics from A to Z.* Weidenfeld & Nicolson. ISBN 0-297-81752-3.
- Schumm, Bruce A. (2004) *Deep Down Things.* Johns Hopkins Univ. Press. Chpt. 4.

Introductory texts

- McMahon, D. (2008). *Quantum Field Theory.* McGraw-Hill. ISBN 978-0-07-154382-8.
- Bogoliubov, N.; Shirkov, D. (1982). *Quantum Fields.* Benjamin-Cummings. ISBN 0-8053-0983-7.
- Frampton, P.H. (2000). *Gauge Field Theories. Frontiers in Physics (2nd ed.). Wiley.*
- Greiner, W; Müller, B. (2000). *Gauge Theory of Weak Interactions.* Springer. ISBN 3-540-67672-4.
- Itzykson, C.; Zuber, J.-B. (1980). *Quantum Field Theory.* McGraw-Hill. ISBN 0-07-032071-3.
- Kane, G.L. (1987). *Modern Elementary Particle Physics.* Perseus Books. ISBN 0-201-11749-5.
- Kleinert, H.; Schulte-Frohlinde, Verena (2001). *Critical Properties of φ^4-Theories.* World Scientific. ISBN 981-02-4658-7.
- Kleinert, H. (2008). *Multivalued Fields in Condensed Matter, Electrodynamics, and Gravitation* (PDF). World Scientific. ISBN 978-981-279-170-2.
- Loudon, R (1983). *The Quantum Theory of Light.* Oxford University Press. ISBN 0-19-851155-8.

- Mandl, F.; Shaw, G. (1993). *Quantum Field Theory*. John Wiley & Sons. ISBN 978-0-471-94186-6.
- Peskin, M.; Schroeder, D. (1995). *An Introduction to Quantum Field Theory*. Westview Press. ISBN 0-201-50397-2.
- Ryder, L.H. (1985). *Quantum Field Theory*. Cambridge University Press. ISBN 0-521-33859-X.
- Schwartz, M.D. (2014). *Quantum Field Theory and the Standard Model*. Cambridge University Press. ISBN 978-1107034730.
- Srednicki, Mark (2007) *Quantum Field Theory*. Cambridge Univ. Press.
- Ynduráin, F.J. (1996). *Relativistic Quantum Mechanics and Introduction to Field Theory* (1st ed.). Springer. ISBN 978-3-540-60453-2.
- Zee, A. (2003). *Quantum Field Theory in a Nutshell*. Princeton University Press. ISBN 0-691-01019-6.

Advanced texts

- Brown, Lowell S. (1994). *Quantum Field Theory*. Cambridge University Press. ISBN 978-0-521-46946-3.
- Bogoliubov, N.; Logunov, A.A.; Oksak, A.I.; Todorov, I.T. (1990). *General Principles of Quantum Field Theory*. Kluwer Academic Publishers. ISBN 978-0-7923-0540-8.
- Weinberg, S. (1995). *The Quantum Theory of Fields* **1–3**. Cambridge University Press.

Articles:

- Gerard 't Hooft (2007) "The Conceptual Basis of Quantum Field Theory" in Butterfield, J., and John Earman, eds., *Philosophy of Physics, Part A*. Elsevier: 661–730.
- Frank Wilczek (1999) "Quantum field theory", *Reviews of Modern Physics* 71: S83–S95. Also doi=10.1103/Rev. Mod. Phys. 71.

7.10 External links

- Hazewinkel, Michiel, ed. (2001), “Quantum field theory”, *Encyclopedia of Mathematics*, Springer, ISBN 978-1-55608-010-4
- Stanford Encyclopedia of Philosophy: "Quantum Field Theory", by Meinard Kuhlmann.
- Siegel, Warren, 2005. *Fields*. A free text, also available from arXiv:hep-th/9912205.
- Quantum Field Theory by P. J. Mulders

Chapter 8

Graviton

This article is about the hypothetical particle. For other uses, see Graviton (disambiguation).

In physics, the **graviton** is a hypothetical elementary particle that mediates the force of gravitation in the framework of quantum field theory. If it exists, the graviton is expected to be massless (because the gravitational force appears to have unlimited range) and must be a spin−2 boson. The spin follows from the fact that the source of gravitation is the stress–energy tensor, a second-rank tensor (compared to electromagnetism's spin-1 photon, the source of which is the four-current, a first-rank tensor). Additionally, it can be shown that any massless spin-2 field would give rise to a force indistinguishable from gravitation, because a massless spin-2 field must couple to (interact with) the stress–energy tensor in the same way that the gravitational field does. Seeing as the graviton is hypothetical, its discovery would unite quantum theory with gravity.[4] This result suggests that, if a massless spin-2 particle is discovered, it must be the graviton, so that the only experimental verification needed for the graviton may simply be the discovery of a massless spin-2 particle.[5]

8.1 Theory

The four other known forces of nature are mediated by elementary particles: electromagnetism by the photon, the strong interaction by the gluons, the Higgs field by the Higgs Boson, and the weak interaction by the W and Z bosons. The hypothesis is that the gravitational interaction is likewise mediated by an – as yet undiscovered – elementary particle, dubbed as *the graviton*. In the classical limit, the theory would reduce to general relativity and conform to Newton's law of gravitation in the weak-field limit.[6][7][8]

8.1.1 Gravitons and renormalization

When describing graviton interactions, the classical theory (i.e., the tree diagrams) and semiclassical corrections (one-loop diagrams) behave normally, but Feynman diagrams with two (or more) loops lead to ultraviolet divergences; that is, infinite results that cannot be removed because the quantized general relativity is not renormalizable, unlike quantum electrodynamics. That is, the usual ways physicists calculate the probability that a particle will emit or absorb a graviton give nonsensical answers and the theory loses its predictive power. These problems, together with some conceptual puzzles, led many physicists to believe that a theory more complete than quantized general relativity must describe the behavior near the Planck scale.

8.1.2 Comparison with other forces

Unlike the force carriers of the other forces, gravitation plays a special role in general relativity in defining the spacetime in which events take place. In some descriptions, matter modifies the 'shape' of spacetime itself, and gravity is a result of this shape, an idea which at first glance may appear hard to match with the idea of a force acting between particles.[9] Because the diffeomorphism invariance of the theory does not allow any particular space-time background to be singled out as the "true" space-time background, general relativity is said to be background independent. In contrast, the Standard Model is *not* background independent, with Minkowski space enjoying a special status as the fixed background space-time.[10] A theory of quantum gravity is needed in order to reconcile these differences.[11] Whether this theory should be background independent is an open question. The answer to this question will determine our understanding of what specific role gravitation plays in the fate of the universe.[12]

8.1.3 Gravitons in speculative theories

String theory predicts the existence of gravitons and their well-defined interactions. A graviton in perturbative string theory is a closed string in a very particular low-energy vibrational state. The scattering of gravitons in string theory can also be computed from the correlation functions in conformal field theory, as dictated by the AdS/CFT correspondence, or from matrix theory.

A feature of gravitons in string theory is that, as closed strings without endpoints, they would not be bound to branes and could move freely between them. If we live on a brane (as hypothesized by brane theories) this "leakage" of gravitons from the brane into higher-dimensional space could explain why gravitation is such a weak force, and gravitons from other branes adjacent to our own could provide a potential explanation for dark matter. See brane cosmology.

A theory by Ahmed Farag Ali and Saurya Das adds quantum mechanical corrections (using Bohm trajectories) to general relativistic geodesics. If gravitons are given a small but non-zero mass, it could explain the cosmological constant without need for dark energy and solve the smallness problem.[13]

8.2 Experimental observation

Unambiguous detection of individual gravitons, though not prohibited by any fundamental law, is impossible with any physically reasonable detector.[14] The reason is the extremely low cross section for the interaction of gravitons with matter. For example, a detector with the mass of Jupiter and 100% efficiency, placed in close orbit around a neutron star, would only be expected to observe one graviton every 10 years, even under the most favorable conditions. It would be impossible to discriminate these events from the background of neutrinos, since the dimensions of the required neutrino shield would ensure collapse into a black hole.[14]

However, experiments to detect gravitational waves, which may be viewed as coherent states of many gravitons, are underway (e.g., LIGO and VIRGO). Although these experiments cannot detect individual gravitons, they might provide information about certain properties of the graviton.[15] For example, if gravitational waves were observed to propagate slower than c (the speed of light in a vacuum), that would imply that the graviton has mass (however, gravitational waves must propagate slower than "c" in a region with non-zero mass density if they are to be detectable).[16] Astronomical observations of the kinematics of galaxies, especially the galaxy rotation problem and modified Newtonian dynamics, might point toward gravitons having non-zero mass.[17]

8.3 Difficulties and outstanding issues

Most theories containing gravitons suffer from severe problems. Attempts to extend the Standard Model or other quantum field theories by adding gravitons run into serious theoretical difficulties at high energies (processes involving energies close to or above the Planck scale) because of infinities arising due to quantum effects (in technical terms, gravitation is nonrenormalizable). Since classical general relativity and quantum mechanics seem to be incompatible at such energies, from a theoretical point of view, this situation is not tenable. One possible solution is to replace particles with strings. String theories are quantum theories of gravity in the sense that they reduce to classical general relativity plus field theory at low energies, but are fully quantum mechanical, contain a graviton, and are believed to be mathematically consistent.[18]

8.4 See also

- Gravitomagnetism
- Gravitational wave
- Planck mass

- Gravitation
- Static forces and virtual-particle exchange
- Multiverse
- Gravitino

8.5 References

[1] G is used to avoid confusion with gluons (symbol g)

[2] Rovelli, C. (2001). "Notes for a brief history of quantum gravity". arXiv:gr-qc/0006061 [gr-qc].

[3] Blokhintsev, D. I.; Gal'perin, F. M. (1934). "Gipoteza neitrino i zakon sokhraneniya energii" [Neutrino hypothesis and conservation of energy]. *Pod Znamenem Marxisma* (in Russian) **6**: 147–157.

[4] Lightman, A. P.; Press, W. H.; Price, R. H.; Teukolsky, S. A. (1975). "Problem 12.16". *Problem book in Relativity and Gravitation*. Princeton University Press. ISBN 0-691-08162-X.

[5] For a comparison of the geometric derivation and the (non-geometric) spin-2 field derivation of general relativity, refer to box 18.1 (and also 17.2.5) of Misner, C. W.; Thorne, K. S.; Wheeler, J. A. (1973). *Gravitation*. W. H. Freeman. ISBN 0-7167-0344-0.

[6] Feynman, R. P.; Morinigo, F. B.; Wagner, W. G.; Hatfield, B. (1995). *Feynman Lectures on Gravitation*. Addison-Wesley. ISBN 0-201-62734-5.

[7] Zee, A. (2003). *Quantum Field Theory in a Nutshell*. Princeton University Press. ISBN 0-691-01019-6.

[8] Randall, L. (2005). *Warped Passages: Unraveling the Universe's Hidden Dimensions*. Ecco Press. ISBN 0-06-053108-8.

[9] See the other articles on General relativity, Gravitational field, Gravitational wave, etc

[10] Colosi, D. et al. (2005). "Background independence in a nutshell: The dynamics of a tetrahedron". *Classical and Quantum Gravity* **22** (14): 2971. arXiv:gr-qc/0408079. Bibcode:2005CQGra..22.2971C. doi:10.1088/0264-9381/22/14/008.

[11] Witten, E. (1993). "Quantum Background Independence In String Theory". arXiv:hep-th/9306122 [hep-th].

[12] Smolin, L. (2005). "The case for background independence". arXiv:hep-th/0507235 [hep-th].

[13] Ali, Ahmed Farang (2014). "Cosmology from quantum potential". *Physical Letters B* **741**: 276–279. arXiv:1404.3093v3. doi:10.1016/j.physletb.2014.12.057.

[14] Rothman, T.; Boughn, S. (2006). "Can Gravitons be Detected?". *Foundations of Physics* **36** (12): 1801–1825. arXiv:gr-qc/0601043. Bibcode:2006FoPh...36.1801R. doi:10.1007/s10701-006-9081-9.

[15] Freeman Dyson (8 October 2013). "Is a graviton detectable?". *International Journal of Modern Physics A* **28** (25): 1330041–1–1330035–14. Bibcode:2013IJMPA..2830041D. doi:10.1142/S0217751X1330041X.

[16] Will, C. M. (1998). "Bounding the mass of the graviton using gravitational-wave observations of inspiralling compact binaries". *Physical Review D* **57** (4): 2061–2068. arXiv:gr-qc/9709011. Bibcode:1998PhRvD..57.2061W. doi:10.1103/PhysRevD.57.2061.

[17] Trippe, S. (2013), "A Simplified Treatment of Gravitational Interaction on Galactic Scales", J. Kor. Astron. Soc. **46**, 41. arXiv:1211.4692

[18] Sokal, A. (July 22, 1996). "Don't Pull the String Yet on Superstring Theory". *The New York Times*. Retrieved March 26, 2010.

8.6 External links

-
- Graviton on *In Our Time* at the BBC. (listen now)

Chapter 9

Quantum gravity

Quantum gravity (**QG**) is a field of theoretical physics that seeks to describe the force of gravity according to the principles of quantum mechanics.

The current understanding of gravity is based on Albert Einstein's general theory of relativity, which is formulated within the framework of classical physics. On the other hand, the nongravitational forces are described within the framework of quantum mechanics, a radically different formalism for describing physical phenomena based on probability.[1] The necessity of a quantum mechanical description of gravity follows from the fact that one cannot consistently couple a classical system to a quantum one.[2]

Although a quantum theory of gravity is needed in order to reconcile general relativity with the principles of quantum mechanics, difficulties arise when one attempts to apply the usual prescriptions of quantum field theory to the force of gravity.[3] From a technical point of view, the problem is that the theory one gets in this way is not renormalizable and therefore cannot be used to make meaningful physical predictions. As a result, theorists have taken up more radical approaches to the problem of quantum gravity, the most popular approaches being string theory and loop quantum gravity.[4] A recent development is the theory of causal fermion systems which gives quantum mechanics, general relativity and quantum field theory as limiting cases.[5][6][7][8][9][10]

Strictly speaking, the aim of quantum gravity is only to describe the quantum behavior of the gravitational field and should not be confused with the objective of unifying all fundamental interactions into a single mathematical framework. Although some quantum gravity theories such as string theory try to unify gravity with the other fundamental forces, others such as loop quantum gravity make no such attempt; instead, they make an effort to quantize the gravitational field while it is kept separate from the other forces. A theory of quantum gravity that is also a grand unification of all known interactions is sometimes referred to as a theory of everything (TOE).

One of the difficulties of quantum gravity is that quantum gravitational effects are only expected to become apparent near the Planck scale, a scale far smaller in distance (equivalently, far larger in energy) than what is currently accessible at high energy particle accelerators. As a result, quantum gravity is a mainly theoretical enterprise, although there are speculations about how quantum gravity effects might be observed in existing experiments.[11]

9.1 Overview

Much of the difficulty in meshing these theories at all energy scales comes from the different assumptions that these theories make on how the universe works. Quantum field theory depends on particle fields embedded in the flat space-time of special relativity. General relativity models gravity as a curvature within space-time that changes as a gravitational mass moves. Historically, the most obvious way of combining the two (such as treating gravity as simply another particle field) ran quickly into what is known as the renormalization problem. In the old-fashioned understanding of renormalization, gravity particles would attract each other and adding together all of the interactions results in many infinite values which cannot easily be cancelled out mathematically to yield sensible, finite results. This is in contrast with quantum electrodynamics where, given that the series still do not converge, the interactions sometimes evaluate to infinite results, but those are few enough in number to be removable via renormalization.

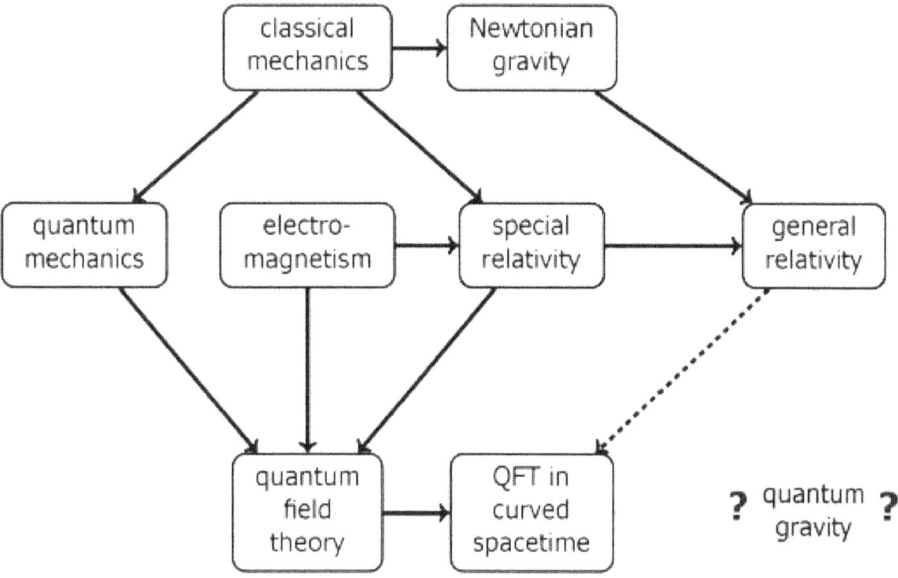

Diagram showing where quantum gravity sits in the hierarchy of physics theories

9.1.1 Effective field theories

Quantum gravity can be treated as an effective field theory. Effective quantum field theories come with some high-energy cutoff, beyond which we do not expect that the theory provides a good description of nature. The "infinities" then become large but finite quantities depending on this finite cutoff scale, and correspond to processes that involve very high energies near the fundamental cutoff. These quantities can then be absorbed into an infinite collection of coupling constants, and at energies well below the fundamental cutoff of the theory, to any desired precision; only a finite number of these coupling constants need to be measured in order to make legitimate quantum-mechanical predictions. This same logic works just as well for the highly successful theory of low-energy pions as for quantum gravity. Indeed, the first quantum-mechanical corrections to graviton-scattering and Newton's law of gravitation have been explicitly computed[12] (although they are so astronomically small that we may never be able to measure them). In fact, gravity is in many ways a much better quantum field theory than the Standard Model, since it appears to be valid all the way up to its cutoff at the Planck scale.

While confirming that quantum mechanics and gravity are indeed consistent at reasonable energies, it is clear that near or above the fundamental cutoff of our effective quantum theory of gravity (the cutoff is generally assumed to be of the order of the Planck scale), a new model of nature will be needed. Specifically, the problem of combining quantum mechanics and gravity becomes an issue only at very high energies, and may well require a totally new kind of model.

9.1.2 Quantum gravity theory for the highest energy scales

The general approach to deriving a quantum gravity theory that is valid at even the highest energy scales is to assume that such a theory will be simple and elegant and, accordingly, to study symmetries and other clues offered by current theories that might suggest ways to combine them into a comprehensive, unified theory. One problem with this approach is that it is unknown whether quantum gravity will actually conform to a simple and elegant theory, as it should resolve the dual conundrums of special relativity with regard to the uniformity of acceleration and gravity, and general relativity with regard to spacetime curvature.

Such a theory is required in order to understand problems involving the combination of very high energy and very small dimensions of space, such as the behavior of black holes, and the origin of the universe.

9.2 Quantum mechanics and general relativity

9.2.1 The graviton

Main article: Graviton

At present, one of the deepest problems in theoretical physics is harmonizing the theory of general relativity, which describes gravitation, and applications to large-scale structures (stars, planets, galaxies), with quantum mechanics, which describes the other three fundamental forces acting on the atomic scale. This problem must be put in the proper context, however. In particular, contrary to the popular claim that quantum mechanics and general relativity are fundamentally incompatible, one can demonstrate that the structure of general relativity essentially follows inevitably from the quantum mechanics of interacting theoretical spin-2 massless particles (called gravitons).[13][14][15][16][17]

While there is no concrete proof of the existence of gravitons, quantized theories of matter may necessitate their existence. Supporting this theory is the observation that all fundamental forces except gravity have one or more known messenger particles, leading researchers to believe that at least one most likely does exist; they have dubbed this hypothetical particle the *graviton*. The predicted find would result in the classification of the graviton as a "force particle" similar to the photon of the electromagnetic field. Many of the accepted notions of a unified theory of physics since the 1970s assume, and to some degree depend upon, the existence of the graviton. These include string theory, superstring theory, M-theory, and loop quantum gravity. Detection of gravitons is thus vital to the validation of various lines of research to unify quantum mechanics and relativity theory.

9.2.2 The dilaton

Main article: Dilaton

The dilaton made its first appearance in Kaluza–Klein theory, a five-dimensional theory that combined gravitation and electromagnetism. Generally, it appears in string theory. More recently, it has appeared in the lower-dimensional many-bodied gravity problem[18] based on the field theoretic approach of Roman Jackiw. The impetus arose from the fact that complete analytical solutions for the metric of a covariant N-body system have proven elusive in general relativity. To simplify the problem, the number of dimensions was lowered to $(1+1)$, i.e. one spatial dimension and one temporal dimension. This model problem, known as $R=T$ theory[19] (as opposed to the general $G=T$ theory) was amenable to exact solutions in terms of a generalization of the Lambert W function. It was also found that the field equation governing the dilaton (derived from differential geometry) was the Schrödinger equation and consequently amenable to quantization.[20]

Thus, one had a theory which combined gravity, quantization and even the electromagnetic interaction, promising ingredients of a fundamental physical theory. It is worth noting that the outcome revealed a previously unknown and already existing *natural link* between general relativity and quantum mechanics. However, this theory needs to be generalized in $(2+1)$ or $(3+1)$ dimensions although, in principle, the field equations are amenable to such generalization as shown with the inclusion of a one-graviton process[21] and yielding the correct Newtonian limit in d dimensions if a dilaton is included. However, it is not yet clear what the fully generalized field equation governing the dilaton in (3+1) dimensions should be. This is further complicated by the fact that gravitons can propagate in $(3+1)$ dimensions and consequently that would imply gravitons and dilatons exist in the real world. Moreover, detection of the dilaton is expected to be even more elusive than the graviton. However, since this approach allows for the combination of gravitational, electromagnetic and quantum effects, their coupling could potentially lead to a means of vindicating the theory, through cosmology and perhaps even *experimentally*.

9.2.3 Nonrenormalizability of gravity

Further information: Renormalization

General relativity, like electromagnetism, is a classical field theory. One might expect that, as with electromagnetism, there should be a corresponding quantum field theory.

However, gravity is perturbatively nonrenormalizable.[22][23] For a quantum field theory to be well-defined according to this understanding of the subject, it must be asymptotically free or asymptotically safe. The theory must be characterized by a choice of *finitely many* parameters, which could, in principle, be set by experiment. For example, in quantum electrodynamics, these parameters are the charge and mass of the electron, as measured at a particular energy scale.

On the other hand, in quantizing gravity, there are *infinitely many independent parameters* (counterterm coefficients) needed to define the theory. For a given choice of those parameters, one could make sense of the theory, but since we can never do infinitely many experiments to fix the values of every parameter, we do not have a meaningful physical theory:

- At low energies, the logic of the renormalization group tells us that, despite the unknown choices of these infinitely many parameters, quantum gravity will reduce to the usual Einstein theory of general relativity.
- On the other hand, if we could probe very high energies where quantum effects take over, then *every one* of the infinitely many unknown parameters would begin to matter, and we could make no predictions at all.

As explained below, there is a way around this problem by treating QG as an effective field theory.

Any meaningful theory of quantum gravity that makes sense and is predictive at all energy scales must have some deep principle that reduces the infinitely many unknown parameters to a finite number that can then be measured.

- One possibility is that normal perturbation theory is not a reliable guide to the renormalizability of the theory, and that there really *is* a UV fixed point for gravity. Since this is a question of non-perturbative quantum field theory, it is difficult to find a reliable answer, but some people still pursue this option.
- Another possibility is that there are new symmetry principles that constrain the parameters and reduce them to a finite set. This is the route taken by string theory, where all of the excitations of the string essentially manifest themselves as new symmetries.

9.2.4 QG as an effective field theory

Main article: Effective field theory

In an effective field theory, all but the first few of the infinite set of parameters in a non-renormalizable theory are suppressed by huge energy scales and hence can be neglected when computing low-energy effects. Thus, at least in the low-energy regime, the model is indeed a predictive quantum field theory.[12] (A very similar situation occurs for the very similar effective field theory of low-energy pions.) Furthermore, many theorists agree that even the Standard Model should really be regarded as an effective field theory as well, with “nonrenormalizable” interactions suppressed by large energy scales and whose effects have consequently not been observed experimentally.

Recent work[12] has shown that by treating general relativity as an effective field theory, one can actually make legitimate predictions for quantum gravity, at least for low-energy phenomena. An example is the well-known calculation of the tiny first-order quantum-mechanical correction to the classical Newtonian gravitational potential between two masses.

9.2.5 Spacetime background dependence

Main article: Background independence

A fundamental lesson of general relativity is that there is no fixed spacetime background, as found in Newtonian mechanics and special relativity; the spacetime geometry is dynamic. While easy to grasp in principle, this is the hardest idea to understand about general relativity, and its consequences are profound and not fully explored, even at the classical level. To a certain extent, general relativity can be seen to be a relational theory,[24] in which the only physically relevant information is the relationship between different events in space-time.

On the other hand, quantum mechanics has depended since its inception on a fixed background (non-dynamic) structure. In the case of quantum mechanics, it is time that is given and not dynamic, just as in Newtonian classical mechanics. In relativistic quantum field theory, just as in classical field theory, Minkowski spacetime is the fixed background of the theory.

String theory

String theory can be seen as a generalization of quantum field theory where instead of point particles, string-like objects propagate in a fixed spacetime background, although the interactions among closed strings give rise to space-time in a dynamical way. Although string theory had its origins in the study of quark confinement and not of quantum gravity, it was soon discovered that the string spectrum contains the graviton, and that "condensation" of certain vibration modes of strings is equivalent to a modification of the original background. In this sense, string perturbation theory exhibits exactly the features one would expect of a perturbation theory that may exhibit a strong dependence on asymptotics (as seen, for example, in the AdS/CFT correspondence) which is a weak form of background dependence.

Background independent theories

Loop quantum gravity is the fruit of an effort to formulate a background-independent quantum theory.

Topological quantum field theory provided an example of background-independent quantum theory, but with no local degrees of freedom, and only finitely many degrees of freedom globally. This is inadequate to describe gravity in 3+1 dimensions, which has local degrees of freedom according to general relativity. In 2+1 dimensions, however, gravity is a topological field theory, and it has been successfully quantized in several different ways, including spin networks.

9.2.6 Semi-classical quantum gravity

Quantum field theory on curved (non-Minkowskian) backgrounds, while not a full quantum theory of gravity, has shown many promising early results. In an analogous way to the development of quantum electrodynamics in the early part of the 20th century (when physicists considered quantum mechanics in classical electromagnetic fields), the consideration of quantum field theory on a curved background has led to predictions such as black hole radiation.

Phenomena such as the Unruh effect, in which particles exist in certain accelerating frames but not in stationary ones, do not pose any difficulty when considered on a curved background (the Unruh effect occurs even in flat Minkowskian backgrounds). The vacuum state is the state with the least energy (and may or may not contain particles). See Quantum field theory in curved spacetime for a more complete discussion.

9.2.7 Points of tension

There are other points of tension between quantum mechanics and general relativity.

- First, classical general relativity breaks down at singularities, and quantum mechanics becomes inconsistent with general relativity in the neighborhood of singularities (however, no one is certain that classical general relativity applies near singularities in the first place).
- Second, it is not clear how to determine the gravitational field of a particle, since under the Heisenberg uncertainty principle of quantum mechanics its location and velocity cannot be known with certainty. The resolution of these points may come from a better understanding of general relativity.[25]
- Third, there is the problem of time in quantum gravity. Time has a different meaning in quantum mechanics and general relativity and hence there are subtle issues to resolve when trying to formulate a theory which combines the two.[26]

9.3 Candidate theories

There are a number of proposed quantum gravity theories.[27] Currently, there is still no complete and consistent quantum theory of gravity, and the candidate models still need to overcome major formal and conceptual problems. They also face the common problem that, as yet, there is no way to put quantum gravity predictions to experimental tests, although there is hope for this to change as future data from cosmological observations and particle physics experiments becomes available.[28][29]

9.3.1 String theory

Main article: String theory

One suggested starting point is ordinary quantum field theories which, after all, are successful in describing the other three basic fundamental forces in the context of the standard model of elementary particle physics. However, while this leads to an acceptable effective (quantum) field theory of gravity at low energies,[30] gravity turns out to be much more problematic at higher energies. Where, for ordinary field theories such as quantum electrodynamics, a technique known as renormalization is an integral part of deriving predictions which take into account higher-energy contributions,[31] gravity turns out to be nonrenormalizable: at high energies, applying the recipes of ordinary quantum field theory yields models that are devoid of all predictive power.[32]

One attempt to overcome these limitations is to replace ordinary quantum field theory, which is based on the classical concept of a point particle, with a quantum theory of one-dimensional extended objects: string theory.[33] At the energies reached in current experiments, these strings are indistinguishable from point-like particles, but, crucially, different modes of oscillation of one and the same type of fundamental string appear as particles with different (electric and other) charges. In this way, string theory promises to be a unified description of all particles and interactions.[34] The theory is successful in that one mode will always correspond to a graviton, the messenger particle of gravity; however, the price to pay are unusual features such as six extra dimensions of space in addition to the usual three for space and one for time.[35]

In what is called the second superstring revolution, it was conjectured that both string theory and a unification of general relativity and supersymmetry known as supergravity[36] form part of a hypothesized eleven-dimensional model known as M-theory, which would constitute a uniquely defined and consistent theory of quantum gravity.[37][38] As presently understood, however, string theory admits a very large number (10^{500} by some estimates) of consistent vacua, comprising the so-called "string landscape". Sorting through this large family of solutions remains a major challenge.

9.3.2 Loop quantum gravity

Main article: Loop quantum gravity

Loop quantum gravity is based first of all on the idea to take seriously the insight of general relativity that spacetime is a dynamical field and therefore is a quantum object. The second idea is that the quantum discreteness that determines the particle-like behavior of other field theories (for instance, the photons of the electromagnetic field) also affects the structure of space.

The main result of loop quantum gravity is the derivation of a granular structure of space at the Planck length. This is derived as follows. In the case of electromagnetism, the quantum operator representing the energy of each frequency of the field has discrete spectrum. Therefore the energy of each frequency is quantized, and the quanta are the photons. In the case of gravity, the operators representing the area and the volume of each surface or space region have discrete spectrum. Therefore area and volume of any portion of space are quantized, and the quanta are elementary quanta of space. It follows that spacetime has an elementary quantum granular structure at the Planck scale, which cuts-off the ultraviolet infinities of quantum field theory.

The quantum state of spacetime is described in the theory by means of a mathematical structure called spin networks. Spin networks were initially introduced by Roger Penrose in abstract form, and later shown by Carlo Rovelli and Lee Smolin to derive naturally from a non perturbative quantization of general relativity. Spin networks do not represent quantum states of a field in spacetime: they represent directly quantum states of spacetime.

The theory is based on the reformulation of general relativity known as Ashtekar variables, which represent geometric gravity using mathematical analogues of electric and magnetic fields.[39][40] In the quantum theory space is represented by a network structure called a spin network, evolving over time in discrete steps.[41][42][43][44]

The dynamics of the theory is today constructed in several versions. One version starts with the canonical quantization of general relativity. The analogue of the Schrödinger equation is a Wheeler–DeWitt equation, which can be defined in the theory.[45] In the covariant, or spinfoam formulation of the theory, the quantum dynamics is obtained via a sum over discrete versions of spacetime, called spinfoams. These represent histories of spin networks.

9.3.3 Scale Relativity

Main article: Scale relativity

Most quantum gravity theories assume quantum laws as a starting point. However, in the framework of scale relativity, this is not needed.[46] The theory is an extension of special and general relativity, including the relativity of scale transformations. It thus takes a geometrical approach to the problem, where quantum phenomena became a manifestation of the fractality of spacetime. This is similar to the geometrical interpretation of gravitation in general relativity, where gravitation become a manifestation of spacetime curvature instead of a force. Although much remains to be developed, validated predictions have already been obtained in physics, astrophysics and cosmology.

9.3.4 Other approaches

There are a number of other approaches to quantum gravity. The approaches differ depending on which features of general relativity and quantum theory are accepted unchanged, and which features are modified.[47][48] Examples include:

- Acoustic metric and other analog models of gravity
- Asymptotic safety in quantum gravity
- Euclidean quantum gravity
- Causal dynamical triangulation[49]
- Causal fermion systems,[5][6][7][8][9][10] giving quantum mechanics, general relativity and quantum field theory as limiting cases.
- Causal sets[50]
- Covariant Feynman path integral approach
- Group field theory[51]
- Wheeler-DeWitt equation
- Geometrodynamics
- Hořava–Lifshitz gravity
- MacDowell–Mansouri action
- Noncommutative geometry.
- Path-integral based models of quantum cosmology[52]
- Regge calculus
- String-nets giving rise to gapless helicity ±2 excitations with no other gapless excitations[53]
- Superfluid vacuum theory a.k.a. theory of BEC vacuum
- Supergravity
- Twistor theory[54]
- Canonical quantum gravity
- E8 Theory

9.4 Weinberg–Witten theorem

In quantum field theory, the Weinberg–Witten theorem places some constraints on theories of composite gravity/emergent gravity. However, recent developments attempt to show that if locality is only approximate and the holographic principle is correct, the Weinberg–Witten theorem would not be valid.

9.5 Experimental tests

As was emphasized above, quantum gravitational effects are extremely weak and therefore difficult to test. For this reason, the possibility of experimentally testing quantum gravity had not received much attention prior to the late 1990s. However, in the past decade, physicists have realized that evidence for quantum gravitational effects can guide the development of the theory. Since theoretical development has been slow, the field of phenomenological quantum gravity, which studies the possibility of experimental tests, has obtained increased attention.[55][56]

The most widely pursued possibilities for quantum gravity phenomenology include violations of Lorentz invariance, imprints of quantum gravitational effects in the cosmic microwave background (in particular its polarization), and decoherence induced by fluctuations in the space-time foam.

The BICEP2 experiment detected what was initially thought to be primordial B-mode polarization caused by gravitational waves in the early universe. If truly primordial, these waves were born as quantum fluctuations in gravity itself. Cosmologist Ken Olum (Tufts University) stated: "I think this is the only observational evidence that we have that actually shows that gravity is quantized....It's probably the only evidence of this that we will ever have."[57]

9.6 See also

9.7 References

[1] Griffiths, David J. (2004). *Introduction to Quantum Mechanics*. Pearson Prentice Hall. OCLC 803860989.

[2] Wald, Robert M. (1984). *General Relativity*. University of Chicago Press. p. 382. OCLC 471881415.

[3] Zee, Anthony (2010). *Quantum Field Theory in a Nutshell* (2nd ed.). Princeton University Press. p. 172. OCLC 659549695.

[4] Penrose, Roger (2007). *The road to reality : a complete guide to the laws of the universe*. Vintage. p. 1017. OCLC 716437154.

[5] F. Finster, J. Kleiner, Causal Fermion Systems as a Candidate for a Unified Physical Theory, http://arxiv.org/abs/1502.03587

[6] F. Finster, The Principle of the Fermionic Projector, hep-th/0001048, hep-th/0202059, hep- th/0210121, AMS/IP Studies in Advanced Mathematics, vol. **35**, American Mathematical Society, Providence, RI, 2006.

[7] F. Finster, A formulation of quantum field theory realizing a sea of interacting Dirac particles, arXiv:0911.2102 [hep-th], Lett. Math. Phys. **97** (2011), no. 2, 165–183.

[8] F. Finster, An action principle for an interacting fermion system and its analysis in the continuum limit, arXiv:0908.1542 [math-ph] (2009).

[9] F. Finster, The continuum limit of a fermion system involving neutrinos: Weak and gravitational interactions, arXiv: 1211.3351 [math-ph] (2012).

[10] F. Finster, Perturbative quantum field theory in the framework of the fermionic projector, arXiv:1310.4121 [math-ph], J. Math. Phys. **55** (2014), no. 4, 042301.

[11] Quantum effects in the early universe might have an observable effect on the structure of the present universe, for example, or gravity might play a role in the unification of the other forces. Cf. the text by Wald cited above.

[12] Donoghue (1995). "Introduction to the Effective Field Theory Description of Gravity". arXiv:gr-qc/9512024. (verify against ISBN 9789810229085)

[13] Kraichnan, R. H. (1955). "Special-Relativistic Derivation of Generally Covariant Gravitation Theory". *Physical Review* **98** (4): 1118–1122. Bibcode:1955PhRv...98.1118K. doi:10.1103/PhysRev.98.1118.

[14] Gupta, S. N. (1954). "Gravitation and Electromagnetism". *Physical Review* **96** (6): 1683–1685. Bibcode:1954PhRv...96.1683G. doi:10.1103/PhysRev.96.1683.

[15] Gupta, S. N. (1957). "Einstein's and Other Theories of Gravitation". *Reviews of Modern Physics* **29** (3): 334–336. Bibcode:1957RvMP...29..334G. doi:10.1103/RevModPhys.29.334.

[16] Gupta, S. N. (1962). "Quantum Theory of Gravitation". *Recent Developments in General Relativity*. Pergamon Press. pp. 251–258.

[17] Deser, S. (1970). "Self-Interaction and Gauge Invariance". *General Relativity and Gravitation* **1**: 9–18. arXiv:gr-qc/0411023. Bibcode:1970GReGr...1....9D. doi:10.1007/BF00759198.

[18] Ohta, Tadayuki; Mann, Robert (1996). "Canonical reduction of two-dimensional gravity for particle dynamics". *Classical and Quantum Gravity* **13** (9): 2585–2602. arXiv:gr-qc/9605004. Bibcode:1996CQGra..13.2585O. doi:10.1088/0264-9381/13/9/022.

[19] Sikkema, A E; Mann, R B (1991). "Gravitation and cosmology in (1+1) dimensions". *Classical and Quantum Gravity* **8**: 219–235. Bibcode:1991CQGra...8..219S. doi:10.1088/0264-9381/8/1/022.

[20] Farrugia; Mann; Scott (2007). "N-body Gravity and the Schroedinger Equation". *Classical and Quantum Gravity* **24** (18): 4647–4659. arXiv:gr-qc/0611144. Bibcode:2007CQGra..24.4647F. doi:10.1088/0264-9381/24/18/006.

[21] Mann, R B; Ohta, T (1997). "Exact solution for the metric and the motion of two bodies in (1+1)-dimensional gravity". *Physical Review D* **55** (8): 4723–4747. arXiv:gr-qc/9611008. Bibcode:1997PhRvD..55.4723M. doi:10.1103/PhysRevD.55.4723.

[22] Feynman, R. P.; Morinigo, F. B.; Wagner, W. G.; Hatfield, B. (1995). *Feynman lectures on gravitation*. Addison-Wesley. ISBN 0-201-62734-5.

[23] Hamber, H. W. (2009). *Quantum Gravitation - The Feynman Path Integral Approach*. Springer Publishing. ISBN 978-3-540-85292-6.

[24] Smolin, Lee (2001). *Three Roads to Quantum Gravity*. Basic Books. pp. 20–25. ISBN 0-465-07835-4. Pages 220–226 are annotated references and guide for further reading.

[25] Hunter Monroe (2005). "Singularity-Free Collapse through Local Inflation". arXiv:astro-ph/0506506.

[26] Edward Anderson (2010). "The Problem of Time in Quantum Gravity". arXiv:1009.2157 [gr-qc]. (also published as chapter 4 of ISBN 9781611229578)

[27] A timeline and overview can be found in Rovelli, Carlo (2000). "Notes for a brief history of quantum gravity". arXiv:gr-qc/0006061. (verify against ISBN 9789812777386)

[28] Ashtekar, Abhay (2007). "Loop Quantum Gravity: Four Recent Advances and a Dozen Frequently Asked Questions". *11th Marcel Grossmann Meeting on Recent Developments in Theoretical and Experimental General Relativity*. p. 126. arXiv:0705.2222. Bibcode:2008mgm..conf..126A. doi:10.1142/9789812834300_0008.

[29] Schwarz, John H. (2007). "String Theory: Progress and Problems". *Progress of Theoretical Physics Supplement* **170**: 214–226. arXiv:hep-th/0702219. Bibcode:2007PThPS.170..214S. doi:10.1143/PTPS.170.214.

[30] Donoghue, John F. (editor) (1995). "Introduction to the Effective Field Theory Description of Gravity". In Cornet, Fernando. *Effective Theories: Proceedings of the Advanced School, Almunecar, Spain, 26 June–1 July 1995*. Singapore: World Scientific. arXiv:gr-qc/9512024. ISBN 981-02-2908-9.

[31] Weinberg, Steven (1996). "Chapters 17–18". *The Quantum Theory of Fields II: Modern Applications*. Cambridge University Press. ISBN 0-521-55002-5.

[32] Goroff, Marc H.; Sagnotti, Augusto; Sagnotti, Augusto (1985). "Quantum gravity at two loops". *Physics Letters B* **160**: 81–86. Bibcode:1985PhLB..160...81G. doi:10.1016/0370-2693(85)91470-4.

[33] An accessible introduction at the undergraduate level can be found in Zwiebach, Barton (2004). *A First Course in String Theory*. Cambridge University Press. ISBN 0-521-83143-1., and more complete overviews in Polchinski, Joseph (1998). *String Theory Vol. I: An Introduction to the Bosonic String*. Cambridge University Press. ISBN 0-521-63303-6. and Polchinski, Joseph (1998b). *String Theory Vol. II: Superstring Theory and Beyond*. Cambridge University Press. ISBN 0-521-63304-4.

[34] Ibanez, L. E. (2000). "The second string (phenomenology) revolution". *Classical & Quantum Gravity* **17** (5): 1117–1128. arXiv:hep-ph/9911499. Bibcode:2000CQGra..17.1117I. doi:10.1088/0264-9381/17/5/321.

[35] For the graviton as part of the string spectrum, e.g. Green, Schwarz & Witten 1987, sec. 2.3 and 5.3; for the extra dimensions, ibid sec. 4.2.

[36] Weinberg, Steven (2000). "Chapter 31". *The Quantum Theory of Fields II: Modern Applications*. Cambridge University Press. ISBN 0-521-55002-5.

[37] Townsend, Paul K. (1996). *Four Lectures on M-Theory*. ICTP Series in Theoretical Physics. p. 385. arXiv:hep-th/9612121. Bibcode:1997hepcbconf..385T.

[38] Duff, Michael (1996). "M-Theory (the Theory Formerly Known as Strings)". *International Journal of Modern Physics A* **11** (32): 5623–5642. arXiv:hep-th/9608117. Bibcode:1996IJMPA..11.5623D. doi:10.1142/S0217751X96002583.

[39] Ashtekar, Abhay (1986). "New variables for classical and quantum gravity". *Physical Review Letters* **57** (18): 2244–2247. Bibcode:1986PhRvL..57.2244A. doi:10.1103/PhysRevLett.57.2244. PMID 10033673.

[40] Ashtekar, Abhay (1987). "New Hamiltonian formulation of general relativity". *Physical Review D* **36** (6): 1587–1602. Bibcode:1987PhRvD..36.1587A. doi:10.1103/PhysRevD.36.1587.

[41] Thiemann, Thomas (2006). "Loop Quantum Gravity: An Inside View". *Approaches to Fundamental Physics*. Lecture Notes in Physics **721**: 185. arXiv:hep-th/0608210. Bibcode:2007LNP...721..185T. doi:10.1007/978-3-540-71117-9_10. ISBN 978-3-540-71115-5.

[42] Rovelli, Carlo (1998). "Loop Quantum Gravity". *Living Reviews in Relativity* **1**. Retrieved 2008-03-13.

[43] Ashtekar, Abhay; Lewandowski, Jerzy (2004). "Background Independent Quantum Gravity: A Status Report". *Classical & Quantum Gravity* **21** (15): R53–R152. arXiv:gr-qc/0404018. Bibcode:2004CQGra..21R..53A. doi:10.1088/0264-9381/21/15/R01.

[44] Thiemann, Thomas (2003). "Lectures on Loop Quantum Gravity". *Lecture Notes in Physics*. Lecture Notes in Physics **631**: 41–135. arXiv:gr-qc/0210094. Bibcode:2003LNP...631...41T. doi:10.1007/978-3-540-45230-0_3. ISBN 978-3-540-40810-9.

[45] Rovelli, Carlo (2004). *Quantum Gravity*. Cambridge University Press. ISBN 0521715962.

[46] Nottale, L. (2011). *Scale Relativity and Fractal Space-Time: A New Approach to Unifying Relativity and Quantum Mechanics*. World Scientific Publishing Company. ISBN 1848166508.;p. 458

[47] Isham, Christopher J. (1994). "Prima facie questions in quantum gravity". In Ehlers, Jürgen; Friedrich, Helmut. *Canonical Gravity: From Classical to Quantum*. Springer. arXiv:gr-qc/9310031. ISBN 3-540-58339-4.

[48] Sorkin, Rafael D. (1997). "Forks in the Road, on the Way to Quantum Gravity". *International Journal of Theoretical Physics* **36** (12): 2759–2781. arXiv:gr-qc/9706002. Bibcode:1997IJTP...36.2759S. doi:10.1007/BF02435709.

[49] Loll, Renate (1998). "Discrete Approaches to Quantum Gravity in Four Dimensions". *Living Reviews in Relativity* **1**: 13. arXiv:gr-qc/9805049. Bibcode:1998LRR.....1...13L. doi:10.12942/lrr-1998-13. Retrieved 2008-03-09.

[50] Sorkin, Rafael D. (2005). "Causal Sets: Discrete Gravity". In Gomberoff, Andres; Marolf, Donald. *Lectures on Quantum Gravity*. Springer. arXiv:gr-qc/0309009. ISBN 0-387-23995-2.

[51] See Daniele Oriti and references therein.

[52] Hawking, Stephen W. (1987). "Quantum cosmology". In Hawking, Stephen W.; Israel, Werner. *300 Years of Gravitation*. Cambridge University Press. pp. 631–651. ISBN 0-521-37976-8.

[53] Wen 2006

[54] See ch. 33 in Penrose 2004 and references therein.

[55] Hossenfelder, Sabine (2011). "Experimental Search for Quantum Gravity". In V. R. Frignanni. *Classical and Quantum Gravity: Theory, Analysis and Applications*. Chapter 5: Nova Publishers. ISBN 978-1-61122-957-8.

[56] Hossenfelder, Sabine (2010-10-17). V. R. Frignanni, ed. "Experimental Search for Quantum Gravity". *Classical and Quantum Gravity: Theory, Analysis and Applications* (Nova Publishers) **5** (2011). arXiv:1010.3420. Bibcode:2010arXiv1010.3420H. |chapter= ignored (help)

[57] Camille Carlisle. "First Direct Evidence of Big Bang Inflation". SkyandTelescope.com. Retrieved March 18, 2014.

9.8 Further reading

- Ahluwalia, D. V. (2002). "Interface of Gravitational and Quantum Realms". *Modern Physics Letters A* **17** (15–17): 1135. arXiv:gr-qc/0205121. Bibcode:2002MPLA...17.1135A. doi:10.1142/S021773230200765X.
- Ashtekar, Abhay (2005). "The winding road to quantum gravity" (PDF). *Current Science* **89**: 2064–2074.
- Carlip, Steven (2001). "Quantum Gravity: a Progress Report". *Reports on Progress in Physics* **64** (8): 885–942. arXiv:gr-qc/0108040. Bibcode:2001RPPh...64..885C. doi:10.1088/0034-4885/64/8/301.

- Herbert W. Hamber (2009). *Quantum Gravitation*. Springer Publishing. doi:10.1007/978-3-540-85293-3. ISBN 978-3-540-85292-6.
- Kiefer, Claus (2007). *Quantum Gravity*. Oxford University Press. ISBN 0-19-921252-X.
- Kiefer, Claus (2005). "Quantum Gravity: General Introduction and Recent Developments". *Annalen der Physik* **15**: 129–148. arXiv:gr-qc/0508120. Bibcode:2006AnP...518..129K. doi:10.1002/andp.200510175.
- Lämmerzahl, Claus, ed. (2003). *Quantum Gravity: From Theory to Experimental Search*. Lecture Notes in Physics. Springer. ISBN 3-540-40810-X.
- Rovelli, Carlo (2004). *Quantum Gravity*. Cambridge University Press. ISBN 0-521-83733-2.
- Trifonov, Vladimir (2008). "GR-friendly description of quantum systems". *International Journal of Theoretical Physics* **47** (2): 492–510. arXiv:math-ph/0702095. Bibcode:2008IJTP...47..492T. doi:10.1007/s10773-007-9474-3.

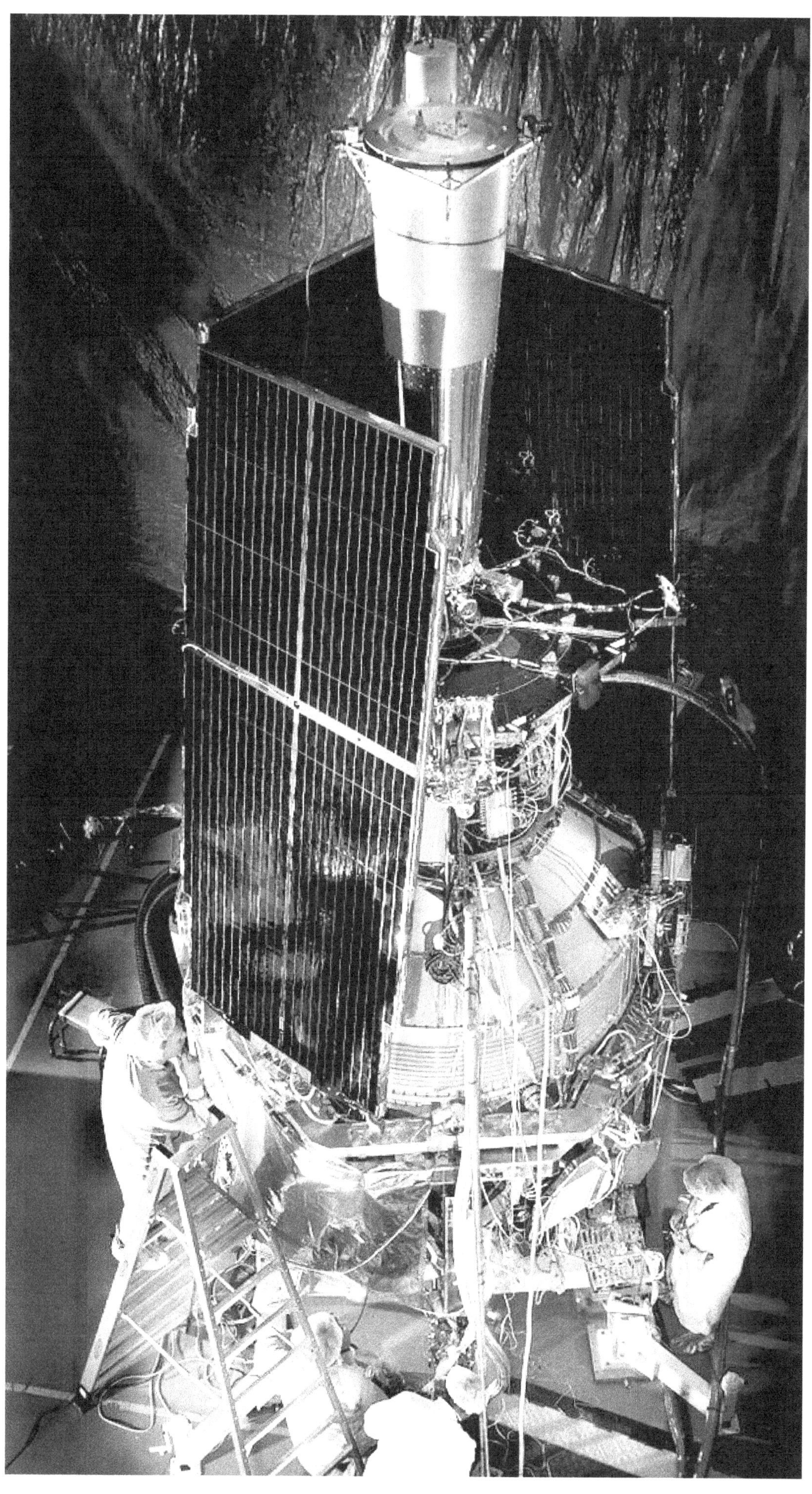

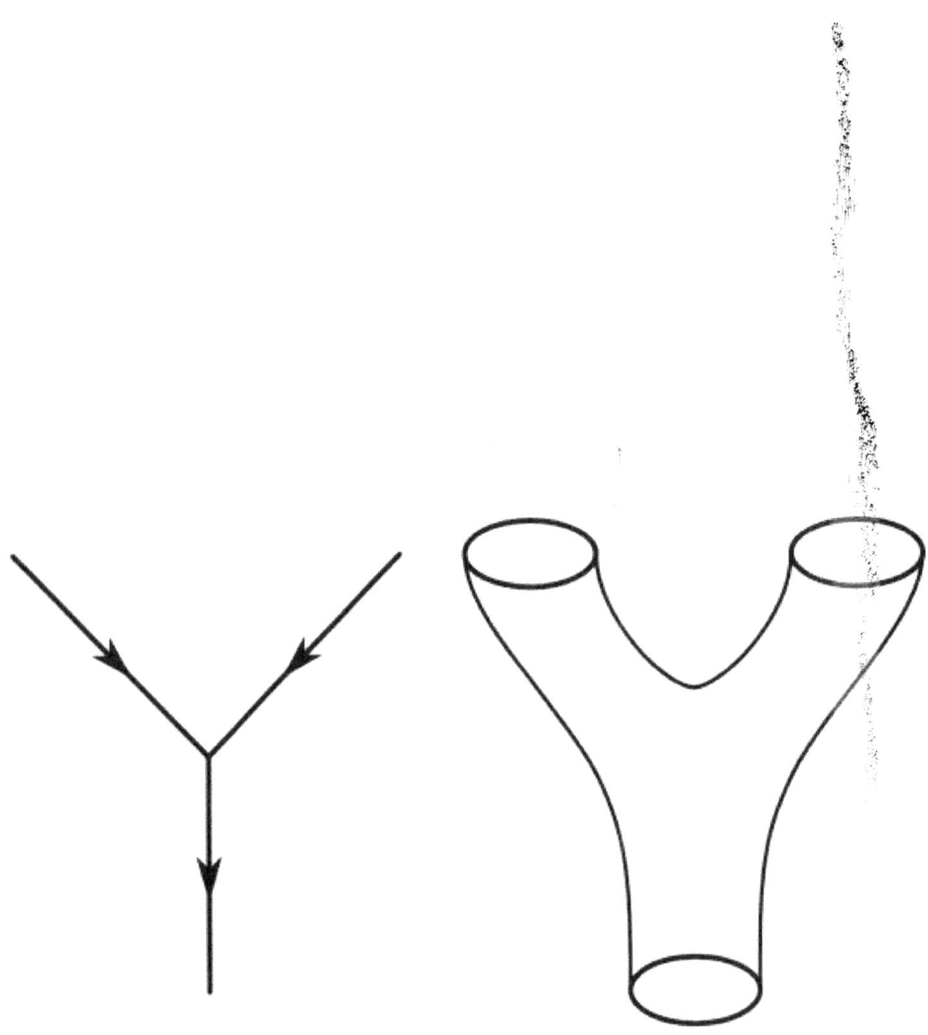

Interaction in the subatomic world: world lines of point-like particles in the Standard Model or a world sheet swept up by closed strings in string theory

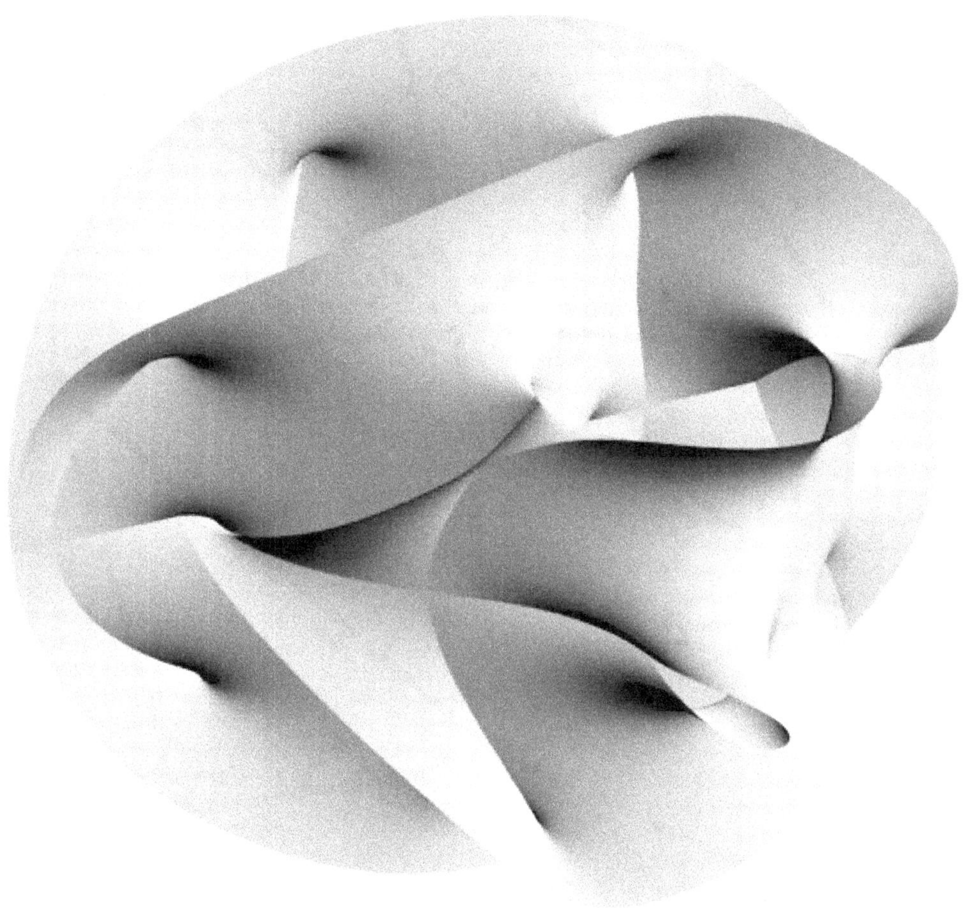

Projection of a Calabi–Yau manifold, one of the ways of compactifying the extra dimensions posited by string theory

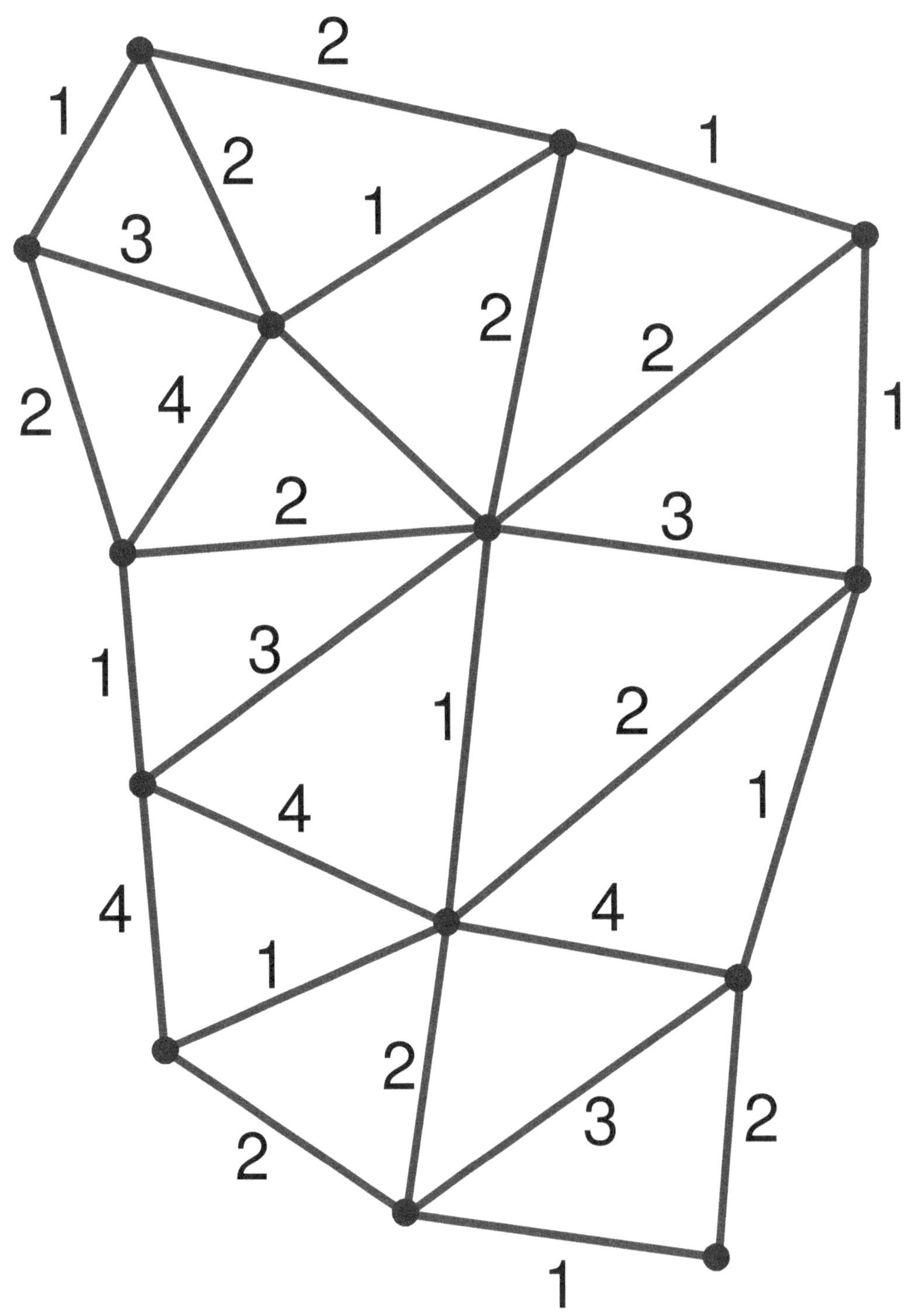

Simple spin network of the type used in loop quantum gravity

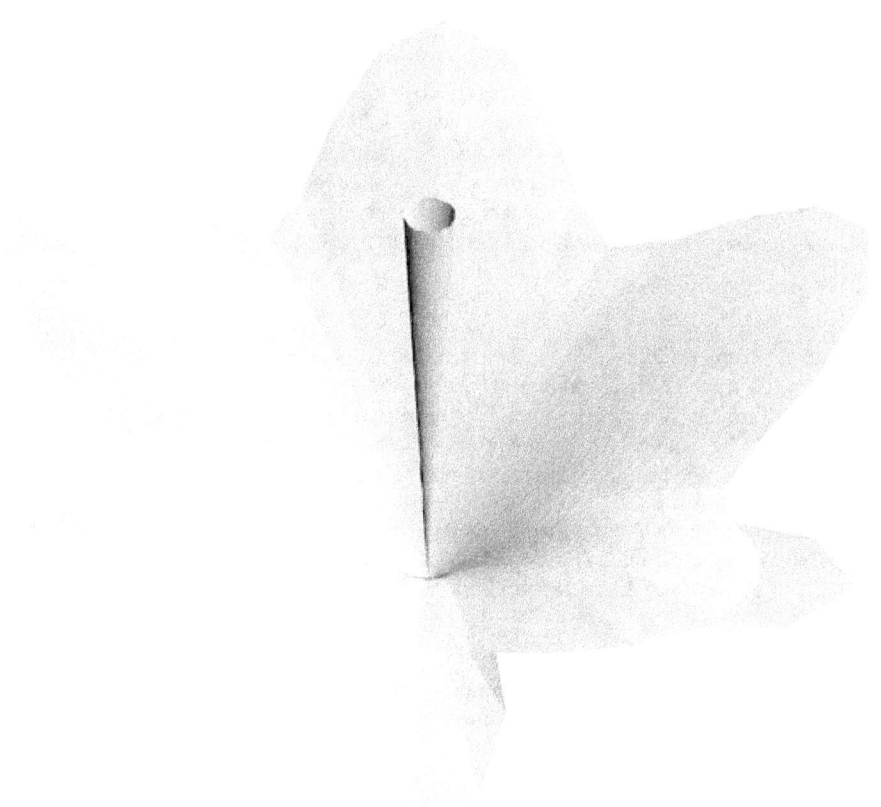

Schrödinger's flower. Morphogenesis of a flower-like structure, solution of a growth process equation that takes the form of a Schrödinger equation under fractal conditions.

Chapter 10

Quantum chromodynamics

In theoretical physics, **quantum chromodynamics** (**QCD**) is the theory of strong interactions, a fundamental force describing the interactions between quarks and gluons which make up hadrons such as the proton, neutron and pion. QCD is a type of quantum field theory called a non-abelian gauge theory with symmetry group SU(3). The QCD analog of electric charge is a property called *color*. Gluons are the force carrier of the theory, like photons are for the electromagnetic force in quantum electrodynamics. The theory is an important part of the Standard Model of particle physics. A huge body of experimental evidence for QCD has been gathered over the years.

QCD enjoys two peculiar properties:

- **Confinement**, which means that the force between quarks does not diminish as they are separated. Because of this, when you do separate a quark from other quarks, the energy in the gluon field is enough to create another quark pair; they are thus forever bound into hadrons such as the proton and the neutron or the pion and kaon. Although analytically unproven, confinement is widely believed to be true because it explains the consistent failure of free quark searches, and it is easy to demonstrate in lattice QCD.
- **Asymptotic freedom**, which means that in very high-energy reactions, quarks and gluons interact very weakly creating a quark–gluon plasma. This prediction of QCD was first discovered in the early 1970s by David Politzer and by Frank Wilczek and David Gross. For this work they were awarded the 2004 Nobel Prize in Physics.

The phase transition temperature between these two properties has been measured by the ALICE experiment to be around 160 MeV. Below this temperature, confinement is dominant, while above it, asymptotic freedom becomes dominant.

10.1 Terminology

The word *quark* was coined by American physicist Murray Gell-Mann (b. 1929) in its present sense. It originally comes from the phrase "Three quarks for Muster Mark" in *Finnegans Wake* by James Joyce. On June 27, 1978, Gell-Mann wrote a private letter to the editor of the *Oxford English Dictionary*, in which he related that he had been influenced by Joyce's words: "The allusion to three quarks seemed perfect." (Originally, only three quarks had been discovered.) Gell-Mann, however, wanted to pronounce the word to rhyme with "fork" rather than with "park", as Joyce seemed to indicate by rhyming words in the vicinity such as *Mark*. Gell-Mann got around that "by supposing that one ingredient of the line 'Three quarks for Muster Mark' was a cry of 'Three quarts for Mister ...' heard in H.C. Earwicker's pub", a plausible suggestion given the complex punning in Joyce's novel.[1]

The three kinds of charge in QCD (as opposed to one in quantum electrodynamics or QED) are usually referred to as "color charge" by loose analogy to the three kinds of color (red, green and blue) perceived by humans. Other than this nomenclature, the quantum parameter "color" is completely unrelated to the everyday, familiar phenomenon of color.

Since the theory of electric charge is dubbed "electrodynamics", the Greek word "chroma" Χρώμα (meaning color) is applied to the theory of color charge, "chromodynamics".

10.2 History

With the invention of bubble chambers and spark chambers in the 1950s, experimental particle physics discovered a large and ever-growing number of particles called hadrons. It seemed that such a large number of particles could not all be fundamental. First, the particles were classified by charge and isospin by Eugene Wigner and Werner Heisenberg; then, in 1953, according to strangeness by Murray Gell-Mann and Kazuhiko Nishijima. To gain greater insight, the hadrons were sorted into groups having similar properties and masses using the *eightfold way*, invented in 1961 by Gell-Mann and Yuval Ne'eman. Gell-Mann and George Zweig, correcting an earlier approach of Shoichi Sakata, went on to propose in 1963 that the structure of the groups could be explained by the existence of three flavors of smaller particles inside the hadrons: the quarks.

Perhaps the first remark that quarks should possess an additional quantum number was made[2] as a short footnote in the preprint of Boris Struminsky[3] in connection with Ω^- hyperon composed of three strange quarks with parallel spins (this situation was peculiar, because since quarks are fermions, such combination is forbidden by the Pauli exclusion principle):

> Three identical quarks cannot form an antisymmetric S-state. In order to realize an antisymmetric orbital S-state, it is necessary for the quark to have an additional quantum number.
> — B. V. Struminsky, *Magnetic moments of barions in the quark model*, JINR-Preprint P-1939, Dubna, Submitted on January 7, 1965

Boris Struminsky was a PhD student of Nikolay Bogolyubov. The problem considered in this preprint was suggested by Nikolay Bogolyubov, who advised Boris Struminsky in this research.[3] In the beginning of 1965, Nikolay Bogolyubov, Boris Struminsky and Albert Tavkhelidze wrote a preprint with a more detailed discussion of the additional quark quantum degree of freedom.[4] This work was also presented by Albert Tavchelidze without obtaining consent of his collaborators for doing so at an international conference in Trieste (Italy), in May 1965.[5][6]

A similar mysterious situation was with the Δ^{++} baryon; in the quark model, it is composed of three up quarks with parallel spins. In 1965, Moo-Young Han with Yoichiro Nambu and Oscar W. Greenberg independently resolved the problem by proposing that quarks possess an additional SU(3) gauge degree of freedom, later called color charge. Han and Nambu noted that quarks might interact via an octet of vector gauge bosons: the gluons.

Since free quark searches consistently failed to turn up any evidence for the new particles, and because an elementary particle back then was *defined* as a particle which could be separated and isolated, Gell-Mann often said that quarks were merely convenient mathematical constructs, not real particles. The meaning of this statement was usually clear in context: He meant quarks are confined, but he also was implying that the strong interactions could probably not be fully described by quantum field theory.

Richard Feynman argued that high energy experiments showed quarks are real particles: he called them *partons* (since they were parts of hadrons). By particles, Feynman meant objects which travel along paths, elementary particles in a field theory.

The difference between Feynman's and Gell-Mann's approaches reflected a deep split in the theoretical physics community. Feynman thought the quarks have a distribution of position or momentum, like any other particle, and he (correctly) believed that the diffusion of parton momentum explained diffractive scattering. Although Gell-Mann believed that certain quark charges could be localized, he was open to the possibility that the quarks themselves could not be localized because space and time break down. This was the more radical approach of S-matrix theory.

James Bjorken proposed that pointlike partons would imply certain relations should hold in deep inelastic scattering of electrons and protons, which were spectacularly verified in experiments at SLAC in 1969. This led physicists to abandon the S-matrix approach for the strong interactions.

The discovery of asymptotic freedom in the strong interactions by David Gross, David Politzer and Frank Wilczek allowed physicists to make precise predictions of the results of many high energy experiments using the quantum field theory technique of perturbation theory. Evidence of gluons was discovered in three-jet events at PETRA in 1979. These experiments became more and more precise, culminating in the verification of perturbative QCD at the level of a few percent at the LEP in CERN.

The other side of asymptotic freedom is confinement. Since the force between color charges does not decrease with distance, it is believed that quarks and gluons can never be liberated from hadrons. This aspect of the theory is verified within lattice QCD computations, but is not mathematically proven. One of the Millennium Prize Problems announced by the Clay Mathematics Institute requires a claimant to produce such a proof. Other aspects of non-perturbative QCD are the exploration of phases of quark matter, including the quark–gluon plasma.

The relation between the short-distance particle limit and the confining long-distance limit is one of the topics recently explored using string theory, the modern form of S-matrix theory.[7][8]

10.3 Theory

10.3.1 Some definitions

Every field theory of particle physics is based on certain symmetries of nature whose existence is deduced from observations. These can be

- local symmetries, that is the symmetry acts independently at each point in spacetime. Each such symmetry is the basis of a gauge theory and requires the introduction of its own gauge bosons.
- global symmetries, which are symmetries whose operations must be simultaneously applied to all points of spacetime.

QCD is a gauge theory of the SU(3) gauge group obtained by taking the color charge to define a local symmetry.

Since the strong interaction does not discriminate between different flavors of quark, QCD has approximate **flavor symmetry**, which is broken by the differing masses of the quarks.

There are additional global symmetries whose definitions require the notion of chirality, discrimination between left and right-handed. If the spin of a particle has a positive projection on its direction of motion then it is called left-handed; otherwise, it is right-handed. Chirality and handedness are not the same, but become approximately equivalent at high energies.

- **Chiral** symmetries involve independent transformations of these two types of particle.
- **Vector** symmetries (also called diagonal symmetries) mean the same transformation is applied on the two chiralities.
- **Axial** symmetries are those in which one transformation is applied on left-handed particles and the inverse on the right-handed particles.

10.3.2 Additional remarks: duality

As mentioned, *asymptotic freedom* means that at large energy – this corresponds also to *short distances* – there is practically no interaction between the particles. This is in contrast – more precisely one would say *dual* – to what one is used to, since usually one connects the absence of interactions with *large* distances. However, as already mentioned in the original paper of Franz Wegner,[9] a solid state theorist who introduced 1971 simple gauge invariant lattice models, the high-temperature behaviour of the *original model*, e.g. the strong decay of correlations at large distances, corresponds to the low-temperature behaviour of the (usually ordered!) *dual model*, namely the asymptotic decay of non-trivial correlations, e.g. short-range deviations from almost perfect arrangements, for short distances. Here, in contrast to Wegner, we have only the dual model, which is that one described in this article.[10]

10.3.3 Symmetry groups

The color group SU(3) corresponds to the local symmetry whose gauging gives rise to QCD. The electric charge labels a representation of the local symmetry group U(1) which is gauged to give QED: this is an abelian group. If one considers a version of QCD with *Nf* flavors of massless quarks, then there is a global (chiral) flavor symmetry group SUL(*Nf*) × SUR(*Nf*) × UB(1) × UA(1). The chiral symmetry is spontaneously broken by the QCD vacuum to the vector (L+R) SUV(*Nf*) with the formation of a chiral condensate. The vector symmetry, UB(1) corresponds to the baryon number of quarks and is an exact symmetry. The axial symmetry UA(1) is exact in the classical theory, but broken in the quantum theory, an occurrence called an anomaly. Gluon field configurations called instantons are closely related to this anomaly.

There are two different types of SU(3) symmetry: there is the symmetry that acts on the different colors of quarks, and this is an exact gauge symmetry mediated by the gluons, and there is also a flavor symmetry which rotates different flavors of quarks to each other, or *flavor SU(3)*. Flavor SU(3) is an approximate symmetry of the vacuum of QCD, and is not a fundamental symmetry at all. It is an accidental consequence of the small mass of the three lightest quarks.

In the QCD vacuum there are vacuum condensates of all the quarks whose mass is less than the QCD scale. This includes the up and down quarks, and to a lesser extent the strange quark, but not any of the others. The vacuum is symmetric under SU(2) isospin rotations of up and down, and to a lesser extent under rotations of up, down and strange, or full flavor group SU(3), and the observed particles make isospin and SU(3) multiplets.

The approximate flavor symmetries do have associated gauge bosons, observed particles like the rho and the omega, but these particles are nothing like the gluons and they are not massless. They are emergent gauge bosons in an approximate string description of QCD.

10.3.4 Lagrangian

The dynamics of the quarks and gluons are controlled by the quantum chromodynamics Lagrangian. The gauge invariant QCD Lagrangian is

where $\psi_i(x)$ is the quark field, a dynamical function of spacetime, in the fundamental representation of the SU(3) gauge group, indexed by $i,\ j,\ \ldots$; $\mathcal{A}^a_\mu(x)$ are the gluon fields, also dynamical functions of spacetime, in the adjoint representation of the SU(3) gauge group, indexed by *a*, *b*,... The γ^μ are Dirac matrices connecting the spinor representation to the vector representation of the Lorentz group.

The symbol $G^a_{\mu\nu}$ represents the gauge invariant gluon field strength tensor, analogous to the electromagnetic field strength tensor, $F^{\mu\nu}$, in quantum electrodynamics. It is given by:[11]

$$G^a_{\mu\nu} = \partial_\mu \mathcal{A}^a_\nu - \partial_\nu \mathcal{A}^a_\mu + g f^{abc} \mathcal{A}^b_\mu \mathcal{A}^c_\nu\,,$$

where *fabc* are the structure constants of SU(3). Note that the rules to move-up or pull-down the *a*, *b*, or *c* indexes are *trivial*, (+, ..., +), so that $f^{abc} = fabc = f^a bc$ whereas for the μ or ν indexes one has the non-trivial *relativistic* rules, corresponding e.g. to the metric signature (+ − − −).

The constants *m* and *g* control the quark mass and coupling constants of the theory, subject to renormalization in the full quantum theory.

An important theoretical notion concerning the final term of the above Lagrangian is the *Wilson loop* variable. This loop variable plays a most important role in discretized forms of the QCD (see lattice QCD), and more generally, it distinguishes confined and deconfined states of a gauge theory. It was introduced by the Nobel prize winner Kenneth G. Wilson and is treated in a separate article.

10.3.5 Fields

Quarks are massive spin-1/2 fermions which carry a color charge whose gauging is the content of QCD. Quarks are represented by Dirac fields in the fundamental representation **3** of the gauge group SU(3). They also carry electric charge (either −1/3 or 2/3) and participate in weak interactions as part of weak isospin doublets. They carry global quantum numbers including the baryon number, which is 1/3 for each quark, hypercharge and one of the flavor quantum numbers.

Gluons are spin-1 bosons which also carry color charges, since they lie in the adjoint representation **8** of SU(3). They have no electric charge, do not participate in the weak interactions, and have no flavor. They lie in the singlet representation **1** of all these symmetry groups.

Every quark has its own antiquark. The charge of each antiquark is exactly the opposite of the corresponding quark.

10.3.6 Dynamics

According to the rules of quantum field theory, and the associated Feynman diagrams, the above theory gives rise to three basic interactions: a quark may emit (or absorb) a gluon, a gluon may emit (or absorb) a gluon, and two gluons

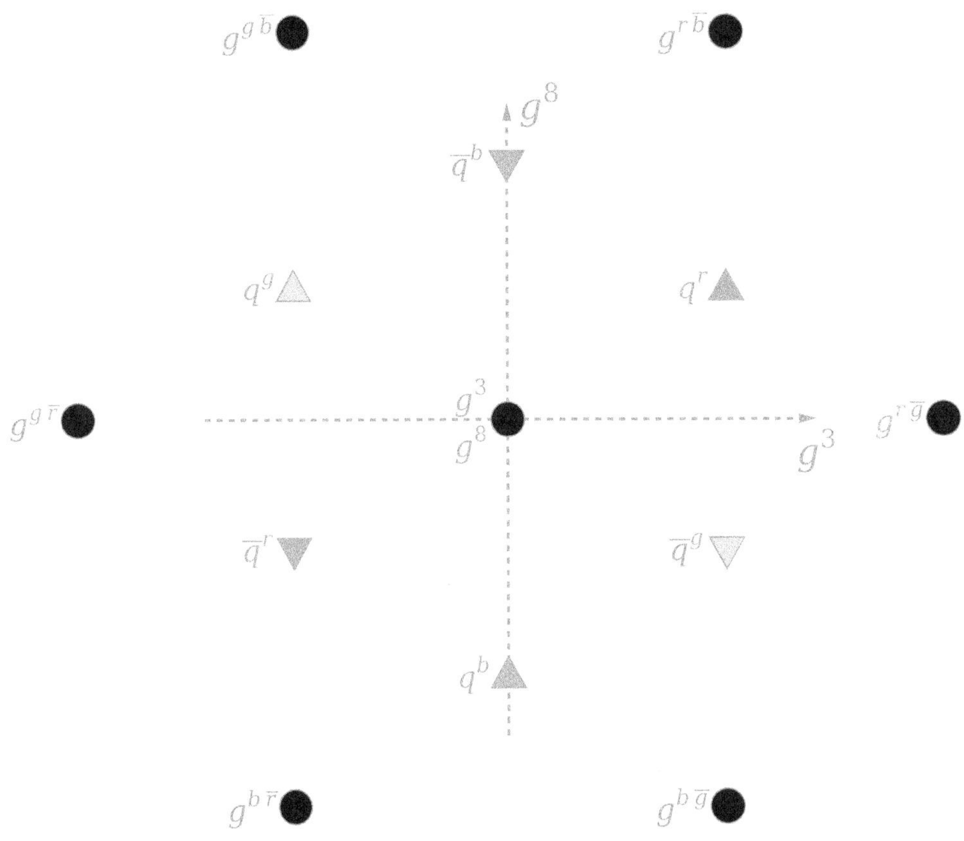

The pattern of strong charges for the three colors of quark, three antiquarks, and eight gluons (with two of zero charge overlapping).

may directly interact. This contrasts with QED, in which only the first kind of interaction occurs, since photons have no charge. Diagrams involving Faddeev–Popov ghosts must be considered too (except in the unitarity gauge).

10.3.7 Area law and confinement

Detailed computations with the above-mentioned Lagrangian[12] show that the effective potential between a quark and its anti-quark in a meson contains a term $\propto r$, which represents some kind of "stiffness" of the interaction between the particle and its anti-particle at large distances, similar to the entropic elasticity of a rubber band (see below). This leads to *confinement* [13] of the quarks to the interior of hadrons, i.e. mesons and nucleons, with typical radii R_c, corresponding to former "Bag models" of the hadrons[14] . The order of magnitude of the "bag radius" is 1 fm (= 10^{-15} m). Moreover, the above-mentioned stiffness is quantitatively related to the so-called "area law" behaviour of the expectation value of the Wilson loop product *PW* of the ordered coupling constants around a closed loop *W*; i.e. $\langle P_W \rangle$ is proportional to the *area* enclosed by the loop. For this behaviour the non-abelian behaviour of the gauge group is essential.

10.4 Methods

Further analysis of the content of the theory is complicated. Various techniques have been developed to work with QCD. Some of them are discussed briefly below.

10.4.1 Perturbative QCD

Main article: Perturbative QCD

This approach is based on asymptotic freedom, which allows perturbation theory to be used accurately in experiments performed at very high energies. Although limited in scope, this approach has resulted in the most precise tests of QCD to date.

10.4.2 Lattice QCD

Main article: Lattice QCD

Among non-perturbative approaches to QCD, the most well established one is lattice QCD. This approach uses a discrete set of spacetime points (called the lattice) to reduce the analytically intractable path integrals of the continuum theory to a very difficult numerical computation which is then carried out on supercomputers like the QCDOC which was constructed for precisely this purpose. While it is a slow and resource-intensive approach, it has wide applicability, giving insight into parts of the theory inaccessible by other means, in particular into the explicit forces acting between quarks and antiquarks in a meson. However, the numerical sign problem makes it difficult to use lattice methods to study QCD at high density and low temperature (e.g. nuclear matter or the interior of neutron stars).

10.4.3 1/N expansion

Main article: 1/N expansion

A well-known approximation scheme, the 1/N expansion, starts from the premise that the number of colors is infinite, and makes a series of corrections to account for the fact that it is not. Until now, it has been the source of qualitative insight rather than a method for quantitative predictions. Modern variants include the AdS/CFT approach.

10.4.4 Effective theories

For specific problems effective theories may be written down which give qualitatively correct results in certain limits. In the best of cases, these may then be obtained as systematic expansions in some parameter of the QCD Lagrangian. One such effective field theory is chiral perturbation theory or ChiPT, which is the QCD effective theory at low energies. More precisely, it is a low energy expansion based on the spontaneous chiral symmetry breaking of QCD, which is an exact symmetry when quark masses are equal to zero, but for the u,d and s quark, which have small mass, it is still a good approximate symmetry. Depending on the number of quarks which are treated as light, one uses either SU(2) ChiPT or SU(3) ChiPT . Other effective theories are heavy quark effective theory (which expands around heavy quark mass near infinity), and soft-collinear effective theory (which expands around large ratios of energy scales). In addition to effective theories, models like the Nambu–Jona-Lasinio model and the chiral model are often used when discussing general features.

10.4.5 QCD sum rules

Main article: QCD sum rules

Based on an Operator product expansion one can derive sets of relations that connect different observables with each other.

10.4.6 Nambu–Jona-Lasinio model

In one of his recent works, Kei-Ichi Kondo derived as a low-energy limit of QCD, a theory linked to the Nambu–Jona-Lasinio model since it is basically a particular non-local version of the Polyakov–Nambu–Jona-Lasinio model.[16] The later being in its local version, nothing but the Nambu–Jona-Lasinio model in which one has included the Polyakov loop effect, in order to describe a 'certain confinement'.

The Nambu–Jona-Lasinio model in itself is, among many other things, used because it is a 'relatively simple' model of chiral symmetry breaking, phenomenon present up to certain conditions (Chiral limit i.e. massless fermions) in QCD itself. In this model, however, there is no confinement. In particular, the energy of an isolated quark in the physical vacuum turns out well defined and finite.

10.5 Experimental tests

The notion of quark flavors was prompted by the necessity of explaining the properties of hadrons during the development of the quark model. The notion of color was necessitated by the puzzle of the Δ++. This has been dealt with in the section on the history of QCD.

The first evidence for quarks as real constituent elements of hadrons was obtained in deep inelastic scattering experiments at SLAC. The first evidence for gluons came in three jet events at PETRA.

Several good quantitative tests of perturbative QCD exist:

- The running of the QCD coupling as deduced from many observations
- Scaling violation in polarized and unpolarized deep inelastic scattering
- Vector boson production at colliders (this includes the Drell-Yan process)
- Jet cross sections in colliders
- Event shape observables at the LEP
- Heavy-quark production in colliders

Quantitative tests of non-perturbative QCD are fewer, because the predictions are harder to make. The best is probably the running of the QCD coupling as probed through lattice computations of heavy-quarkonium spectra. There is a recent claim about the mass of the heavy meson B_c . Other non-perturbative tests are currently at the level of 5% at best. Continuing work on masses and form factors of hadrons and their weak matrix elements are promising candidates for future quantitative tests. The whole subject of quark matter and the quark–gluon plasma is a non-perturbative test bed for QCD which still remains to be properly exploited.

One qualitative prediction of QCD is that there exist composite particles made solely of gluons called glueballs that have not yet been definitively observed experimentally. A definitive observation of a glueball with the properties predicted by QCD would strongly confirm the theory. In principle, if glueballs could be definitively ruled out, this would be a serious experimental blow to QCD. But, as of 2013, scientists are unable to confirm or deny the existence of glueballs definitively, despite the fact that particle accelerators have sufficient energy to generate them.

10.6 Cross-relations to solid state physics

There are unexpected cross-relations to solid state physics. For example, the notion of gauge invariance forms the basis of the well-known Mattis spin glasses,[17] which are systems with the usual spin degrees of freedom $s_i = \pm 1$ for i =1,...,N, with the special fixed "random" couplings $J_{i,k} = \epsilon_i \, J_0 \, \epsilon_k$. Here the ε_i and ε_k quantities can independently and "randomly" take the values ±1, which corresponds to a most-simple gauge transformation ($s_i \to s_i \cdot \epsilon_i \quad J_{i,k} \to \epsilon_i J_{i,k} \epsilon_k \quad s_k \to s_k \cdot \epsilon_k$). This means that thermodynamic expectation values of measurable quantities, e.g. of the energy $\mathcal{H} := -\sum s_i \, J_{i,k} \, s_k$, are invariant.

However, here the *coupling degrees of freedom* $J_{i,k}$, which in the QCD correspond to the *gluons*, are "frozen" to fixed values (quenching). In contrast, in the QCD they "fluctuate" (annealing), and through the large number of gauge degrees of freedom the entropy plays an important role (see below).

For positive J_0 the thermodynamics of the Mattis spin glass corresponds in fact simply to a "ferromagnet in disguise", just because these systems have no "frustration" at all. This term is a basic measure in spin glass theory.[18] Quantitatively it is identical with the loop product $P_W := J_{i,k}J_{k,l}...J_{n,m}J_{m,i}$ along a closed loop *W*. However, for a Mattis spin glass – in contrast to "genuine" spin glasses – the quantity *PW* never becomes negative.

The basic notion "frustration" of the spin-glass is actually similar to the Wilson loop quantity of the QCD. The only difference is again that in the QCD one is dealing with SU(3) matrices, and that one is dealing with a "fluctuating" quantity. Energetically, perfect absence of frustration should be non-favorable and atypical for a spin glass, which means that one should add the loop product to the Hamiltonian, by some kind of term representing a "punishment". In the QCD the Wilson loop is essential for the Lagrangian rightaway.

The relation between the QCD and "disordered magnetic systems" (the spin glasses belong to them) were additionally stressed in a paper by Fradkin, Huberman und Shenker,[19] which also stresses the notion of duality.

A further analogy consists in the already mentioned similarity to polymer physics, where, analogously to Wilson Loops, so-called "entangled nets" appear, which are important for the formation of the entropy-elasticity (force proportional to the length) of a rubber band. The non-abelian character of the SU(3) corresponds thereby to the non-trivial "chemical links", which glue different loop segments together, and "asymptotic freedom" means in the polymer analogy simply the fact that in the short-wave limit, i.e. for $0 \leftarrow \lambda_w \ll R_c$ (where *Rc* is a characteristic correlation length for the glued loops, corresponding to the above-mentioned "bag radius", while λ_w is the wavelength of an excitation) any non-trivial correlation vanishes totally, as if the system had crystallized.[20]

There is also a correspondence between confinement in QCD – the fact that the color field is only different from zero in the interior of hadrons – and the behaviour of the usual magnetic field in the theory of type-II superconductors: there the magnetism is confined to the interiour of the Abrikosov flux-line lattice,[21] i.e., the London penetration depth λ of that theory is analogous to the confinement radius *Rc* of quantum chromodynamics. Mathematically, this correspondendence is supported by the second term, $\propto gG^a_\mu \bar{\psi}_i \gamma^\mu T^a_{ij} \psi_j$, on the r.h.s. of the Lagrangian.

10.7 See also

- For overviews, see Standard Model, its field theoretical formulation, strong interactions, quarks and gluons, hadrons, confinement, QCD matter, or quark–gluon plasma.
- For details, see gauge theory, quantization procedure including BRST quantization and Faddeev–Popov ghosts. A more general category is quantum field theory.
- For techniques, see Lattice QCD, 1/N expansion, perturbative QCD, Soft-collinear effective theory, heavy quark effective theory, chiral models, and the Nambu and Jona-Lasinio model.
- For experiments, see quark search experiments, deep inelastic scattering, jet physics, quark–gluon plasma.
- Symmetry in quantum mechanics

10.8 References

[1] Gell-Mann, Murray (1995). *The Quark and the Jaguar*. Owl Books. ISBN 978-0-8050-7253-2.

[2] Fyodor Tkachov (2009). "A contribution to the history of quarks: Boris Struminsky's 1965 JINR publication". arXiv:0904.0343 [physics.hist-ph].

[3] B. V. Struminsky, Magnetic moments of barions in the quark model. JINR-Preprint P-1939, Dubna, Russia. Submitted on January 7, 1965.

[4] N. Bogolubov, B. Struminsky, A. Tavkhelidze. On composite models in the theory of elementary particles. JINR Preprint D-1968, Dubna 1965.

[5] A. Tavkhelidze. Proc. Seminar on High Energy Physics and Elementary Particles, Trieste, 1965, Vienna IAEA, 1965, p. 763.

[6] V. A. Matveev and A. N. Tavkhelidze (INR, RAS, Moscow) The quantum number color, colored quarks and QCD (Dedicated to the 40th Anniversary of the Discovery of the Quantum Number Color). Report presented at the 99th Session of the JINR Scientific Council, Dubna, 19–20 January 2006.

[7] J. Polchinski, M. Strassler (2002). "Hard Scattering and Gauge/String duality". *Physical Review Letters* **88** (3): 31601. arXiv:hep-th/0109174. Bibcode:2002PhRvL..88c1601P. doi:10.1103/PhysRevLett.88.031601. PMID 11801052.

[8] Brower, Richard C.; Mathur, Samir D.; Chung-I Tan (2000). "Glueball Spectrum for QCD from AdS Supergravity Duality". *Nuclear Physics B* **587**: 249–276. arXiv:hep-th/0003115. Bibcode:2000NuPhB.587..249B. doi:10.1016/S0550-3213(00)00435-1.

[9] F. Wegner, *Duality in Generalized Ising Models and Phase Transitions without Local Order Parameter*, J. Math. Phys. **12** (1971) 2259–2272.

Reprinted in Claudio Rebbi (ed.), *Lattice Gauge Theories and Monte Carlo Simulations*, World Scientific, Singapore (1983), p. 60–73. Abstract:

[10] Perhaps one can guess that in the "original" model mainly the quarks would fluctuate, whereas in the present one, the "dual" model, mainly the gluons do.

[11] M. Eidemüller, H.G. Dosch, M. Jamin (1999). "The field strength correlator from QCD sum rules". *Nucl.Phys.Proc.Suppl.86:421–425,2000* (Heidelberg, Germany). arXiv:hep-ph/9908318.

[12] See all standard textbooks on the QCD, e.g., those noted above

[13] Only at extremely large pressures and or temperatures, e.g. for $T \cong 5 \cdot 10^{12}$ K or larger, *confinement* gives way to a quark–gluon plasma.

[14] Kenneth A. Johnson, "The bag model of quark confinement", Scientific American, July 1979

[15] M. Cardoso et al., "Lattice QCD computation of the colour fields for the static hybrid quark–gluon–antiquark system, and microscopic study of the Casimir scaling", Phys. Rev. D 81, 034504 (2010)).

[16] Kei-Ichi Kondo (2010). "Toward a first-principle derivation of confinement and chiral-symmetry-breaking crossover transitions in QCD". *Physical Review D* **82** (6): 065024. arXiv:1005.0314v2. Bibcode:2010PhRvD..82f5024K. doi:10.1103/PhysRevD.82.065024.

[17] D.C. Mattis, Phys. Lett. 56a (1976) 421

[18] J. Vanninemus and G. Toulouse, J. Phys. C 10 (1977) 537

[19] E. Fradkin, B.A. Huberman, S. Shenker, *Gauge Symmetries in random magnetic systems*, Phys. Rev. B 18 (1978) 4783–4794,

[20] A. Bergmann, A. Owen, "Dielectric relaxation spectroscopy of poly[(R)−3-Hydroxybutyrate] (PHD) during crystallization", Polymer International 53 (7) (2004) 863–868,

[21] Mathematically, the flux-line lattices are described by Emil Artin's braid group, which is nonabelian, since one braid can wind around another one.

10.9 Further reading

- Greiner, Walter;Schäfer, Andreas (1994). *Quantum Chromodynamics*. Springer. ISBN 0-387-57103-5.
- Halzen, Francis; Martin, Alan (1984). *Quarks & Leptons: An Introductory Course in Modern Particle Physics*. John Wiley & Sons. ISBN 0-471-88741-2.
- Creutz, Michael (1985). *Quarks, Gluons and Lattices*. Cambridge University Press. ISBN 978-0-521-31535-7.

10.10 External links

- Particle data group
- The millennium prize for proving confinement
- Ab Initio Determination of Light Hadron Masses
- Andreas S Kronfeld *The Weight of the World Is Quantum Chromodynamics*
- Andreas S Kronfeld *Quantum chromodynamics with advanced computing*
- Standard model gets right answer
- Quantum Chromodynamics

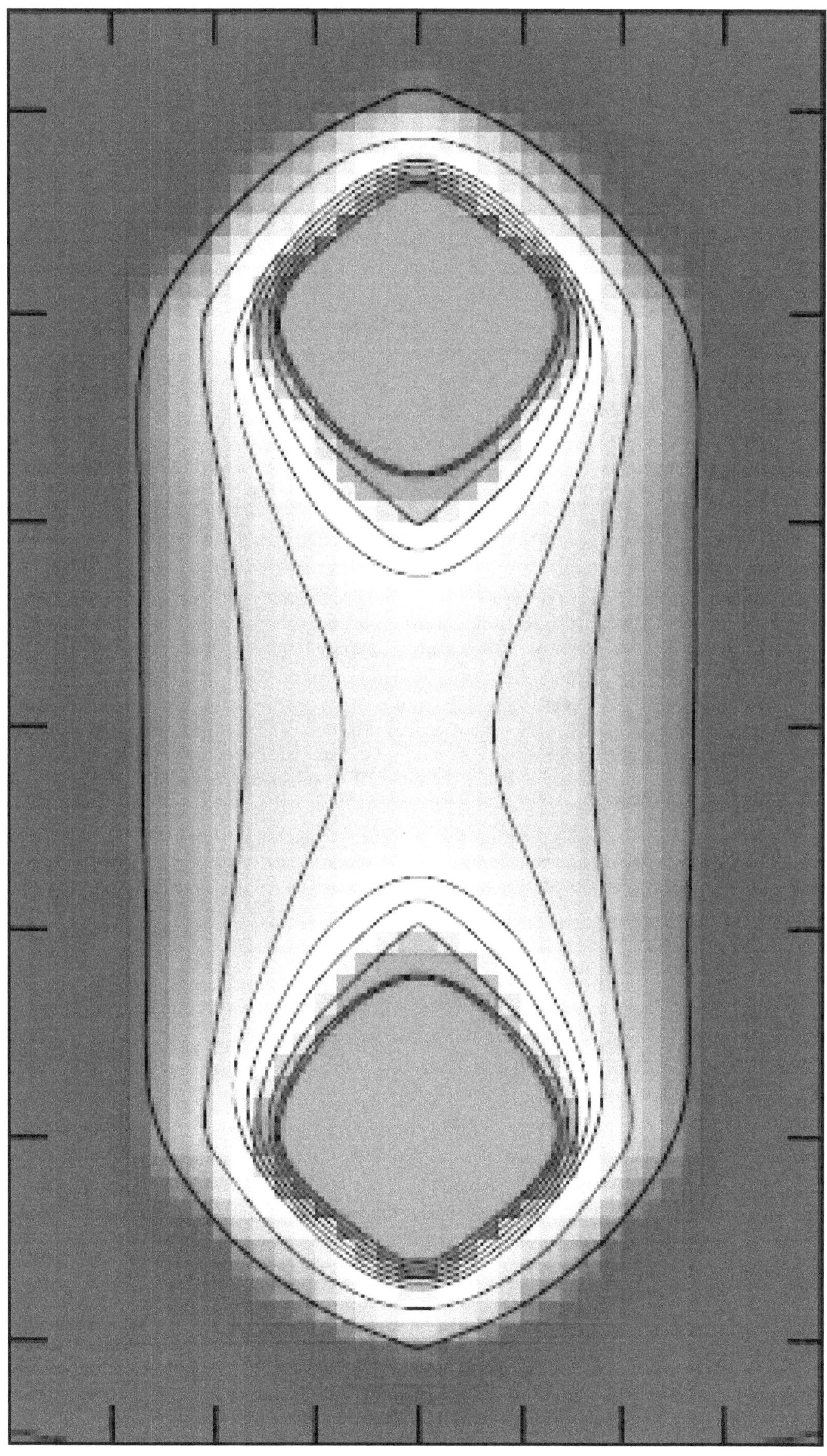

A quark and an antiquark (red color) are glued together (green color) to form a meson (result of a lattice QCD simulation by M. Cardoso et al [15])

Chapter 11

Brane

For other uses, see Brane (disambiguation).

In string theory and related theories such as supergravity theories, a **brane** is a physical object that generalizes the notion of a point particle to higher dimensions. For example, a point particle can be viewed as a brane of dimension zero, while a string can be viewed as a brane of dimension one. It is also possible to consider higher-dimensional branes. In dimension p, these are called p-branes. The word brane comes from the word "membrane" which refers to a two-dimensional brane.[1]

Branes are dynamical objects which can propagate through spacetime according to the rules of quantum mechanics. They have mass and can have other attributes such as charge. A p-brane sweeps out a $(p+1)$-dimensional volume in spacetime called its *worldvolume*. Physicists often study fields analogous to the electromagnetic field which live on the worldvolume of a brane.[2]

In string theory, D-branes are an important class of branes that arise when one considers open strings. As an open string propagates through spacetime, its endpoints are required to lie on a D-brane. The letter "D" in D-brane refers to a certain mathematical condition on the system known as the Dirichlet boundary condition. The study of D-branes in string theory has led to important results such as the AdS/CFT correspondence, which has shed light on many problems in quantum field theory.

Branes are also frequently studied from a purely mathematical point of view since they are related to subjects such as homological mirror symmetry and noncommutative geometry. Mathematically, branes may be represented as objects of certain categories, such as the derived category of coherent sheaves on a Calabi–Yau manifold, or the Fukaya category.

11.1 D-branes

Main article: D-brane

In string theory, a string may be open (forming a segment with two endpoints) or closed (forming a closed loop). D-branes are an important class of branes that arise when one considers open strings. As an open string propagates through spacetime, its endpoints are required to lie on a D-brane. The letter "D" in D-brane refers to a condition that it satisfies, the Dirichlet boundary condition.[3]

One crucial point about D-branes is that the dynamics on the D-brane worldvolume is described by a gauge theory, a kind of highly symmetric physical theory which is also used to describe the behavior of elementary particles in the standard model of particle physics. This connection has led to many important insights into gauge theory. For example, it led to the discovery of the AdS/CFT correspondence, a theoretical tool that physicists use to translate difficult problems in gauge theory into more mathematically tractable problems in string theory.[4]

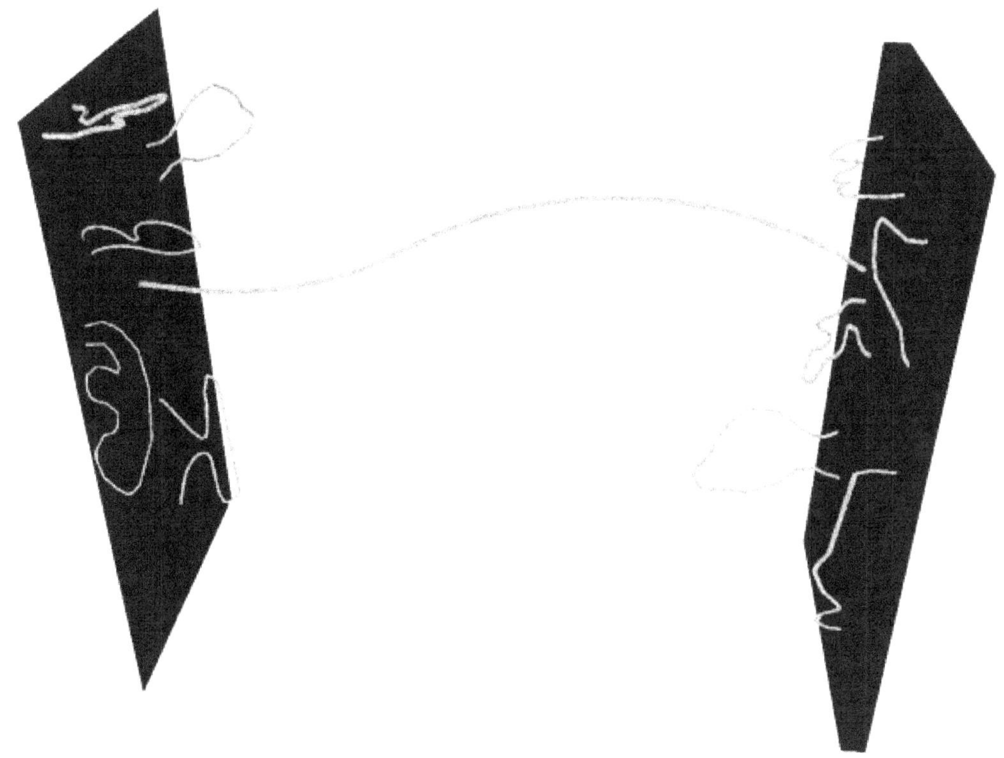

Open strings attached to a pair of D-branes

11.2 Mathematical viewpoint

Mathematically, branes can be described using the notion of a category.[5] This is a mathematical structure consisting of *objects*, and for any pair of objects, a set of *morphisms* between them. In most examples, the objects are mathematical structures (such as sets, vector spaces, or topological spaces) and the morphisms are functions between these structures.[6] One can also consider categories where the objects are D-branes and the morphisms between two branes α and β are states of open strings stretched between α and β .[7]

In one version of string theory known as the topological B-model, the D-branes are complex submanifolds of certain six-dimensional shapes called Calabi–Yau manifolds, together with additional data that arise physically from having charges at the endpoints of strings.[8] Intuitively, one can think of a submanifold as a surface embedded inside of a Calabi–Yau manifold, although submanifolds can also exist in dimensions different from two.[9] In mathematical language, the category having these branes as its objects is known as the derived category of coherent sheaves on the Calabi–Yau.[10] In another version of string theory called the topological A-model, the D-branes can again be viewed as submanifolds of a Calabi–Yau manifold. Roughly speaking, they are what mathematicians call special Lagrangian submanifolds.[11] This means among other things that they have half the dimension of the space in which they sit, and they are length-, area-, or volume-minimizing.[12] The category having these branes as its objects is called the Fukaya category.[13]

The derived category of coherent sheaves is constructed using tools from complex geometry, a branch of mathematics that describes geometric curves in algebraic terms and solves geometric problems using algebraic equations.[14] On the other hand, the Fukaya category is constructed using symplectic geometry, a branch of mathematics that arose from studies of classical physics. Symplectic geometry studies spaces equipped with a symplectic form, a mathematical tool that can be used to compute area in two-dimensional examples.[15]

The homological mirror symmetry conjecture of Maxim Kontsevich states that the derived category of coherent sheaves on one Calabi–Yau manifold is equivalent in a certain sense to the Fukaya category of a completely different Calabi–Yau manifold.[16] This equivalence provides an unexpected bridge between two branches of geometry, namely complex and symplectic geometry.[17]

A cross section of a Calabi–Yau manifold

11.3 See also

- Black brane
- Brane cosmology
- M2-brane
- M5-brane
- NS5-brane

11.4 Notes

[1] Moore 2005, p. 214

[2] Moore 2005, p. 214

[3] Moore 2005, p. 215

[4] Moore 2005, p. 215

[5] Aspinwall et al. 2009

[6] A basic reference on category theory is Mac Lane 1998.

[7] Zaslow 2008, p. 536

[8] Zaslow 2008, p. 536

[9] Yau and Nadis 2010, p. 165

[10] Aspinwal et al. 2009, p. 575

[11] Aspinwal et al. 2009, p. 575

[12] Yau and Nadis 2010, p. 175

[13] Aspinwal et al. 2009, p. 575

[14] Yau and Nadis 2010, pp. 180–1

[15] Zaslow 2008, p. 531

[16] Aspinwall et al. 2009, p. 616

[17] Yau and Nadis 2010, p. 181

11.5 References

- Aspinwall, Paul; Bridgeland, Tom; Craw, Alastair; Douglas, Michael; Gross, Mark; Kapustin, Anton; Moore, Gregory; Segal, Graeme; Szendröi, Balázs; Wilson, P.M.H., eds. (2009). *Dirichlet Branes and Mirror Symmetry*. American Mathematical Society. ISBN 978-0-8218-3848-8.

- Mac Lane, Saunders (1998). *Categories for the Working Mathematician*. ISBN 978-0-387-98403-2.

- Moore, Gregory (2005). “What is ... a Brane?" (PDF). *Notices of the AMS* **52**: 214. Retrieved June 2013.

- Yau, Shing-Tung; Nadis, Steve (2010). *The Shape of Inner Space: String Theory and the Geometry of the Universe's Hidden Dimensions*. Basic Books. ISBN 978-0-465-02023-2.

- Zaslow, Eric (2008). “Mirror Symmetry”. In Gowers, Timothy. *The Princeton Companion to Mathematics*. ISBN 978-0-691-11880-2.

Chapter 12

D-brane

In string theory, **D-branes** are a class of extended objects upon which open strings can end with Dirichlet boundary conditions, after which they are named. D-branes were discovered by Dai, Leigh and Polchinski, and independently by Hořava in 1989. In 1995, Polchinski identified D-branes with black p-brane solutions of supergravity, a discovery that triggered the Second Superstring Revolution and led to both holographic and M-theory dualities.

D-branes are typically classified by their spatial dimension, which is indicated by a number written after the *D*. A D0-brane is a single point, a D1-brane is a line (sometimes called a "D-string"), a D2-brane is a plane, and a D25-brane fills the highest-dimensional space considered in bosonic string theory. There are also instantonic D(-1)-branes, which are localized in both space and time.

12.1 Theoretical background

The equations of motion of string theory require that the endpoints of an open string (a string with endpoints) satisfy one of two types of boundary conditions: The Neumann boundary condition, corresponding to free endpoints moving through spacetime at the speed of light, or the Dirichlet boundary conditions, which pin the string endpoint. Each coordinate of the string must satisfy one or the other of these conditions. There can also exist strings with mixed boundary conditions, where the two endpoints satisfy NN, DD, ND and DN boundary conditions. If p spatial dimensions satisfy the Neumann boundary condition, then the string endpoint is confined to move within a p-dimensional hyperplane. This hyperplane provides one description of a Dp-brane.

Although rigid in the limit of zero coupling, the spectrum of open strings ending on a D-brane contains modes associated with its fluctuations, implying that D-branes are dynamical objects. When N D-branes are nearly coincident, the spectrum of strings stretching between them becomes very rich. One set of modes produce a non-abelian gauge theory on the world-volume. Another set of modes is an $N \times N$ dimensional matrix for each transverse dimension of the brane. If these matrices commute, they may be diagonalized, and the eigenvalues define the position of the N D-branes in space. More generally, the branes are described by non-commutative geometry, which allows exotic behavior such as the Myers effect, in which a collection of Dp-branes expand into a D(p+2)-brane.

Tachyon condensation is a central concept in this field. Ashoke Sen has argued that in Type IIB string theory, tachyon condensation allows (in the absence of Neveu-Schwarz 3-form flux) an arbitrary D-brane configuration to be obtained from a stack of D9 and anti D9-branes. Edward Witten has shown that such configurations will be classified by the K-theory of the spacetime. Tachyon condensation is still very poorly understood. This is due to the lack of an exact string field theory that would describe the off-shell evolution of the tachyon.

12.2 Braneworld cosmology

This has implications for physical cosmology. Because string theory implies that the Universe has more dimensions than we expect—26 for bosonic string theories and 10 for superstring theories—we have to find a reason why the extra dimensions are not apparent. One possibility would be that the visible Universe is in fact a very large D-brane extending over three spatial dimensions. Material objects, made of open strings, are bound to the D-brane, and cannot move "at right angles to reality" to explore the Universe outside the brane. This scenario is called a brane cosmology.

The force of gravity is *not* due to open strings; the gravitons which carry gravitational forces are vibrational states of *closed* strings. Because closed strings do not have to be attached to D-branes, gravitational effects could depend upon the extra dimensions at right angles to the brane.

12.3 D-brane scattering

When two D-branes approach each other the interaction is captured by the one loop annulus amplitude of strings between the two branes. The scenario of two parallel branes approaching each other at a constant velocity can be mapped to the problem of two stationary branes that are rotated relative to each other by some angle. The annulus amplitude yields singularities that correspond to the on-shell production of open strings stretched between the two branes. This is true irrespective of the charge of the D-branes. At non-relativistic scattering velocities the open strings may be described by a low-energy effective action that contains two complex scalar fields that are coupled via a term $\phi^2\chi^2$. Thus, as the field ϕ (separation of the branes) changes, the mass of the field χ changes. This induces open string production and as a result the two scattering branes will be trapped.

12.4 Gauge theories

The arrangement of D-branes constricts the types of string states which can exist in a system. For example, if we have two parallel D2-branes, we can easily imagine strings stretching from brane 1 to brane 2 or vice versa. (In most theories, strings are *oriented* objects: each one carries an "arrow" defining a direction along its length.) The open strings permissible in this situation then fall into two categories, or "sectors": those originating on brane 1 and terminating on brane 2, and those originating on brane 2 and terminating on brane 1. Symbolically, we say we have the [1 2] and the [2 1] sectors. In addition, a string may begin and end on the same brane, giving [1 1] and [2 2] sectors. (The numbers inside the brackets are called *Chan-Paton indices*, but they are really just labels identifying the branes.) A string in either the [1 2] or the [2 1] sector has a minimum length: it cannot be shorter than the separation between the branes. All strings have some tension, against which one must pull to lengthen the object; this pull does work on the string, adding to its energy. Because string theories are by nature relativistic, adding energy to a string is equivalent to adding mass, by Einstein's relation $E = mc^2$. Therefore, the separation between D-branes controls the minimum mass open strings may have.

Furthermore, affixing a string's endpoint to a brane influences the way the string can move and vibrate. Because particle states "emerge" from the string theory as the different vibrational states the string can experience, the arrangement of D-branes controls the types of particles present in the theory. The simplest case is the [1 1] sector for a Dp-brane, that is to say the strings which begin and end on any particular D-brane of p dimensions. Examining the consequences of the Nambu-Goto action (and applying the rules of quantum mechanics to quantize the string), one finds that among the spectrum of particles is one resembling the photon, the fundamental quantum of the electromagnetic field. The resemblance is precise: a p-dimensional version of the electromagnetic field, obeying a p-dimensional analogue of Maxwell's equations, exists on every Dp-brane.

In this sense, then, one can say that string theory "predicts" electromagnetism: D-branes are a necessary part of the theory if we permit open strings to exist, and all D-branes carry an electromagnetic field on their volume.

Other particle states originate from strings beginning and ending on the same D-brane. Some correspond to massless particles like the photon; also in this group are a set of massless scalar particles. If a Dp-brane is embedded in a spacetime of d spatial dimensions, the brane carries (in addition to its Maxwell field) a set of d - p massless scalars (particles which do not have polarizations like the photons making up light). Intriguingly, there are just as many massless scalars as there are directions perpendicular to the brane; the *geometry* of the brane arrangement is closely related to the *quantum field theory* of the particles existing on it. In fact, these massless scalars are Goldstone excitations of the brane, corresponding to the different ways the symmetry of empty space can be broken. Placing a D-brane in a universe breaks the symmetry among locations, because it defines a particular place, assigning a special meaning to a particular location along each of the d - p directions perpendicular to the brane.

The quantum version of Maxwell's electromagnetism is only one kind of gauge theory, a **U**(1) gauge theory where the gauge group is made of unitary matrices of order 1. D-branes can be used to generate gauge theories of higher order, in the following way:

Consider a group of N separate Dp-branes, arranged in parallel for simplicity. The branes are labeled 1,2,...,N for convenience. Open strings in this system exist in one of many sectors: the strings beginning and ending on some brane

i give that brane a Maxwell field and some massless scalar fields on its volume. The strings stretching from brane i to another brane j have more intriguing properties. For starters, it is worthwhile to ask which sectors of strings can interact with one another. One straightforward mechanism for a string interaction is for two strings to join endpoints (or, conversely, for one string to "split down the middle" and make two "daughter" strings). Since endpoints are restricted to lie on D-branes, it is evident that a [1 2] string may interact with a [2 3] string, but not with a [3 4] or a [4 17] one. The masses of these strings will be influenced by the separation between the branes, as discussed above, so for simplicity's sake we can imagine the branes squeezed closer and closer together, until they lie atop one another. If we regard two overlapping branes as distinct objects, then we still have all the sectors we had before, but without the effects due to the brane separations.

The zero-mass states in the open-string particle spectrum for a system of N coincident D-branes yields a set of interacting quantum fields which is exactly a $\mathbf{U}(N)$ gauge theory. (The string theory does contain other interactions, but they are only detectable at very high energies.) Gauge theories were not invented starting with bosonic or fermionic strings; they originated from a different area of physics, and have become quite useful in their own right. If nothing else, the relation between D-brane geometry and gauge theory offers a useful pedagogical tool for explaining gauge interactions, even if string theory fails to be the "theory of everything".

12.5 Black holes

Another important use of D-branes has been in the study of black holes. Since the 1970s, scientists have debated the problem of black holes having entropy. Consider, as a thought experiment, dropping an amount of hot gas into a black hole. Since the gas cannot escape from the hole's gravitational pull, its entropy would seem to have vanished from the universe. In order to maintain the second law of thermodynamics, one must postulate that the black hole gained whatever entropy the infalling gas originally had. Attempting to apply quantum mechanics to the study of black holes, Stephen Hawking discovered that a hole should emit energy with the characteristic spectrum of thermal radiation. The characteristic temperature of this Hawking radiation is given by

$$T_{\rm H} = \frac{\hbar c^3}{8\pi G M k_B} \qquad (\approx \frac{1.227 \times 10^{23}\ kg}{M}\ K)$$

where G is Newton's gravitational constant, M is the black hole's mass and kB is Boltzmann's constant.

Using this expression for the Hawking temperature, and assuming that a zero-mass black hole has zero entropy, one can use thermodynamic arguments to derive the "Bekenstein entropy":

$$S_{\rm B} = \frac{k_B 4\pi G}{\hbar c} M^2.$$

The Bekenstein entropy is proportional to the black hole mass squared; because the Schwarzschild radius is proportional to the mass, the Bekenstein entropy is proportional to the black hole's *surface area.* In fact,

$$S_{\rm B} = \frac{A k_B}{4 l_{\rm P}^2},$$

where $l_{\rm P}$ is the Planck length.

The concept of black hole entropy poses some interesting conundra. In an ordinary situation, a system has entropy when a large number of different "microstates" can satisfy the same macroscopic condition. For example, given a box full of gas, many different arrangements of the gas atoms can have the same total energy. However, a black hole was believed to be a featureless object (in John Wheeler's catchphrase, "Black holes have no hair"). What, then, are the "degrees of freedom" which can give rise to black hole entropy?

String theorists have constructed models in which a black hole is a very long (and hence very massive) string. This model gives rough agreement with the expected entropy of a Schwarzschild black hole, but an exact proof has yet to be found one way or the other. The chief difficulty is that it is relatively easy to count the degrees of freedom quantum strings possess *if they do not interact with one another.* This is analogous to the ideal gas studied in introductory thermodynamics: the easiest situation to model is when the gas atoms do not have interactions among themselves.

Developing the kinetic theory of gases in the case where the gas atoms or molecules experience inter-particle forces (like the van der Waals force) is more difficult. However, a world without interactions is an uninteresting place: most significantly for the black hole problem, gravity is an interaction, and so if the "string coupling" is turned off, no black hole could ever arise. Therefore, calculating black hole entropy requires working in a regime where string interactions exist.

Extending the simpler case of non-interacting strings to the regime where a black hole could exist requires supersymmetry. In certain cases, the entropy calculation done for zero string coupling remains valid when the strings interact. The challenge for a string theorist is to devise a situation in which a black hole can exist which does not "break" supersymmetry. In recent years, this has been done by building black holes out of D-branes. Calculating the entropies of these hypothetical holes gives results which agree with the expected Bekenstein entropy. Unfortunately, the cases studied so far all involve higher-dimensional spaces — D5-branes in nine-dimensional space, for example. They do not directly apply to the familiar case, the Schwarzschild black holes observed in our own universe.

12.6 History

Dirichlet boundary conditions and D-branes had a long "pre-history" before their full significance was recognized. Mixed Dirichlet/Neumann boundary conditions were first considered by Warren Siegel in 1976 as a means of lowering the critical dimension of open string theory from 26 or 10 to 4 (Siegel also cites unpublished work by Halpern, and a 1974 paper by Chodos and Thorn, but a reading of the latter paper shows that it is actually concerned with linear dilation backgrounds, not Dirichlet boundary conditions). This paper, though prescient, was little-noted in its time (a 1985 parody by Siegel, "The Super-g String," contains an almost dead-on description of braneworlds). Dirichlet conditions for all coordinates including Euclidean time (defining what are now known as D-instantons) were introduced by Michael Green in 1977 as a means of introducing point-like structure into string theory, in an attempt to construct a string theory of the strong interaction. String compactifications studied by Harvey and Minahan, Ishibashi and Onogi, and Pradisi and Sagnotti in 1987-89 also employed Dirichlet boundary conditions.

The fact that T-duality interchanges the usual Neumann boundary conditions with Dirichlet boundary conditions was discovered independently by Horava and by Dai, Leigh, and Polchinski in 1989; this result implies that such boundary conditions must necessarily appear in regions of the moduli space of any open string theory. The Dai et al. paper also notes that the locus of the Dirichlet boundary conditions is dynamical, and coins the term Dirichlet-brane (D-brane) for the resulting object (this paper also coins orientifold for another object that arises under string T-duality). A 1989 paper by Leigh showed that D-brane dynamics are governed by the Dirac-Born-Infeld action. D-instantons were extensively studied by Green in the early 1990s, and were shown by Polchinski in 1994 to produce the $e^{-1/g}$ nonperturbative string effects anticipated by Shenker. In 1995 Polchinski showed that D-branes are the sources of electric and magnetic Ramond–Ramond fields that are required by string duality, leading to rapid progress in the nonperturbative understanding of string theory.

12.7 See also

- Bogomol'nyi–Prasad–Sommerfield bound
- M-theory

12.8 References

- Bachas, C. P. "Lectures on D-branes" (1998). arXiv:hep-th/9806199.
- Giveon, A. and Kutasov, D. "Brane dynamics and gauge theory", *Rev. Mod. Phys.* **71**, 983 (1999). arXiv:hep-th/9802067.
- Hashimoto, Koji, *D-Brane: Superstrings and New Perspective of Our World.* Springer (2012). ISBN 978-3-642-23573-3
- Johnson, Clifford (2003). *D-branes.* Cambridge: Cambridge University Press. ISBN 0-521-80912-6.
- Polchinski, Joseph, *TASI Lectures on D-branes*, arXiv:hep-th/9611050. Lectures given at TASI '96.

- Polchinski, Joseph, *Phys. Rev. Lett.* **75**, 4724 (1995). An article which established D-branes' significance in string theory.
- Zwiebach, Barton. *A First Course in String Theory.* Cambridge University Press (2004). ISBN 0-521-83143-1.

Chapter 13

Perturbation theory

This article is about perturbation theory as a general mathematical method. For perturbation theory as applied to quantum mechanics, see Perturbation theory (quantum mechanics).

Perturbation theory comprises mathematical methods for finding an approximate solution to a problem, by starting from the exact solution of a related problem. A critical feature of the technique is a middle step that breaks the problem into "solvable" and "perturbation" parts.[1] Perturbation theory is applicable if the problem at hand cannot be solved exactly, but can be formulated by adding a "small" term to the mathematical description of the exactly solvable problem.

Perturbation theory leads to an expression for the desired solution in terms of a formal power series in some "small" parameter – known as a **perturbation series** – that quantifies the deviation from the exactly solvable problem. The leading term in this power series is the solution of the exactly solvable problem, while further terms describe the deviation in the solution, due to the deviation from the initial problem. Formally, we have for the approximation to the full solution A, a series in the small parameter (here called ε), like the following:

$$A = A_0 + \varepsilon^1 A_1 + \varepsilon^2 A_2 + \cdots$$

In this example, A_0 would be the known solution to the exactly solvable initial problem and A_1, A_2, ... represent the **higher-order terms** which may be found iteratively by some systematic procedure. For small ε these higher-order terms in the series become successively smaller.

An approximate "perturbation solution" is obtained by truncating the series, usually by keeping only the first two terms, the initial solution and the "first-order" perturbation correction

$$A \approx A_0 + \varepsilon A_1 \ .$$

13.1 General description

Perturbation theory is closely related to methods used in numerical analysis. The earliest use of what would now be called *perturbation theory* was to deal with the otherwise unsolvable mathematical problems of celestial mechanics: for example the orbit of the Moon, which moves noticeably differently from a simple Keplerian ellipse because of the competing gravitation of the Earth and the Sun.

Perturbation methods start with a simplified form of the original problem, which is *simple enough* to be solved exactly. In celestial mechanics, this is usually a Keplerian ellipse. Under non relativistic gravity, an ellipse is exactly correct when there are only two gravitating bodies (say, the Earth and the Moon) but not quite correct when there are three or more objects (say, the Earth, Moon, Sun, and the rest of the solar system).

The solved, but simplified problem is then *"perturbed"* to make the conditions that the perturbed solution actually satisfies closer to the real problem, such as including the gravitational attraction of a third body (the Sun). The "conditions" are a formula (or several) that represent reality, often something arising from a physical law like Newton's second law, the force-acceleration equation,

$$\mathbf{F} = m\mathbf{a} .$$

In the case of the example, the force **F** is calculated based on the number of gravitationally relevant bodies; the acceleration **a** is obtained, using calculus, from the path of the Moon in its orbit. Both of these come in two forms: approximate values for force and acceleration, which result from simplifications, and hypothetical exact values for force and acceleration, which would require the complete answer to calculate.

The slight changes that result from accommodating the perturbation, which themselves may have been simplified yet again, are used as corrections to the approximate solution. Because of simplifications introduced along every step of the way, the corrections are never perfect, and the conditions met by the corrected solution do not perfectly match the equation demanded by reality. However, even only one cycle of corrections often provides an excellent approximate answer to what the real solution should be.

There is no requirement to stop at only one cycle of corrections. A partially corrected solution can be re-used as the new starting point for yet another cycle of perturbations and corrections. In principle, cycles of finding increasingly better corrections could go on indefinitely. In practice, one typically stops at one or two cycles of corrections. The usual difficulty with the method is that the corrections progressively make the new solutions very much more complicated, so each cycle is much more difficult to manage than the previous cycle of corrections. Isaac Newton is reported to have said, regarding the problem of the Moon's orbit, that *"It causeth my head to ache."*[2]

This general procedure is a widely used mathematical tool in advanced sciences and engineering: start with a simplified problem and gradually add corrections that make the formula that the corrected problem matches closer and closer to the formula that represents reality. It is the natural extension to mathematical functions of the "guess, check, and fix" method used by older civilisations to compute certain numbers, such as square roots.

13.2 Examples

Examples for the "mathematical description" are: an algebraic equation, a differential equation (e.g., the equations of motion in celestial mechanics or a wave equation), a free energy (in statistical mechanics), a Hamiltonian operator (in quantum mechanics).

Examples for the kind of solution to be found perturbatively: the solution of the equation (e.g., the trajectory of a particle), the statistical average of some physical quantity (e.g., average magnetization), the ground state energy of a quantum mechanical problem.

Examples for the exactly solvable problems to start with: linear equations, including linear equations of motion (harmonic oscillator, linear wave equation), statistical or quantum-mechanical systems of non-interacting particles (or in general, Hamiltonians or free energies containing only terms quadratic in all degrees of freedom).

Examples of "perturbations" to deal with: Nonlinear contributions to the equations of motion, interactions between particles, terms of higher powers in the Hamiltonian/Free Energy.

For physical problems involving interactions between particles, the terms of the perturbation series may be displayed (and manipulated) using Feynman diagrams.

13.3 History

Perturbation theory was first proposed for the solution of problems in celestial mechanics, in the context of the motions of planets in the solar system. Since the planets are very remote from each other, and since their mass is small as compared to the mass of the Sun, the gravitational forces between the planets can be neglected, and the planetary motion is considered, to a first approximation, as taking place along Kepler's orbits, which are defined by the equations of the two-body problem, the two bodies being the planet and the Sun.[3]

Since astronomic data came to be known with much greater accuracy, it became necessary to consider how the motion of a planet around the Sun is affected by other planets. This was the origin of the three-body problem; thus, in studying the system Moon–Earth–Sun the mass ratio between the Moon and the Earth was chosen as the small parameter. Lagrange and Laplace were the first to advance the view that the constants which describe the motion of a planet around the Sun are "perturbed" , as it were, by the motion of other planets and vary as a function of time; hence the name "perturbation theory" .[3]

Perturbation theory was investigated by the classical scholars — Laplace, Poisson, Gauss — as a result of which the computations could be performed with a very high accuracy. The discovery of the planet Neptune in 1848 by John Couch Adams and Urbain Le Verrier, based on the deviations in motion of the planet Uranus, represented a triumph of perturbation theory.[3]

The development of basic perturbation theory for differential equations was fairly complete by the middle of the 19th century. It was at that time that Charles-Eugène Delaunay was studying the perturbative expansion for the Earth-Moon-Sun system, and discovered the so-called "problem of small denominators". Here, the denominator appearing in the n-th term of the perturbative expansion could become arbitrarily small, causing the n-th correction to be as large or larger than the first-order correction. At the turn of the 20th century, this problem led Henri Poincaré to make one of the first deductions of the existence of chaos, or what is poetically called the "butterfly effect": that even a very small perturbation can have a very large effect on a system.

Perturbation theory saw a particularly dramatic expansion and evolution with the arrival of quantum mechanics. Although perturbation theory was used in the semi-classical theory of the Bohr atom, the calculations were monstrously complicated, and subject to somewhat ambiguous interpretation. The discovery of Heisenberg's matrix mechanics allowed a vast simplification of the application of perturbation theory. Notable examples are the Stark effect and the Zeeman effect, which have a simple enough theory to be included in standard undergraduate textbooks in quantum mechanics. Other early applications include the fine structure and the hyperfine structure in the hydrogen atom.

In modern times, perturbation theory underlies much of quantum chemistry and quantum field theory. In chemistry, perturbation theory was used to obtain the first solutions for the helium atom.

In the middle of the 20th century, Richard Feynman realized that the perturbative expansion could be given a dramatic and beautiful graphical representation in terms of what are now called Feynman diagrams. Although originally applied only in quantum field theory, such diagrams now find increasing use in any area where perturbative expansions are studied.

A partial resolution of the small-divisor problem was given by the statement of the KAM theorem in 1954. Developed by Andrey Kolmogorov, Vladimir Arnold and Jürgen Moser, this theorem stated the conditions under which a system of partial differential equations will have only mildly chaotic behaviour under small perturbations.

In the late 20th century, broad dissatisfaction with perturbation theory in the quantum physics community, including not only the difficulty of going beyond second order in the expansion, but also questions about whether the perturbative expansion is even convergent, has led to a strong interest in the area of non-perturbative analysis, that is, the study of exactly solvable models. The prototypical model is the Korteweg–de Vries equation, a highly non-linear equation for which the interesting solutions, the solitons, cannot be reached by perturbation theory, even if the perturbations were carried out to infinite order. Much of the theoretical work in non-perturbative analysis goes under the name of quantum groups and non-commutative geometry.

13.4 Perturbation orders

The standard exposition of perturbation theory is given in terms of the order to which the perturbation is carried out: **first-order perturbation theory** or **second-order perturbation theory**, and whether the perturbed states are degenerate (that is, singular), in which case extra care must be taken, and the theory is slightly more elaborate.

13.5 First-order non-singular perturbation theory

This section develops, in simple terms,[4] the general theory for the perturbative solution to a differential equation to the first order. To keep the exposition simple, a crucial assumption is made: that the solutions to the unperturbed system are not *degenerate*, so that the perturbation series can be inverted. There are ways of dealing with the degenerate (or *singular*) case; these require extra care.

Suppose one wants to solve a differential equation of the form

$$Dg(x) = \lambda g(x)\,,$$

where D is some specific differential operator, and λ is an eigenvalue. Many problems involving ordinary or partial differential equations can be cast in this form.

It is presumed that the differential operator can be written in the form

$$D = D^{(0)} + \varepsilon D^{(1)}$$

where ε is presumed to be small, and that, furthermore, the complete set of solutions for $D^{(0)}$ are known.

That is, one has a set of solutions $f_n^{(0)}(x)$, labelled by some arbitrary index n, such that

$$D^{(0)} f_n^{(0)}(x) = \lambda_n^{(0)} f_n^{(0)}(x).$$

Furthermore, one assumes that the set of solutions $\{f_n^{(0)}(x)\}$ form an orthonormal set,

$$\int f_m^{(0)}(x) f_n^{(0)}(x)\, dx = \delta_{mn}$$

with δmn the Kronecker delta function.

To zeroth order, one expects that the solutions $g(x)$ are then somehow "close" to one of the unperturbed solutions $f_n^{(0)}(x)$. That is,

$$g(x) = f_n^{(0)}(x) + \mathcal{O}(\varepsilon)$$

and

$$\lambda = \lambda_n^{(0)} + \mathcal{O}(\varepsilon).$$

where $\mathcal{O}$ denotes the relative size, in big-O notation, of the perturbation.

To solve this problem, one assumes that the solution $g(x)$ can be written as a linear combination of the $f_n^{(0)}(x)$,

$$g(x) = \sum_m c_m f_m^{(0)}(x)$$

with all of the constants $c_m = \mathcal{O}(\varepsilon)$ except for n, where $c_n = \mathcal{O}(1)$.

Substituting this last expansion into the differential equation, taking the inner product of the result with $f_n^{(0)}(x)$, and making use of orthogonality, one obtains

$$c_n \lambda_n^{(0)} + \varepsilon \sum_m c_m \int f_n^{(0)}(x) D^{(1)} f_m^{(0)}(x)\, dx = \lambda c_n\,.$$

This can be trivially rewritten as a simple linear algebra problem of finding the eigenvalue of a matrix, where

$$\sum_m A_{nm} c_m = \lambda c_n$$

where the matrix elements Anm are given by

$$A_{nm} = \delta_{nm}\lambda_n^{(0)} + \varepsilon \int f_n^{(0)}(x) D^{(1)} f_m^{(0)}(x)\, dx \,.$$

Rather than solving this full matrix equation, one notes that, of all the *cm* in the linear equation, only one, namely *cn*, is not small. Thus, to the *first order* in ε, the linear equation may be solved trivially as

$$\lambda = \lambda_n^{(0)} + \varepsilon \int f_n^{(0)}(x) D^{(1)} f_n^{(0)}(x)\, dx$$

since all of the other terms in the linear equation are of order $\mathcal{O}(\varepsilon^2)$. The above gives the solution of the eigenvalue to first order in perturbation theory.

The function *g*(*x*) to first order is obtained through similar reasoning. Substituting

$$g(x) = f_n^{(0)}(x) + \varepsilon f_n^{(1)}(x)$$

so that

$$\left(D^{(0)} + \varepsilon D^{(1)}\right)\left(f_n^{(0)}(x) + \varepsilon f_n^{(1)}(x)\right) = \left(\lambda_n^{(0)} + \varepsilon \lambda_n^{(1)}\right)\left(f_n^{(0)}(x) + \varepsilon f_n^{(1)}(x)\right)$$

gives an equation for $f_n^{(1)}(x)$.

It may be solved integrating with the partition of unity

$$\delta(x-y) = \sum_n f_n^{(0)}(x) f_n^{(0)}(y)$$

to give

$$f_n^{(1)}(x) = \sum_{m\,(\neq n)} \frac{f_m^{(0)}(x)}{\lambda_n^{(0)} - \lambda_m^{(0)}} \int f_m^{(0)}(y) D^{(1)} f_n^{(0)}(y)\, dy$$

which finally gives the exact solution to the perturbed differential equation to first order in the perturbation ε.

Several observations may be made about the form of this solution. First, the sum over functions with differences of eigenvalues in the denominator evokes the resolvent in Fredholm theory. This is no accident; the resolvent acts essentially as a kind of Green's function or propagator, passing the perturbation along. Higher-order perturbations resemble this form, with an additional sum over a resolvent appearing at each order.

The form of this solution also illustrates the idea behind the small-divisor problem. If, for whatever reason, two eigenvalues are close, so that the difference $\lambda_n^{(0)} - \lambda_m^{(0)}$ becomes small, the corresponding term in the above sum will become disproportionately large. In particular, if this happens in higher-order terms, the higher-order perturbation may become as large or larger in magnitude than the first-order perturbation. Such a situation calls into question the validity of utilizing a perturbative analysis to begin with, which can be understood to be a fairly catastrophic situation; it is frequently encountered in chaotic dynamical systems, and requires the development of techniques other than perturbation theory to solve the problem.

Curiously, the situation is not at all bad if two or more eigenvalues are *exactly equal*. This case is referred to as singular or degenerate perturbation theory, addressed below. The degeneracy of eigenvalues indicates that the unperturbed system has some sort of symmetry, and that the generators of that symmetry commute with the unperturbed differential operator. Typically, the perturbing term does not possess the symmetry, and so the full solutions do not, either; one says that the perturbation *lifts* or *breaks* the degeneracy. In this case, the perturbation can still be performed, as in following sections; however, care must be taken to work in a basis for the unperturbed states, so that these map one-to-one to the perturbed states, rather than being a mixture.

13.6 Perturbation theory of degenerate states

One may note that a problem occurs in the above first order perturbation theory when two or more eigenfunctions of the unperturbed system correspond to the same eigenvalue, i.e. when the eigenvalue equation becomes

$$D^{(0)}f^{(0)}_{n,i}(x) = \lambda^{(0)}_n f^{(0)}_{n,i}(x) ,$$

and the index i labels *several states with the same eigenvalue* $\lambda^{(0)}_n$. The expression for the eigenfunctions which has energy differences in the denominators becomes infinite. In that case, degenerate perturbation theory must be applied.

The degeneracy must first be removed for higher order perturbation theory. First, consider the eigenfunction which is a linear combination of eigenfunctions with the same eigenvalue only,

$$g(x) = \sum_k c_{n,k} f^{(0)}_{n,k}(x) ,$$

which, again from the orthogonality of $f^{(0)}_{n,k}$, leads to the following equation,

$$c_{n,i}\lambda^{(0)}_n + \varepsilon \sum_k c_{n,k} \int f^{(0)}_{n,i}(x) D^{(1)} f^{(0)}_{n,k}(x)\, dx = \lambda c_{n,i}$$

for each n.

As for the majority of low quantum numbers n, i changes over a *small range of integers*, so often the later equation can be solved analytically as an at most 4 × 4 matrix equation. Once the degeneracy is removed, the first and any order of the above perturbation theory may be further applied relying on the new eigenfunctions.

13.7 Example of second-order singular perturbation theory

Main article: Singular perturbation

Consider the following equation for the unknown variable x:

$$x = 1 + \varepsilon x^5 .$$

For the initial problem with $\varepsilon = 0$, the solution is $x_0 = 1$. For small ε the lowest-order approximation may be found by inserting the ansatz

$$x = x_0 + \varepsilon x_1 + \cdots$$

into the equation and demanding the equation to be fulfilled up to terms that involve powers of ε higher than the first. This yields $x_1 = 1$. In the same way, the higher orders may be found. However, even in this simple example it may be observed that for (arbitrarily) small positive ε there are four other solutions to the equation (with very large magnitude). The reason we don't find these solutions in the above perturbation method is because these solutions diverge when $\varepsilon \to 0$ while the ansatz assumes regular behavior in this limit.

The four additional solutions can be found using the methods of **singular perturbation theory**. In this case this works as follows. Since the four solutions diverge at $\varepsilon = 0$, it makes sense to rescale x. We put

$$x = y\varepsilon^{-\nu}$$

such that in terms of y the solutions stay finite. This means that we need to choose the exponent ν to match the rate at which the solutions diverge. In terms of y the equation reads:

$$\varepsilon^{-\nu}y = 1 + \varepsilon^{1-5\nu}y^5$$

The 'right' value for ν is obtained when the exponent of ε in the prefactor of the term proportional to y is equal to the exponent of ε in the prefactor of the term proportional to y^5, i.e. when $\nu = 1/4$. This is called 'significant degeneration'. If we choose ν larger, then the four solutions will collapse to zero in terms of y and they will become degenerate with the solution we found above. If we choose ν smaller, then the four solutions will still diverge to infinity.

Putting $\nu = 1/4$ in the above equation yields:

$$y = \varepsilon^{\frac{1}{4}} + y^5$$

This equation can be solved using ordinary perturbation theory in the same way as regular expansion for x was obtained. Since the expansion parameter is now $\varepsilon^{1/4}$ we put:

$$y = y_0 + \varepsilon^{\frac{1}{4}}y_1 + \varepsilon^{\frac{1}{2}}y_2 + \cdots$$

There are five solutions for y_0: $\{0, \pm 1, \pm i\}$. We must disregard the solution $y = 0$ since it corresponds to the original regular solution which appears to be at zero for $\varepsilon = 0$, because in the limit $\varepsilon \to 0$ we are rescaling by an infinite amount. The next term is $y_1 = -1/4$. In terms of x the four solutions are thus given as:

$$x = \varepsilon^{-\frac{1}{4}}\left[y_0 - \tfrac{1}{4}\varepsilon^{\frac{1}{4}} + \cdots\right]$$

13.8 Example of degenerate perturbation theory – Stark effect in resonant rotating wave

Let us consider a hydrogen atom rotating with a constant angular frequency ω in an electric field. The Hamiltonian is given by:

$$H = H_0 + \varepsilon x$$

where the unperturbed Hamiltonian is

$$H_0 = \frac{\mathbf{p}^2}{2} - \frac{1}{r} - \omega L_z,$$

and Lz is the operator for the z-component of angular momentum: $Lz = i\partial/\partial\varphi$. The perturbation εx can be seen as the strength of the applied electric field multiplied by one of the space coordinates (This calculation is in atomic units, so that every quantity is dimensionless).

The eigenvalues of H_0 are

$$E_{n,m} = -\tfrac{1}{2}n^2 - m\omega$$

For the lowest energy eigenstates of Hydrogen $|n, l, m\rangle$, $|1, 0, 0\rangle$ and $|2, 1, 1\rangle$ in the resonance $E_{2,1} - E_{1,0} = 0$ their energies are therefore equal $E_{1,0} = E_{2,1} = -1/2$, while the eigenstates are different.

The eigenvalue equation for the Hamiltonian takes the form

$$\begin{bmatrix} E_{1,0} & \varepsilon d \\ \varepsilon d & E_{1,0} \end{bmatrix}\begin{bmatrix} a \\ b \end{bmatrix} = E\begin{bmatrix} a \\ b \end{bmatrix}$$

where

$d = \frac{128}{243} a_0$

which leads to the quadratic equation which can be readily solved

$$(E_{1,0} - E)^2 - d^2 \varepsilon^2 = 0$$

with the solution

$$|\chi 1\rangle = \frac{1}{\sqrt{2}}(|1,0,0\rangle + |2,1,1\rangle)$$
$$E(1) = E_{1,0} + d\varepsilon$$

$$|\chi 2\rangle = \frac{1}{\sqrt{2}}(|1,0,0\rangle - |2,1,1\rangle)$$
$$E(2) = E_{1,0} - d\varepsilon$$

These states are the Stark states in the rotating frame, they are Trojan (higher eigenvalue) and anti-Trojan wavepackets.

13.9 Commentary

Both regular and singular perturbation theory are frequently used in physics and engineering. Regular perturbation theory may only be used to find those solutions of a problem that evolve smoothly out of the initial solution when changing the parameter (that are "adiabatically connected" to the initial solution). A well-known example from physics where regular perturbation theory fails is in fluid dynamics when one treats the viscosity as a small parameter. Close to a boundary, the fluid velocity goes to zero, even for very small viscosity (the no-slip condition). For zero viscosity, it is not possible to impose this boundary condition and a regular perturbative expansion amounts to an expansion about an unrealistic physical solution. Singular perturbation theory can, however, be applied here and this amounts to 'zooming in' at the boundaries (using the method of matched asymptotic expansions).

Perturbation theory can fail when the system can transition to a different "phase" of matter, with a qualitatively different behaviour, that cannot be modelled by the physical formulas put into the perturbation theory (e.g., a solid crystal melting into a liquid). In some cases, this failure manifests itself by divergent behavior of the perturbation series. Such divergent series can sometimes be resummed using techniques such as Borel resummation.

Perturbation techniques can be also used to find approximate solutions to non-linear differential equations. Examples of techniques used to find approximate solutions to these types of problems are the Lindstedt–Poincaré technique and the method of multiple time scales.

There is absolutely no guarantee that perturbative methods result in a convergent solution. In fact, asymptotic series are the norm.

13.10 Perturbation theory in chemistry

Many of the ab initio quantum chemistry methods use perturbation theory directly or are closely related methods. Implicit perturbation theory[5] works with the complete Hamiltonian from the very beginning and never specifies a perturbation operator as such. Møller–Plesset perturbation theory uses the difference between the Hartree–Fock Hamiltonian and the exact non-relativistic Hamiltonian as the perturbation. The zero-order energy is the sum of orbital energies. The first-order energy is the Hartree–Fock energy and electron correlation is included at second-order or higher. Calculations to second, third or fourth order are very common and the code is included in most ab initio quantum chemistry programs. A related but more accurate method is the coupled cluster method.

13.11 See also

- Cosmological perturbation theory
- Dynamic nuclear polarisation
- Alternative approach to perturbation theory[6]
- Eigenvalue perturbation
- Interval FEM
- Orders of approximation
- Structural stability
- Lyapunov stability

13.12 References

[1] William E. Wiesel (2010). *Modern Astrodynamics*. Ohio: Aphelion Press. p. 107. ISBN 978-145378-1470.

[2] Cropper, William H. (2004), *Great Physicists: The Life and Times of Leading Physicists from Galileo to Hawking*, Oxford University Press, p. 34, ISBN 978-0-19-517324-6.

[3] Perturbation theory. N. N. Bogolyubov, jr. (originator), Encyclopedia of Mathematics. URL: http://www.encyclopediaofmath.org/index.php?title=Perturbation_theory&oldid=11676

[4] • Sakurai, J.J., and Napolitano, J. (1964,2011). *Modern quantum mechanics* (2nd ed.), Addison Wesley ISBN 978-0-8053-8291-4 . Chapter 5

[5] King, Matcha (1976). "Theory of the Chemical Bond". *JACS* **98** (12): 3415–3420. doi:10.1021/ja00428a004.

[6] Martínez-Carranza, J.; Soto-Eguibar, F.; Moya-Cessa, H. (2012). "Alternative analysis to perturbation theory in quantum mechanics". *The European Physical Journal D* **66**. doi:10.1140/epjd/e2011-20654-5.

13.13 External links

- Introduction to regular perturbation theory by Eric Vanden-Eijnden (PDF)
- Perturbation Method of Multiple Scales

Chapter 14

Gauge theory

For a more accessible and less technical introduction to this topic, see Introduction to gauge theory.

In physics, a **gauge theory** is a type of field theory in which the Lagrangian is invariant under a continuous group of local transformations.

The term *gauge* refers to redundant degrees of freedom in the Lagrangian. The transformations between possible gauges, called *gauge transformations*, form a Lie group—referred to as the *symmetry group* or the *gauge group* of the theory. Associated with any Lie group is the Lie algebra of group generators. For each group generator there necessarily arises a corresponding vector field called the *gauge field*. Gauge fields are included in the Lagrangian to ensure its invariance under the local group transformations (called *gauge invariance*). When such a theory is quantized, the quanta of the gauge fields are called *gauge bosons*. If the symmetry group is non-commutative, the gauge theory is referred to as *non-abelian*, the usual example being the Yang–Mills theory.

Many powerful theories in physics are described by Lagrangians that are invariant under some symmetry transformation groups. When they are invariant under a transformation identically performed at *every* point in the space in which the physical processes occur, they are said to have a global symmetry. The requirement of local symmetry, the cornerstone of gauge theories, is a stricter constraint. In fact, a global symmetry is just a local symmetry whose group's parameters are fixed in space-time.

Gauge theories are important as the successful field theories explaining the dynamics of elementary particles. Quantum electrodynamics is an abelian gauge theory with the symmetry group U(1) and has one gauge field, the electromagnetic four-potential, with the photon being the gauge boson. The Standard Model is a non-abelian gauge theory with the symmetry group U(1)×SU(2)×SU(3) and has a total of twelve gauge bosons: the photon, three weak bosons and eight gluons.

Gauge theories are also important in explaining gravitation in the theory of general relativity. Its case is somewhat unique in that the gauge field is a tensor, the Lanczos tensor. Theories of quantum gravity, beginning with gauge gravitation theory, also postulate the existence of a gauge boson known as the graviton. Gauge symmetries can be viewed as analogues of the principle of general covariance of general relativity in which the coordinate system can be chosen freely under arbitrary diffeomorphisms of spacetime. Both gauge invariance and diffeomorphism invariance reflect a redundancy in the description of the system. An alternative theory of gravitation, gauge theory gravity, replaces the principle of general covariance with a true gauge principle with new gauge fields.

Historically, these ideas were first stated in the context of classical electromagnetism and later in general relativity. However, the modern importance of gauge symmetries appeared first in the relativistic quantum mechanics of electrons – quantum electrodynamics, elaborated on below. Today, gauge theories are useful in condensed matter, nuclear and high energy physics among other subfields.

14.1 History and importance

The earliest field theory having a gauge symmetry was Maxwell's formulation of electrodynamics in 1864. The importance of this symmetry remained unnoticed in the earliest formulations. Similarly unnoticed, Hilbert had derived the Einstein field equations by postulating the invariance of the action under a general coordinate transformation.

Later Hermann Weyl, in an attempt to unify general relativity and electromagnetism, conjectured that *Eichinvarianz* or invariance under the change of scale (or "gauge") might also be a local symmetry of general relativity. After the development of quantum mechanics, Weyl, Vladimir Fock and Fritz London modified gauge by replacing the scale factor with a complex quantity and turned the scale transformation into a change of phase, which is a U(1) gauge symmetry. This explained the electromagnetic field effect on the wave function of a charged quantum mechanical particle. This was the first widely recognised gauge theory, popularised by Pauli in the 1940s.[1]

In 1954, attempting to resolve some of the great confusion in elementary particle physics, Chen Ning Yang and Robert Mills introduced **non-abelian gauge theories** as models to understand the strong interaction holding together nucleons in atomic nuclei. (Ronald Shaw, working under Abdus Salam, independently introduced the same notion in his doctoral thesis.) Generalizing the gauge invariance of electromagnetism, they attempted to construct a theory based on the action of the (non-abelian) SU(2) symmetry group on the isospin doublet of protons and neutrons. This is similar to the action of the U(1) group on the spinor fields of quantum electrodynamics. In particle physics the emphasis was on using **quantized gauge theories**.

This idea later found application in the quantum field theory of the weak force, and its unification with electromagnetism in the electroweak theory. Gauge theories became even more attractive when it was realized that non-abelian gauge theories reproduced a feature called asymptotic freedom. Asymptotic freedom was believed to be an important characteristic of strong interactions. This motivated searching for a strong force gauge theory. This theory, now known as quantum chromodynamics, is a gauge theory with the action of the SU(3) group on the color triplet of quarks. The Standard Model unifies the description of electromagnetism, weak interactions and strong interactions in the language of gauge theory.

In the 1970s, Sir Michael Atiyah began studying the mathematics of solutions to the classical Yang–Mills equations. In 1983, Atiyah's student Simon Donaldson built on this work to show that the differentiable classification of smooth 4-manifolds is very different from their classification up to homeomorphism. Michael Freedman used Donaldson's work to exhibit exotic $\mathbf{R}^4$s, that is, exotic differentiable structures on Euclidean 4-dimensional space. This led to an increasing interest in gauge theory for its own sake, independent of its successes in fundamental physics. In 1994, Edward Witten and Nathan Seiberg invented gauge-theoretic techniques based on supersymmetry that enabled the calculation of certain topological invariants (the Seiberg–Witten invariants). These contributions to mathematics from gauge theory have led to a renewed interest in this area.

The importance of gauge theories in physics is exemplified in the tremendous success of the mathematical formalism in providing a unified framework to describe the quantum field theories of electromagnetism, the weak force and the strong force. This theory, known as the Standard Model, accurately describes experimental predictions regarding three of the four fundamental forces of nature, and is a gauge theory with the gauge group SU(3) × SU(2) × U(1). Modern theories like string theory, as well as general relativity, are, in one way or another, gauge theories.

See Pickering[2] for more about the history of gauge and quantum field theories.

14.2 Description

14.2.1 Global and local symmetries

In physics, the mathematical description of any physical situation usually contains excess degrees of freedom; the same physical situation is equally well described by many equivalent mathematical configurations. For instance, in Newtonian dynamics, if two configurations are related by a Galilean transformation (an inertial change of reference frame) they represent the same physical situation. These transformations form a group of "symmetries" of the theory, and a physical situation corresponds not to an individual mathematical configuration but to a class of configurations related to one another by this symmetry group.

This idea can be generalized to include local as well as global symmetries, analogous to much more abstract "changes of coordinates" in a situation where there is no preferred "inertial" coordinate system that covers the entire physical system. A gauge theory is a mathematical model that has symmetries of this kind, together with a set of techniques for making physical predictions consistent with the symmetries of the model.

14.2.2 Example of global symmetry

When a quantity occurring in the mathematical configuration is not just a number but has some geometrical significance, such as a velocity or an axis of rotation, its representation as numbers arranged in a vector or matrix is also changed by a coordinate transformation. For instance, if one description of a pattern of fluid flow states that the fluid velocity in the neighborhood of (x=1, y=0) is 1 m/s in the positive x direction, then a description of the same situation in which the coordinate system has been rotated clockwise by 90 degrees states that the fluid velocity in the neighborhood of (x=0, y=1) is 1 m/s in the positive y direction. The coordinate transformation has affected both the coordinate system used to identify the *location* of the measurement and the basis in which its *value* is expressed. As long as this transformation is performed globally (affecting the coordinate basis in the same way at every point), the effect on values that represent the *rate of change* of some quantity along some path in space and time as it passes through point P is the same as the effect on values that are truly local to P.

14.2.3 Use of fiber bundles to describe local symmetries

In order to adequately describe physical situations in more complex theories, it is often necessary to introduce a "coordinate basis" for some of the objects of the theory that do not have this simple relationship to the coordinates used to label points in space and time. (In mathematical terms, the theory involves a fiber bundle in which the fiber at each point of the base space consists of possible coordinate bases for use when describing the values of objects at that point.) In order to spell out a mathematical configuration, one must choose a particular coordinate basis at each point (a *local section* of the fiber bundle) and express the values of the objects of the theory (usually "fields" in the physicist's sense) using this basis. Two such mathematical configurations are equivalent (describe the same physical situation) if they are related by a transformation of this abstract coordinate basis (a change of local section, or *gauge transformation*).

In most gauge theories, the set of possible transformations of the abstract gauge basis at an individual point in space and time is a finite-dimensional Lie group. The simplest such group is U(1), which appears in the modern formulation of quantum electrodynamics (QED) via its use of complex numbers. QED is generally regarded as the first, and simplest, physical gauge theory. The set of possible gauge transformations of the entire configuration of a given gauge theory also forms a group, the *gauge group* of the theory. An element of the gauge group can be parameterized by a smoothly varying function from the points of spacetime to the (finite-dimensional) Lie group, such that the value of the function and its derivatives at each point represents the action of the gauge transformation on the fiber over that point.

A gauge transformation with constant parameter at every point in space and time is analogous to a rigid rotation of the geometric coordinate system; it represents a global symmetry of the gauge representation. As in the case of a rigid rotation, this gauge transformation affects expressions that represent the rate of change along a path of some gauge-dependent quantity in the same way as those that represent a truly local quantity. A gauge transformation whose parameter is *not* a constant function is referred to as a local symmetry; its effect on expressions that involve a derivative is qualitatively different from that on expressions that don't. (This is analogous to a non-inertial change of reference frame, which can produce a Coriolis effect.)

14.2.4 Gauge fields

The "gauge covariant" version of a gauge theory accounts for this effect by introducing a gauge field (in mathematical language, an Ehresmann connection) and formulating all rates of change in terms of the covariant derivative with respect to this connection. The gauge field becomes an essential part of the description of a mathematical configuration. A configuration in which the gauge field can be eliminated by a gauge transformation has the property that its field strength (in mathematical language, its curvature) is zero everywhere; a gauge theory is *not* limited to these configurations. In other words, the distinguishing characteristic of a gauge theory is that the gauge field does not merely compensate for a poor choice of coordinate system; there is generally no gauge transformation that makes the gauge field vanish.

When analyzing the dynamics of a gauge theory, the gauge field must be treated as a dynamical variable, similarly to other objects in the description of a physical situation. In addition to its interaction with other objects via the covariant derivative, the gauge field typically contributes energy in the form of a "self-energy" term. One can obtain the equations for the gauge theory by:

- starting from a naïve ansatz without the gauge field (in which the derivatives appear in a "bare" form);

- listing those global symmetries of the theory that can be characterized by a continuous parameter (generally an abstract equivalent of a rotation angle);
- computing the correction terms that result from allowing the symmetry parameter to vary from place to place; and
- reinterpreting these correction terms as couplings to one or more gauge fields, and giving these fields appropriate self-energy terms and dynamical behavior.

This is the sense in which a gauge theory "extends" a global symmetry to a local symmetry, and closely resembles the historical development of the gauge theory of gravity known as general relativity.

14.2.5 Physical experiments

Gauge theories are used to model the results of physical experiments, essentially by:

- limiting the universe of possible configurations to those consistent with the information used to set up the experiment, and then
- computing the probability distribution of the possible outcomes that the experiment is designed to measure.

The mathematical descriptions of the "setup information" and the "possible measurement outcomes" (loosely speaking, the "boundary conditions" of the experiment) are generally not expressible without reference to a particular coordinate system, including a choice of gauge. (If nothing else, one assumes that the experiment has been adequately isolated from "external" influence, which is itself a gauge-dependent statement.) Mishandling gauge dependence in boundary conditions is a frequent source of anomalies in gauge theory calculations, and gauge theories can be broadly classified by their approaches to anomaly avoidance.

14.2.6 Continuum theories

The two gauge theories mentioned above (continuum electrodynamics and general relativity) are examples of continuum field theories. The techniques of calculation in a continuum theory implicitly assume that:

- given a completely fixed choice of gauge, the boundary conditions of an individual configuration can in principle be completely described;
- given a completely fixed gauge and a complete set of boundary conditions, the principle of least action determines a unique mathematical configuration (and therefore a unique physical situation) consistent with these bounds;
- the likelihood of possible measurement outcomes can be determined by:
  - establishing a probability distribution over all physical situations determined by boundary conditions that are consistent with the setup information,
  - establishing a probability distribution of measurement outcomes for each possible physical situation, and
  - convolving these two probability distributions to get a distribution of possible measurement outcomes consistent with the setup information; and
- fixing the gauge introduces no anomalies in the calculation, due either to gauge dependence in describing partial information about boundary conditions or to incompleteness of the theory.

These assumptions are close enough to be valid across a wide range of energy scales and experimental conditions, to allow these theories to make accurate predictions about almost all of the phenomena encountered in daily life, from light, heat, and electricity to eclipses and spaceflight. They fail only at the smallest and largest scales (due to omissions in the theories themselves) and when the mathematical techniques themselves break down (most notably in the case of turbulence and other chaotic phenomena).

14.2.7 Quantum field theories

Other than these classical continuum field theories, the most widely known gauge theories are quantum field theories, including quantum electrodynamics and the Standard Model of elementary particle physics. The starting point of a quantum field theory is much like that of its continuum analog: a gauge-covariant action integral that characterizes "allowable" physical situations according to the principle of least action. However, continuum and quantum theories differ significantly in how they handle the excess degrees of freedom represented by gauge transformations. Continuum theories, and most pedagogical treatments of the simplest quantum field theories, use a gauge fixing prescription to reduce the orbit of mathematical configurations that represent a given physical situation to a smaller orbit related by a smaller gauge group (the global symmetry group, or perhaps even the trivial group).

More sophisticated quantum field theories, in particular those that involve a non-abelian gauge group, break the gauge symmetry within the techniques of perturbation theory by introducing additional fields (the Faddeev–Popov ghosts) and counterterms motivated by anomaly cancellation, in an approach known as BRST quantization. While these concerns are in one sense highly technical, they are also closely related to the nature of measurement, the limits on knowledge of a physical situation, and the interactions between incompletely specified experimental conditions and incompletely understood physical theory . The mathematical techniques that have been developed in order to make gauge theories tractable have found many other applications, from solid-state physics and crystallography to low-dimensional topology.

14.3 Classical gauge theory

14.3.1 Classical electromagnetism

Historically, the first example of gauge symmetry discovered was classical electromagnetism. In electrostatics, one can either discuss the electric field, **E**, or its corresponding electric potential, *V*. Knowledge of one makes it possible to find the other, except that potentials differing by a constant, $V \rightarrow V + C$, correspond to the same electric field. This is because the electric field relates to *changes* in the potential from one point in space to another, and the constant *C* would cancel out when subtracting to find the change in potential. In terms of vector calculus, the electric field is the gradient of the potential, $\mathbf{E} = -\nabla V$. Generalizing from static electricity to electromagnetism, we have a second potential, the vector potential **A**, with

$$\mathbf{E} = -\nabla V - \frac{\partial \mathbf{A}}{\partial t}$$
$$\mathbf{B} = \nabla \times \mathbf{A}$$

The general gauge transformations now become not just $V \rightarrow V + C$ but

$$\mathbf{A} \rightarrow \mathbf{A} + \nabla f$$
$$V \rightarrow V - \frac{\partial f}{\partial t}$$

where *f* is any function that depends on position and time. The fields remain the same under the gauge transformation, and therefore Maxwell's equations are still satisfied. That is, Maxwell's equations have a gauge symmetry.

14.3.2 An example: Scalar O(*n*) gauge theory

The remainder of this section requires some familiarity with classical or quantum field theory, and the use of Lagrangians.

Definitions in this section: gauge group, gauge field, interaction Lagrangian, gauge boson.

The following illustrates how local gauge invariance can be "motivated" heuristically starting from global symmetry properties, and how it leads to an interaction between originally non-interacting fields.

Consider a set of n non-interacting real scalar fields, with equal masses m. This system is described by an action that is the sum of the (usual) action for each scalar field φ_i

$$\mathcal{S} = \int \mathrm{d}^4x \sum_{i=1}^{n} \left[\frac{1}{2}\partial_\mu\varphi_i\partial^\mu\varphi_i - \frac{1}{2}m^2\varphi_i^2\right]$$

The Lagrangian (density) can be compactly written as

$$\mathcal{L} = \frac{1}{2}(\partial_\mu\Phi)^T\partial^\mu\Phi - \frac{1}{2}m^2\Phi^T\Phi$$

by introducing a vector of fields

$$\Phi = (\varphi_1, \varphi_2, \ldots, \varphi_n)^T$$

The term ∂_μ is Einstein notation for the partial derivative of Φ in each of the four dimensions. It is now transparent that the Lagrangian is invariant under the transformation

$$\Phi \mapsto \Phi' = G\Phi$$

whenever G is a *constant* matrix belonging to the n-by-n orthogonal group O(n). This is seen to preserve the Lagrangian, since the derivative of Φ transforms identically to Φ and both quantities appear inside dot products in the Lagrangian (orthogonal transformations preserve the dot product).

$$(\partial_\mu\Phi) \mapsto (\partial_\mu\Phi)' = G\partial_\mu\Phi$$

This characterizes the *global* symmetry of this particular Lagrangian, and the symmetry group is often called the **gauge group**; the mathematical term is **structure group**, especially in the theory of G-structures. Incidentally, Noether's theorem implies that invariance under this group of transformations leads to the conservation of the *currents*

$$J_\mu^a = i\partial_\mu\Phi^T T^a\Phi$$

where the T^a matrices are generators of the SO(n) group. There is one conserved current for every generator.

Now, demanding that this Lagrangian should have *local* O(n)-invariance requires that the G matrices (which were earlier constant) should be allowed to become functions of the space-time coordinates x.

Unfortunately, the G matrices do not "pass through" the derivatives, when $G = G(x)$,

$$\partial_\mu(G\Phi) \neq G(\partial_\mu\Phi)$$

The failure of the derivative to commute with "G" introduces an additional term (in keeping with the product rule), which spoils the invariance of the Lagrangian. In order to rectify this we define a new derivative operator such that the derivative of Φ again transforms identically with Φ

$$(D_\mu\Phi)' = GD_\mu\Phi$$

This new "derivative" is called a (gauge) covariant derivative and takes the form

$$D_\mu = \partial_\mu + igA_\mu$$

Where g is called the coupling constant; a quantity defining the strength of an interaction. After a simple calculation we can see that the **gauge field** $A(x)$ must transform as follows

$$A'_\mu = GA_\mu G^{-1} + \frac{i}{g}(\partial_\mu G)G^{-1}$$

The gauge field is an element of the Lie algebra, and can therefore be expanded as

$$A_\mu = \sum_a A_\mu^a T^a$$

There are therefore as many gauge fields as there are generators of the Lie algebra.

Finally, we now have a *locally gauge invariant* Lagrangian

$$\mathcal{L}_{\text{loc}} = \frac{1}{2}(D_\mu \Phi)^T D^\mu \Phi - \frac{1}{2} m^2 \Phi^T \Phi$$

Pauli uses the term *gauge transformation of the first type* to mean the transformation of Φ , while the compensating transformation in A is called a *gauge transformation of the second type.*

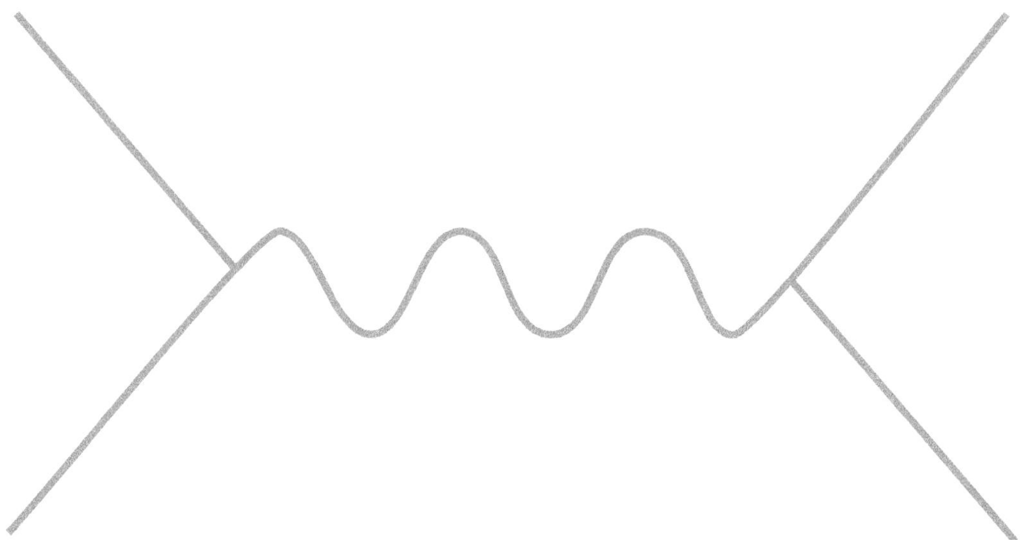

Feynman diagram of scalar bosons interacting via a gauge boson

The difference between this Lagrangian and the original *globally gauge-invariant* Lagrangian is seen to be the **interaction Lagrangian**

$$\mathcal{L}_{\text{int}} = i\frac{g}{2}\Phi^T A_\mu^T \partial^\mu \Phi + i\frac{g}{2}(\partial_\mu \Phi)^T A^\mu \Phi - \frac{g^2}{2}(A_\mu \Phi)^T A^\mu \Phi$$

This term introduces interactions between the n scalar fields just as a consequence of the demand for local gauge invariance. However, to make this interaction physical and not completely arbitrary, the mediator $A(x)$ needs to propagate in space. That is dealt with in the next section by adding yet another term, $\mathcal{L}_{\text{gf}}$, to the Lagrangian. In the quantized version of the obtained classical field theory, the quanta of the gauge field $A(x)$ are called gauge bosons. The interpretation of the interaction Lagrangian in quantum field theory is of scalar bosons interacting by the exchange of these gauge bosons.

14.3.3 The Yang–Mills Lagrangian for the gauge field

Main article: Yang–Mills theory

The picture of a classical gauge theory developed in the previous section is almost complete, except for the fact that to define the covariant derivatives *D*, one needs to know the value of the gauge field $A(x)$ at all space-time points. Instead of manually specifying the values of this field, it can be given as the solution to a field equation. Further requiring that the Lagrangian that generates this field equation is locally gauge invariant as well, one possible form for the gauge field Lagrangian is (conventionally) written as

$$\mathcal{L}_{\text{gf}} = -\frac{1}{2}\operatorname{Tr}(F^{\mu\nu}F_{\mu\nu})$$

with

$$F_{\mu\nu} = \frac{1}{ig}[D_\mu, D_\nu]$$

and the trace being taken over the vector space of the fields. This is called the **Yang–Mills action**. Other gauge invariant actions also exist (e.g., nonlinear electrodynamics, Born–Infeld action, Chern–Simons model, theta term, etc.).

Note that in this Lagrangian term there is no field whose transformation counterweighs the one of A . Invariance of this term under gauge transformations is a particular case of *a priori* classical (geometrical) symmetry. This symmetry must be restricted in order to perform quantization, the procedure being denominated gauge fixing, but even after restriction, gauge transformations may be possible.[3]

The complete Lagrangian for the gauge theory is now

$$\mathcal{L} = \mathcal{L}_{\text{loc}} + \mathcal{L}_{\text{gf}} = \mathcal{L}_{\text{global}} + \mathcal{L}_{\text{int}} + \mathcal{L}_{\text{gf}}$$

14.3.4 An example: Electrodynamics

As a simple application of the formalism developed in the previous sections, consider the case of electrodynamics, with only the electron field. The bare-bones action that generates the electron field's Dirac equation is

$$\mathcal{S} = \int \bar{\psi}(i\hbar c\,\gamma^\mu\partial_\mu - mc^2)\psi\,\mathrm{d}^4x$$

The global symmetry for this system is

$$\psi \mapsto e^{i\theta}\psi$$

The gauge group here is U(1), just rotations of the phase angle of the field, with the particular rotation determined by the constant θ.

"Localising" this symmetry implies the replacement of θ by $\theta(x)$. An appropriate covariant derivative is then

$$D_\mu = \partial_\mu - i\frac{e}{\hbar}A_\mu$$

Identifying the "charge" e (not to be confused with the mathematical constant e in the symmetry description) with the usual electric charge (this is the origin of the usage of the term in gauge theories), and the gauge field $A(x)$ with the four-vector potential of electromagnetic field results in an interaction Lagrangian

$$\mathcal{L}_{\text{int}} = \frac{e}{\hbar}\bar{\psi}(x)\gamma^\mu\psi(x)A_\mu(x) = J^\mu(x)A_\mu(x)$$

where $J^\mu(x)$ is the usual four vector electric current density. The gauge principle is therefore seen to naturally introduce the so-called minimal coupling of the electromagnetic field to the electron field.

Adding a Lagrangian for the gauge field $A_\mu(x)$ in terms of the field strength tensor exactly as in electrodynamics, one obtains the Lagrangian used as the starting point in quantum electrodynamics.

$$\mathcal{L}_{\text{QED}} = \bar{\psi}(i\hbar c\,\gamma^\mu D_\mu - mc^2)\psi - \frac{1}{4\mu_0}F_{\mu\nu}F^{\mu\nu}$$

See also: Dirac equation, Maxwell's equations, Quantum electrodynamics

14.4 Mathematical formalism

Gauge theories are usually discussed in the language of differential geometry. Mathematically, a *gauge* is just a choice of a (local) section of some principal bundle. A **gauge transformation** is just a transformation between two such sections.

Although gauge theory is dominated by the study of connections (primarily because it's mainly studied by high-energy physicists), the idea of a connection is not central to gauge theory in general. In fact, a result in general gauge theory shows that affine representations (i.e., affine modules) of the gauge transformations can be classified as sections of a jet bundle satisfying certain properties. There are representations that transform covariantly pointwise (called by physicists gauge transformations of the first kind), representations that transform as a connection form (called by physicists gauge transformations of the second kind, an affine representation)—and other more general representations, such as the B field in BF theory. There are more general nonlinear representations (realizations), but these are extremely complicated. Still, nonlinear sigma models transform nonlinearly, so there are applications.

If there is a principal bundle *P* whose base space is space or spacetime and structure group is a Lie group, then the sections of *P* form a principal homogeneous space of the group of gauge transformations.

Connections (gauge connection) define this principal bundle, yielding a covariant derivative ∇ in each associated vector bundle. If a local frame is chosen (a local basis of sections), then this covariant derivative is represented by the connection form *A*, a Lie algebra-valued 1-form, which is called the **gauge potential** in physics. This is evidently not an intrinsic but a frame-dependent quantity. The curvature form *F*, a Lie algebra-valued 2-form that is an intrinsic quantity, is constructed from a connection form by

$$\mathbf{F} = \mathrm{d}\mathbf{A} + \mathbf{A} \wedge \mathbf{A}$$

where d stands for the exterior derivative and $\wedge$ stands for the wedge product. ($\mathbf{A}$ is an element of the vector space spanned by the generators T^a , and so the components of $\mathbf{A}$ do not commute with one another. Hence the wedge product $\mathbf{A} \wedge \mathbf{A}$ does not vanish.)

Infinitesimal gauge transformations form a Lie algebra, which is characterized by a smooth Lie-algebra-valued scalar, ε. Under such an infinitesimal gauge transformation,

$$\delta_\varepsilon \mathbf{A} = [\varepsilon, \mathbf{A}] - \mathrm{d}\varepsilon$$

where $[\cdot, \cdot]$ is the Lie bracket.

One nice thing is that if $\delta_\varepsilon X = \varepsilon X$, then $\delta_\varepsilon DX = \varepsilon DX$ where D is the covariant derivative

$$DX \stackrel{\text{def}}{=} \mathrm{d}X + \mathbf{A}X$$

Also, $\delta_\varepsilon \mathbf{F} = \varepsilon \mathbf{F}$, which means $\mathbf{F}$ transforms covariantly.

Not all gauge transformations can be generated by infinitesimal gauge transformations in general. An example is when the base manifold is a compact manifold without boundary such that the homotopy class of mappings from that manifold to the Lie group is nontrivial. See instanton for an example.

The *Yang–Mills action* is now given by

$$\frac{1}{4g^2}\int \mathrm{Tr}[*F \wedge F]$$

where * stands for the Hodge dual and the integral is defined as in differential geometry.

A quantity which is **gauge-invariant** (i.e., invariant under gauge transformations) is the Wilson loop, which is defined over any closed path, γ, as follows:

$$\chi^{(\rho)}\left(\mathcal{P}\left\{e^{\int_\gamma A}\right\}\right)$$

where χ is the character of a complex representation ρ and $\mathcal{P}$ represents the path-ordered operator.

14.5 Quantization of gauge theories

Main article: Quantum gauge theory

Gauge theories may be quantized by specialization of methods which are applicable to any quantum field theory. However, because of the subtleties imposed by the gauge constraints (see section on Mathematical formalism, above) there are many technical problems to be solved which do not arise in other field theories. At the same time, the richer structure of gauge theories allows simplification of some computations: for example Ward identities connect different renormalization constants.

14.5.1 Methods and aims

The first gauge theory quantized was quantum electrodynamics (QED). The first methods developed for this involved gauge fixing and then applying canonical quantization. The Gupta–Bleuler method was also developed to handle this problem. Non-abelian gauge theories are now handled by a variety of means. Methods for quantization are covered in the article on quantization.

The main point to quantization is to be able to compute quantum amplitudes for various processes allowed by the theory. Technically, they reduce to the computations of certain correlation functions in the vacuum state. This involves a renormalization of the theory.

When the running coupling of the theory is small enough, then all required quantities may be computed in perturbation theory. Quantization schemes intended to simplify such computations (such as canonical quantization) may be called **perturbative quantization schemes**. At present some of these methods lead to the most precise experimental tests of gauge theories.

However, in most gauge theories, there are many interesting questions which are non-perturbative. Quantization schemes suited to these problems (such as lattice gauge theory) may be called **non-perturbative quantization schemes**. Precise computations in such schemes often require supercomputing, and are therefore less well-developed currently than other schemes.

14.5.2 Anomalies

Some of the symmetries of the classical theory are then seen not to hold in the quantum theory; a phenomenon called an **anomaly**. Among the most well known are:

- The scale anomaly, which gives rise to a *running coupling constant*. In QED this gives rise to the phenomenon of the Landau pole. In Quantum Chromodynamics (QCD) this leads to asymptotic freedom.

- The chiral anomaly in either chiral or vector field theories with fermions. This has close connection with topology through the notion of instantons. In QCD this anomaly causes the decay of a pion to two photons.
- The gauge anomaly, which must cancel in any consistent physical theory. In the electroweak theory this cancellation requires an equal number of quarks and leptons.

14.6 Pure gauge

A pure gauge is the set of field configurations obtained by a gauge transformation on the null-field configuration, i.e., a gauge-transform of zero. So it is a particular "gauge orbit" in the field configuration's space.

Thus, in the abelian case, where $A_\mu(x) \to A'_\mu(x) = A_\mu(x) + \partial_\mu f(x)$, the pure gauge is just the set of field configurations $A'_\mu(x) = \partial_\mu f(x)$ for all $f(x)$.

14.7 See also

14.8 References

[1] Wolfgang Pauli (1941) "Relativistic Field Theories of Elementary Particles," *Rev. Mod. Phys.* **13**: 203–32.

[2] Pickering, A. (1984). *Constructing Quarks*. University of Chicago Press. ISBN 0-226-66799-5.

[3] Sakurai, *Advanced Quantum Mechanics*, sect 1–4

14.9 Bibliography

General readers

- Schumm, Bruce (2004) *Deep Down Things*. Johns Hopkins University Press. Esp. chpt. 8. A serious attempt by a physicist to explain gauge theory and the Standard Model with little formal mathematics.

Texts

- Bromley, D.A. (2000). *Gauge Theory of Weak Interactions*. Springer. ISBN 3-540-67672-4.
- Cheng, T.-P.; Li, L.-F. (1983). *Gauge Theory of Elementary Particle Physics*. Oxford University Press. ISBN 0-19-851961-3.
- Frampton, P. (2008). *Gauge Field Theories* (3rd ed.). Wiley-VCH.
- Kane, G.L. (1987). *Modern Elementary Particle Physics*. Perseus Books. ISBN 0-201-11749-5.

Articles

- Becchi, C. (1997). "Introduction to Gauge Theories". p. 5211. arXiv:hep-ph/9705211. Bibcode:1997hep.ph....5211B.
- Gross, D. (1992). "Gauge theory – Past, Present and Future" (PDF). Retrieved 2009-04-23.
- Jackson, J.D. (2002). "From Lorenz to Coulomb and other explicit gauge transformations". *Am.J.Phys* **70** (9): 917–928. arXiv:physics/0204034. Bibcode:2002AmJPh..70..917J. doi:10.1119/1.1491265.
- Svetlichny, George (1999). "Preparation for Gauge Theory". p. 2027. arXiv:math-ph/9902027. Bibcode:1999math.ph...2027

14.10 External links

- Hazewinkel, Michiel, ed. (2001), "Gauge transformation", *Encyclopedia of Mathematics*, Springer, ISBN 978-1-55608-010-4
- Yang–Mills equations on DispersiveWiki
- Gauge theories on Scholarpedia

Chapter 15

Group theory

This article covers advanced notions. For basic topics, see Group (mathematics).
For group theory in social sciences, see social group.
In mathematics and abstract algebra, **group theory** studies the algebraic structures known as groups. The concept of a group is central to abstract algebra: other well-known algebraic structures, such as rings, fields, and vector spaces, can all be seen as groups endowed with additional operations and axioms. Groups recur throughout mathematics, and the methods of group theory have influenced many parts of algebra. Linear algebraic groups and Lie groups are two branches of group theory that have experienced advances and have become subject areas in their own right.

Various physical systems, such as crystals and the hydrogen atom, can be modelled by symmetry groups. Thus group theory and the closely related representation theory have many important applications in physics, chemistry, and materials science. Group theory is also central to public key cryptography.

One of the most important mathematical achievements of the 20th century[1] was the collaborative effort, taking up more than 10,000 journal pages and mostly published between 1960 and 1980, that culminated in a complete classification of finite simple groups.

15.1 History

Main article: History of group theory

Group theory has three main historical sources: number theory, the theory of algebraic equations, and geometry. The number-theoretic strand was begun by Leonhard Euler, and developed by Gauss's work on modular arithmetic and additive and multiplicative groups related to quadratic fields. Early results about permutation groups were obtained by Lagrange, Ruffini, and Abel in their quest for general solutions of polynomial equations of high degree. Évariste Galois coined the term "group" and established a connection, now known as Galois theory, between the nascent theory of groups and field theory. In geometry, groups first became important in projective geometry and, later, non-Euclidean geometry. Felix Klein's Erlangen program proclaimed group theory to be the organizing principle of geometry.

Galois, in the 1830s, was the first to employ groups to determine the solvability of polynomial equations. Arthur Cayley and Augustin Louis Cauchy pushed these investigations further by creating the theory of permutation groups. The second historical source for groups stems from geometrical situations. In an attempt to come to grips with possible geometries (such as euclidean, hyperbolic or projective geometry) using group theory, Felix Klein initiated the Erlangen programme. Sophus Lie, in 1884, started using groups (now called Lie groups) attached to analytic problems. Thirdly, groups were, at first implicitly and later explicitly, used in algebraic number theory.

The different scope of these early sources resulted in different notions of groups. The theory of groups was unified starting around 1880. Since then, the impact of group theory has been ever growing, giving rise to the birth of abstract algebra in the early 20th century, representation theory, and many more influential spin-off domains. The classification of finite simple groups is a vast body of work from the mid 20th century, classifying all the finite simple groups.

The popular puzzle Rubik's cube invented in 1974 by Ernő Rubik has been used as an illustration of permutation groups.

15.2 Main classes of groups

Main articles: Group (mathematics) and Glossary of group theory

The range of groups being considered has gradually expanded from finite permutation groups and special examples of matrix groups to abstract groups that may be specified through a presentation by generators and relations.

15.2.1 Permutation groups

The first class of groups to undergo a systematic study was permutation groups. Given any set X and a collection G of bijections of X into itself (known as *permutations*) that is closed under compositions and inverses, G is a group acting on X. If X consists of n elements and G consists of *all* permutations, G is the symmetric group S_n; in general, any permutation group G is a subgroup of the symmetric group of X. An early construction due to Cayley exhibited any group as a permutation group, acting on itself ($X = G$) by means of the left regular representation.

In many cases, the structure of a permutation group can be studied using the properties of its action on the corresponding set. For example, in this way one proves that for $n \geq 5$, the alternating group An is simple, i.e. does not admit any proper normal subgroups. This fact plays a key role in the impossibility of solving a general algebraic equation of degree $n' \geq 5$ *in radicals*.

15.2.2 Matrix groups

The next important class of groups is given by *matrix groups*, or linear groups. Here G is a set consisting of invertible matrices of given order n over a field K that is closed under the products and inverses. Such a group acts on the n-dimensional vector space K^n by linear transformations. This action makes matrix groups conceptually similar to permutation groups, and the geometry of the action may be usefully exploited to establish properties of the group G.

15.2.3 Transformation groups

Permutation groups and matrix groups are special cases of transformation groups: groups that act on a certain space X preserving its inherent structure. In the case of permutation groups, X is a set; for matrix groups, X is a vector space. The concept of a transformation group is closely related with the concept of a symmetry group: transformation groups frequently consist of *all* transformations that preserve a certain structure.

The theory of transformation groups forms a bridge connecting group theory with differential geometry. A long line of research, originating with Lie and Klein, considers group actions on manifolds by homeomorphisms or diffeomorphisms. The groups themselves may be discrete or continuous.

15.2.4 Abstract groups

Most groups considered in the first stage of the development of group theory were "concrete", having been realized through numbers, permutations, or matrices. It was not until the late nineteenth century that the idea of an abstract group as a set with operations satisfying a certain system of axioms began to take hold. A typical way of specifying an abstract group is through a presentation by *generators and relations*,

$$G = \langle S|R \rangle.$$

A significant source of abstract groups is given by the construction of a *factor group*, or quotient group, G/H, of a group G by a normal subgroup H. Class groups of algebraic number fields were among the earliest examples of factor groups, of much interest in number theory. If a group G is a permutation group on a set X, the factor group G/H is no longer acting on X; but the idea of an abstract group permits one not to worry about this discrepancy.

The change of perspective from concrete to abstract groups makes it natural to consider properties of groups that are independent of a particular realization, or in modern language, invariant under isomorphism, as well as the classes of group with a given such property: finite groups, periodic groups, simple groups, solvable groups, and so on. Rather than exploring properties of an individual group, one seeks to establish results that apply to a whole class of groups. The new paradigm was of paramount importance for the development of mathematics: it foreshadowed the creation of abstract algebra in the works of Hilbert, Emil Artin, Emmy Noether, and mathematicians of their school.

15.2.5 Topological and algebraic groups

An important elaboration of the concept of a group occurs if G is endowed with additional structure, notably, of a topological space, differentiable manifold, or algebraic variety. If the group operations m (multiplication) and i (inversion),

$$m : G \times G \to G, (g, h) \mapsto gh, \quad i : G \to G, g \mapsto g^{-1},$$

are compatible with this structure, i.e. are continuous, smooth or regular (in the sense of algebraic geometry) maps, then G becomes a topological group, a Lie group, or an algebraic group.[2]

The presence of extra structure relates these types of groups with other mathematical disciplines and means that more tools are available in their study. Topological groups form a natural domain for abstract harmonic analysis, whereas Lie groups (frequently realized as transformation groups) are the mainstays of differential geometry and unitary representation theory. Certain classification questions that cannot be solved in general can be approached and resolved for special subclasses of groups. Thus, compact connected Lie groups have been completely classified. There is a fruitful relation between infinite abstract groups and topological groups: whenever a group Γ can be realized as a lattice in a topological group G, the geometry and analysis pertaining to G yield important results about Γ. A comparatively recent trend in the theory of finite groups exploits their connections with compact topological groups (profinite groups): for example, a single p-adic analytic group G has a family of quotients which are finite p-groups of various orders, and properties of G translate into the properties of its finite quotients.

15.3 Branches of group theory

15.3.1 Finite group theory

Main article: Finite group

During the twentieth century, mathematicians investigated some aspects of the theory of finite groups in great depth, especially the local theory of finite groups and the theory of solvable and nilpotent groups. As a consequence, the complete classification of finite simple groups was achieved, meaning that all those simple groups from which all finite groups can be built are now known.

During the second half of the twentieth century, mathematicians such as Chevalley and Steinberg also increased our understanding of finite analogs of classical groups, and other related groups. One such family of groups is the family of general linear groups over finite fields. Finite groups often occur when considering symmetry of mathematical or physical objects, when those objects admit just a finite number of structure-preserving transformations. The theory of Lie groups, which may be viewed as dealing with "continuous symmetry", is strongly influenced by the associated Weyl groups. These are finite groups generated by reflections which act on a finite-dimensional Euclidean space. The properties of finite groups can thus play a role in subjects such as theoretical physics and chemistry.

15.3.2 Representation of groups

Main article: Representation theory

Saying that a group G *acts* on a set X means that every element of G defines a bijective map on the set X in a way compatible with the group structure. When X has more structure, it is useful to restrict this notion further: a representation of G on a vector space V is a group homomorphism:

$$\varrho : G \to \mathrm{GL}(V),$$

where $\mathrm{GL}(V)$ consists of the invertible linear transformations of V. In other words, to every group element g is assigned an automorphism $\varrho(g)$ such that $\varrho(g) \circ \varrho(h) = \varrho(gh)$ for any h in G.

This definition can be understood in two directions, both of which give rise to whole new domains of mathematics.[3] On the one hand, it may yield new information about the group G: often, the group operation in G is abstractly given, but via ϱ, it corresponds to the multiplication of matrices, which is very explicit.[4] On the other hand, given a well-understood group acting on a complicated object, this simplifies the study of the object in question. For example, if G is finite, it is known that V above decomposes into irreducible parts. These parts in turn are much more easily manageable than the whole V (via Schur's lemma).

Given a group G, representation theory then asks what representations of G exist. There are several settings, and the employed methods and obtained results are rather different in every case: representation theory of finite groups and representations of Lie groups are two main subdomains of the theory. The totality of representations is governed by the group's characters. For example, Fourier polynomials can be interpreted as the characters of U(1), the group of complex numbers of absolute value *1*, acting on the L^2-space of periodic functions.

15.3.3 Lie theory

Main article: Lie group

A Lie group is a group that is also a differentiable manifold, with the property that the group operations are compatible with the smooth structure. Lie groups are named after Sophus Lie, who laid the foundations of the theory of continuous transformation groups. The term *groupes de Lie* first appeared in French in 1893 in the thesis of Lie's student Arthur Tresse, page 3.[5]

Lie groups represent the best-developed theory of continuous symmetry of mathematical objects and structures, which makes them indispensable tools for many parts of contemporary mathematics, as well as for modern theoretical physics. They provide a natural framework for analysing the continuous symmetries of differential equations (differential Galois theory), in much the same way as permutation groups are used in Galois theory for analysing the discrete symmetries of algebraic equations. An extension of Galois theory to the case of continuous symmetry groups was one of Lie's principal motivations.

15.3.4 Combinatorial and geometric group theory

Main article: Geometric group theory

Groups can be described in different ways. Finite groups can be described by writing down the group table consisting of all possible multiplications $g \bullet h$. A more compact way of defining a group is by *generators and relations*, also called the *presentation* of a group. Given any set F of generators $\{gi\} i \in I$, the free group generated by F subjects onto the group G. The kernel of this map is called subgroup of relations, generated by some subset D. The presentation is usually denoted by $\langle F \mid D \rangle$. For example, the group $\mathbf{Z} = \langle a \mid \rangle$ can be generated by one element a (equal to +1 or −1) and no relations, because $n \cdot 1$ never equals 0 unless n is zero. A string consisting of generator symbols and their inverses is called a *word*.

Combinatorial group theory studies groups from the perspective of generators and relations.[6] It is particularly useful where finiteness assumptions are satisfied, for example finitely generated groups, or finitely presented groups (i.e. in addition the relations are finite). The area makes use of the connection of graphs via their fundamental groups. For example, one can show that every subgroup of a free group is free.

There are several natural questions arising from giving a group by its presentation. The *word problem* asks whether two words are effectively the same group element. By relating the problem to Turing machines, one can show that there is in general no algorithm solving this task. Another, generally harder, algorithmically insoluble problem is the group isomorphism problem, which asks whether two groups given by different presentations are actually isomorphic. For example the additive group $\mathbf{Z}$ of integers can also be presented by

$$\langle x, y \mid xyxyx = e \rangle;$$

it may not be obvious that these groups are isomorphic.[7]

Geometric group theory attacks these problems from a geometric viewpoint, either by viewing groups as geometric objects, or by finding suitable geometric objects a group acts on.[8] The first idea is made precise by means of the Cayley graph, whose vertices correspond to group elements and edges correspond to right multiplication in the group. Given two elements, one constructs the word metric given by the length of the minimal path between the elements. A theorem of Milnor and Svarc then says that given a group G acting in a reasonable manner on a metric space X, for example a compact manifold, then G is quasi-isometric (i.e. looks similar from a distance) to the space X.

15.4 Connection of groups and symmetry

Main article: Symmetry group

Given a structured object X of any sort, a symmetry is a mapping of the object onto itself which preserves the structure. This occurs in many cases, for example

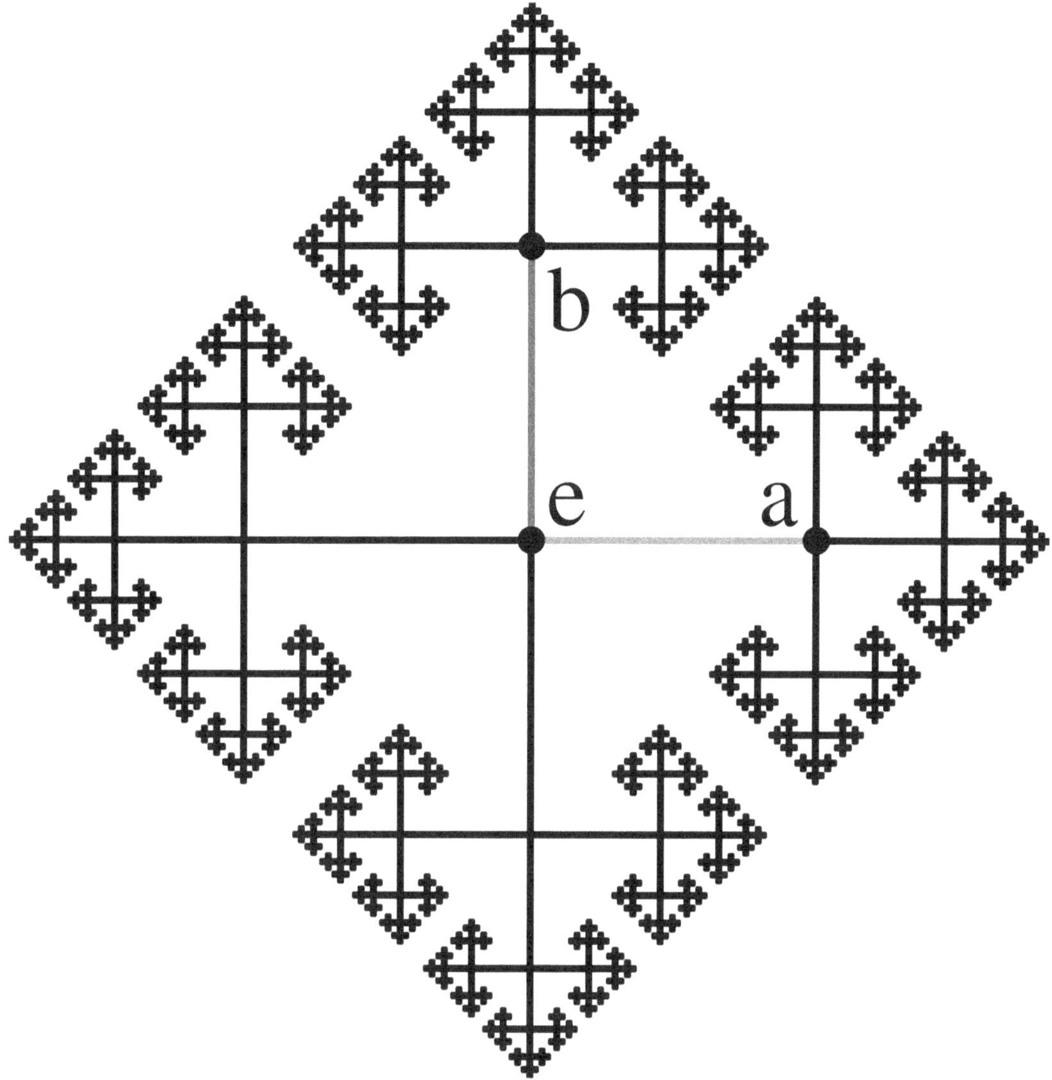

The Cayley graph of ⟨x, y | ⟩, the free group of rank 2.

1. If X is a set with no additional structure, a symmetry is a bijective map from the set to itself, giving rise to permutation groups.
2. If the object X is a set of points in the plane with its metric structure or any other metric space, a symmetry is a bijection of the set to itself which preserves the distance between each pair of points (an isometry). The corresponding group is called isometry group of X.
3. If instead angles are preserved, one speaks of conformal maps. Conformal maps give rise to Kleinian groups, for example.
4. Symmetries are not restricted to geometrical objects, but include algebraic objects as well. For instance, the equation

   $$x^2 - 3 = 0$$

   has the two solutions $+\sqrt{3}$, and $-\sqrt{3}$. In this case, the group that exchanges the two roots is the Galois group belonging to the equation. Every polynomial equation in one variable has a Galois group, that is a certain permutation group on its roots.

The axioms of a group formalize the essential aspects of symmetry. Symmetries form a group: they are closed because if you take a symmetry of an object, and then apply another symmetry, the result will still be a symmetry. The identity keeping the object fixed is always a symmetry of an object. Existence of inverses is guaranteed by undoing the symmetry and the associativity comes from the fact that symmetries are functions on a space, and composition of functions are associative.

Frucht's theorem says that every group is the symmetry group of some graph. So every abstract group is actually the symmetries of some explicit object.

The saying of "preserving the structure" of an object can be made precise by working in a category. Maps preserving the structure are then the morphisms, and the symmetry group is the automorphism group of the object in question.

15.5 Applications of group theory

Applications of group theory abound. Almost all structures in abstract algebra are special cases of groups. Rings, for example, can be viewed as abelian groups (corresponding to addition) together with a second operation (corresponding to multiplication). Therefore, group theoretic arguments underlie large parts of the theory of those entities.

15.5.1 Galois theory

Main article: Galois theory

Galois theory uses groups to describe the symmetries of the roots of a polynomial (or more precisely the automorphisms of the algebras generated by these roots). The fundamental theorem of Galois theory provides a link between algebraic field extensions and group theory. It gives an effective criterion for the solvability of polynomial equations in terms of the solvability of the corresponding Galois group. For example, S_5, the symmetric group in 5 elements, is not solvable which implies that the general quintic equation cannot be solved by radicals in the way equations of lower degree can. The theory, being one of the historical roots of group theory, is still fruitfully applied to yield new results in areas such as class field theory.

15.5.2 Algebraic topology

Main article: Algebraic topology

Algebraic topology is another domain which prominently associates groups to the objects the theory is interested in. There, groups are used to describe certain invariants of topological spaces. They are called "invariants" because they are defined in such a way that they do not change if the space is subjected to some deformation. For example, the fundamental group "counts" how many paths in the space are essentially different. The Poincaré conjecture, proved in 2002/2003 by Grigori Perelman, is a prominent application of this idea. The influence is not unidirectional, though. For example, algebraic topology makes use of Eilenberg–MacLane spaces which are spaces with prescribed homotopy groups. Similarly algebraic K-theory relies in a way on classifying spaces of groups. Finally, the name of the torsion subgroup of an infinite group shows the legacy of topology in group theory.

15.5.3 Algebraic geometry and cryptography

Main articles: Algebraic geometry and Cryptography

Algebraic geometry and cryptography likewise uses group theory in many ways. Abelian varieties have been introduced above. The presence of the group operation yields additional information which makes these varieties particularly accessible. They also often serve as a test for new conjectures.[9] The one-dimensional case, namely elliptic curves is studied in particular detail. They are both theoretically and practically intriguing.[10] Very large groups of prime order constructed in Elliptic-Curve Cryptography serve for public key cryptography. Cryptographical methods of this kind benefit from the flexibility of the geometric objects, hence their group structures, together with the complicated structure of these groups, which make the discrete logarithm very hard to calculate. One of the earliest

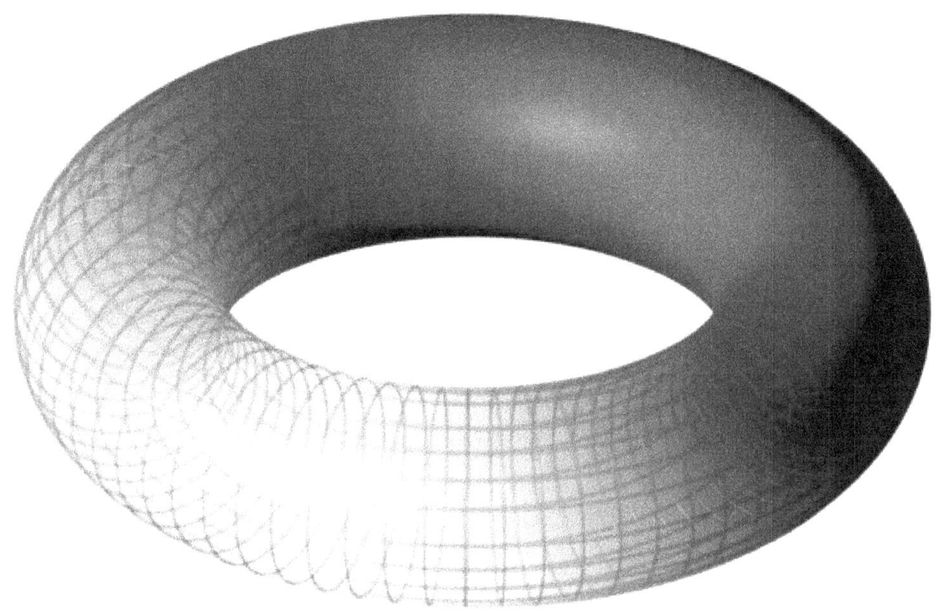

A torus. Its abelian group structure is induced from the map $\mathbf{C} \to \mathbf{C}/\mathbf{Z} + \tau\mathbf{Z}$, *where* τ *is a parameter living in the upper half plane.*

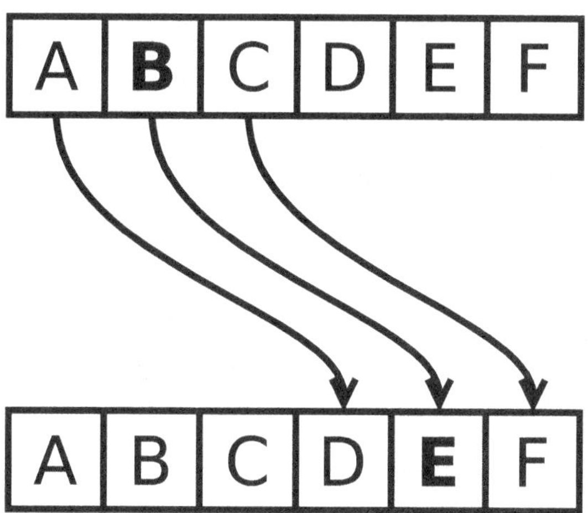

The cyclic group $\mathbf{Z}_{26}$ *underlies Caesar's cipher.*

encryption protocols, Caesar's cipher, may also be interpreted as a (very easy) group operation. In another direction, toric varieties are algebraic varieties acted on by a torus. Toroidal embeddings have recently led to advances in algebraic geometry, in particular resolution of singularities.[11]

15.5.4 Algebraic number theory

Main article: Algebraic number theory

Algebraic number theory is a special case of group theory, thereby following the rules of the latter. For example, Euler's product formula

$$\sum_{n\geq 1} \frac{1}{n^s} = \prod_{p \text{ prime}} \frac{1}{1-p^{-s}}$$

captures the fact that any integer decomposes in a unique way into primes. The failure of this statement for more general rings gives rise to class groups and regular primes, which feature in Kummer's treatment of Fermat's Last Theorem.

15.5.5 Harmonic analysis

Main article: Harmonic analysis

Analysis on Lie groups and certain other groups is called harmonic analysis. Haar measures, that is, integrals invariant under the translation in a Lie group, are used for pattern recognition and other image processing techniques.[12]

15.5.6 Combinatorics

In combinatorics, the notion of permutation group and the concept of group action are often used to simplify the counting of a set of objects; see in particular Burnside's lemma.

15.5.7 Music

The presence of the 12-periodicity in the circle of fifths yields applications of elementary group theory in musical set theory.

15.5.8 Physics

In physics, groups are important because they describe the symmetries which the laws of physics seem to obey. According to Noether's theorem, every continuous symmetry of a physical system corresponds to a conservation law of the system. Physicists are very interested in group representations, especially of Lie groups, since these representations often point the way to the "possible" physical theories. Examples of the use of groups in physics include the Standard Model, gauge theory, the Lorentz group, and the Poincaré group.

15.5.9 Chemistry and materials science

In chemistry and materials science, groups are used to classify crystal structures, regular polyhedra, and the symmetries of molecules. The assigned point groups can then be used to determine physical properties (such as polarity and chirality), spectroscopic properties (particularly useful for Raman spectroscopy and infrared spectroscopy), and to construct molecular orbitals.

Molecular symmetry is responsible for many physical and spectroscopic properties of compounds and provides relevant information about how chemical reactions occur. In order to assign a point group for any given molecule, it is necessary to find the set of symmetry operations present on it. The symmetry operation is an action, such as a rotation around an axis or a reflection through a mirror plane. In other words, it is an operation that moves the molecule such that it is indistinguishable from the original configuration. In group theory, the rotation axes and mirror planes are called "symmetry elements". These elements can be a point, line or plane with respect to which the symmetry operation is carried out. The symmetry operations of a molecule determine the specific point group for this molecule.

In chemistry, there are five important symmetry operations. The identity operation (E) consists of leaving the molecule as it is. This is equivalent to any number of full rotations around any axis. This is a symmetry of all molecules, whereas the symmetry group of a chiral molecule consists of only the identity operation. Rotation around an axis (Cn) consists of rotating the molecule around a specific axis by a specific angle. For example, if a water molecule rotates 180° around the axis that passes through the oxygen atom and between the hydrogen atoms, it is in the same configuration as it started. In this case, $n = 2$, since applying it twice produces the identity operation. Other symmetry operations are: reflection, inversion and improper rotation (rotation followed by reflection).[13]

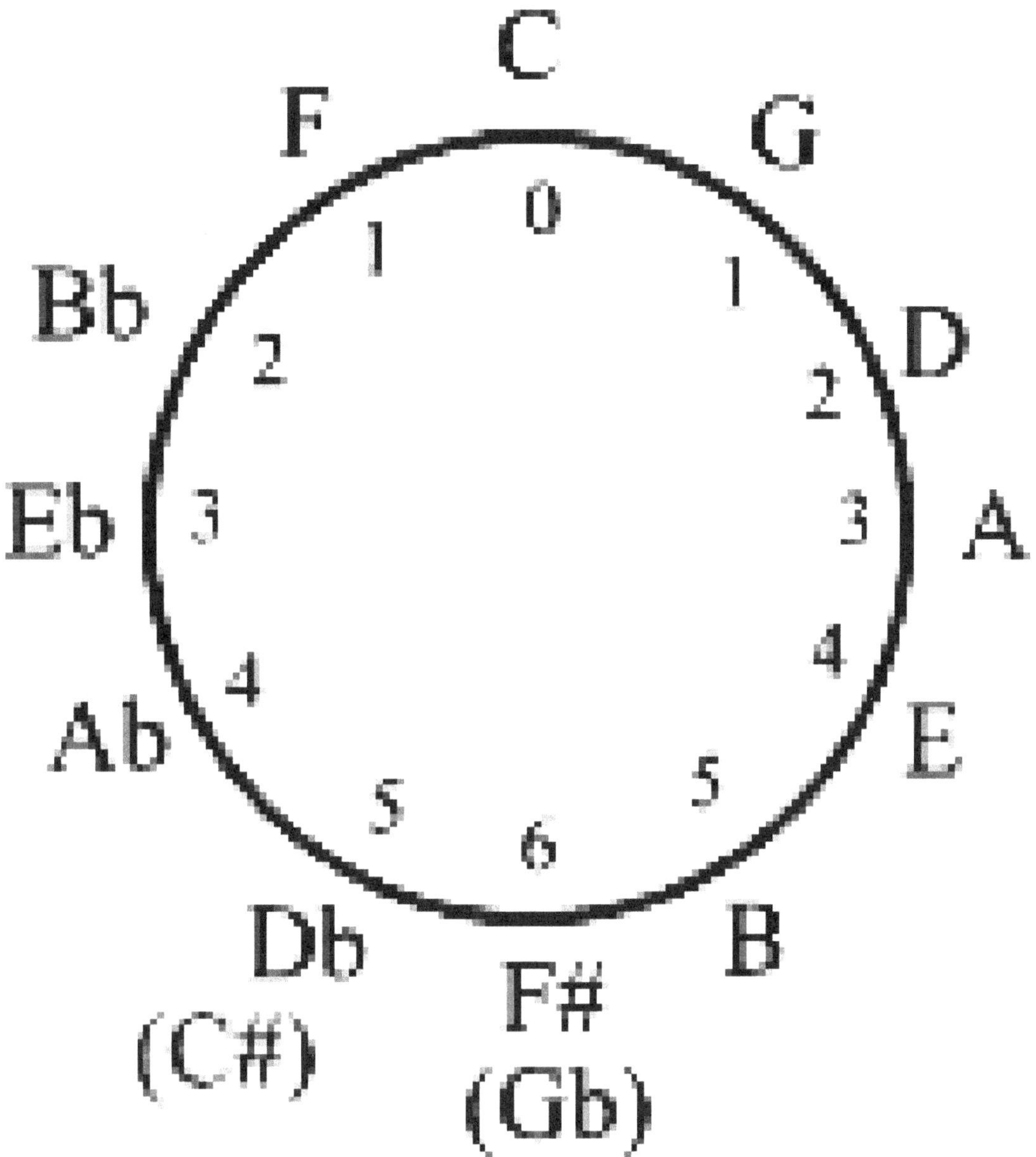

The circle of fifths may be endowed with a cyclic group structure

15.6 See also

- Group (mathematics)
- Glossary of group theory
- List of group theory topics

15.7 Notes

[1] • Elwes, Richard, "An enormous theorem: the classification of finite simple groups," *Plus Magazine*, Issue 41, December 2006.

[2] This process of imposing extra structure has been formalized through the notion of a group object in a suitable category. Thus Lie groups are group objects in the category of differentiable manifolds and affine algebraic groups are group objects in the category of affine algebraic varieties.

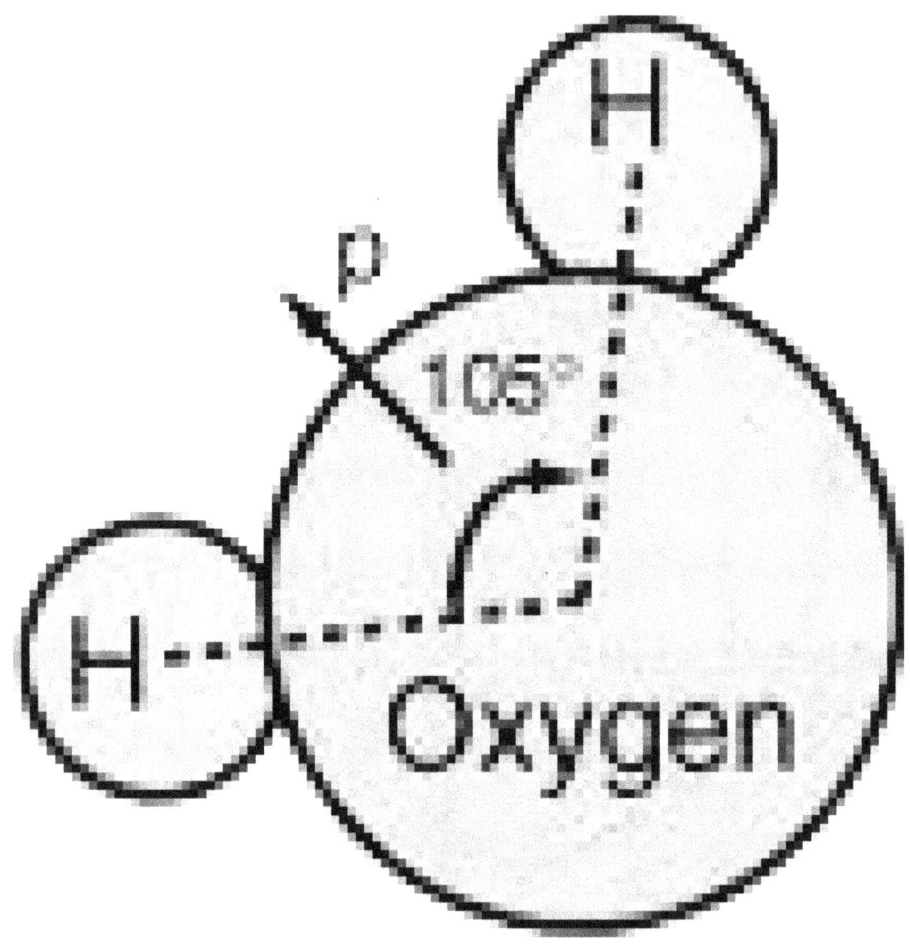

Water molecule with symmetry axis

[3] Such as group cohomology or equivariant K-theory.

[4] In particular, if the representation is faithful.

[5] Arthur Tresse (1893). "Sur les invariants différentiels des groupes continus de transformations". *Acta Mathematica* **18**: 1–88. doi:10.1007/bf02418270.

[6] Schupp & Lyndon 2001

[7] Writing $z = xy$, one has $G = \langle z, y \mid z^3 = y\rangle = \langle z\rangle$.

[8] La Harpe 2000

[9] For example the Hodge conjecture (in certain cases).

[10] See the Birch-Swinnerton-Dyer conjecture, one of the millennium problems

[11] Abramovich, Dan; Karu, Kalle; Matsuki, Kenji; Wlodarczyk, Jaroslaw (2002), "Torification and factorization of birational maps", *Journal of the American Mathematical Society* **15** (3): 531–572, doi:10.1090/S0894-0347-02-00396-X, MR 1896232

[12] Lenz, Reiner (1990), *Group theoretical methods in image processing*, Lecture Notes in Computer Science **413**, Berlin, New York: Springer-Verlag, doi:10.1007/3-540-52290-5, ISBN 978-0-387-52290-6

[13] Shriver, D.F.; Atkins, P.W. Química Inorgânica, 3ª ed., Porto Alegre, Bookman, 2003.

15.8 References

- Borel, Armand (1991), *Linear algebraic groups*, Graduate Texts in Mathematics **126** (2nd ed.), Berlin, New York: Springer-Verlag, ISBN 978-0-387-97370-8, MR 1102012
- Carter, Nathan C. (2009), *Visual group theory*, Classroom Resource Materials Series, Mathematical Association of America, ISBN 978-0-88385-757-1, MR 2504193
- Cannon, John J. (1969), "Computers in group theory: A survey", *Communications of the Association for Computing Machinery* **12**: 3–12, doi:10.1145/362835.362837, MR 0290613
- Frucht, R. (1939), "Herstellung von Graphen mit vorgegebener abstrakter Gruppe", *Compositio Mathematica* **6**: 239–50, ISSN 0010-437X
- Golubitsky, Martin; Stewart, Ian (2006), "Nonlinear dynamics of networks: the groupoid formalism", *Bull. Amer. Math. Soc. (N.S.)* **43** (03): 305–364, doi:10.1090/S0273-0979-06-01108-6, MR 2223010 Shows the advantage of generalising from group to groupoid.
- Judson, Thomas W. (1997), *Abstract Algebra: Theory and Applications* An introductory undergraduate text in the spirit of texts by Gallian or Herstein, covering groups, rings, integral domains, fields and Galois theory. Free downloadable PDF with open-source GFDL license.
- Kleiner, Israel (1986), "The evolution of group theory: a brief survey", *Mathematics Magazine* **59** (4): 195–215, doi:10.2307/2690312, ISSN 0025-570X, JSTOR 2690312, MR 863090
- La Harpe, Pierre de (2000), *Topics in geometric group theory*, University of Chicago Press, ISBN 978-0-226-31721-2
- Livio, M. (2005), *The Equation That Couldn't Be Solved: How Mathematical Genius Discovered the Language of Symmetry*, Simon & Schuster, ISBN 0-7432-5820-7 Conveys the practical value of group theory by explaining how it points to symmetries in physics and other sciences.
- Mumford, David (1970), *Abelian varieties*, Oxford University Press, ISBN 978-0-19-560528-0, OCLC 138290
- Ronan M., 2006. *Symmetry and the Monster*. Oxford University Press. ISBN 0-19-280722-6. For lay readers. Describes the quest to find the basic building blocks for finite groups.
- Rotman, Joseph (1994), *An introduction to the theory of groups*, New York: Springer-Verlag, ISBN 0-387-94285-8 A standard contemporary reference.
- Schupp, Paul E.; Lyndon, Roger C. (2001), *Combinatorial group theory*, Berlin, New York: Springer-Verlag, ISBN 978-3-540-41158-1
- Scott, W. R. (1987) [1964], *Group Theory*, New York: Dover, ISBN 0-486-65377-3 Inexpensive and fairly readable, but somewhat dated in emphasis, style, and notation.
- Shatz, Stephen S. (1972), *Profinite groups, arithmetic, and geometry*, Princeton University Press, ISBN 978-0-691-08017-8, MR 0347778
- Weibel, Charles A. (1994), *An introduction to homological algebra*, Cambridge Studies in Advanced Mathematics **38**, Cambridge University Press, ISBN 978-0-521-55987-4, OCLC 36131259, MR 1269324

15.9 External links

- History of the abstract group concept
- Higher dimensional group theory This presents a view of group theory as level one of a theory which extends in all dimensions, and has applications in homotopy theory and to higher dimensional nonabelian methods for local-to-global problems.
- Plus teacher and student package: Group Theory This package brings together all the articles on group theory from *Plus*, the online mathematics magazine produced by the Millennium Mathematics Project at the University of Cambridge, exploring applications and recent breakthroughs, and giving explicit definitions and examples of groups.

- US Naval Academy group theory guide A general introduction to group theory with exercises written by Tony Gaglione.

Chapter 16

Feynman diagram

For a less technical version, see this article on the Simple English Wikipedia.

In theoretical physics, **Feynman diagrams** are pictorial representations of the mathematical expressions describing the behavior of subatomic particles. The scheme is named for its inventor, American physicist Richard Feynman, and was first introduced in 1948. The interaction of sub-atomic particles can be complex and difficult to understand intuitively, and the Feynman diagrams allow for a simple visualization of what would otherwise be a rather arcane and abstract formula. As David Kaiser writes, "since the middle of the 20th century, theoretical physicists have increasingly turned to this tool to help them undertake critical calculations", and as such "Feynman diagrams have revolutionized nearly every aspect of theoretical physics".[1] While the diagrams are applied primarily to quantum field theory, they can also be used in other fields, such as solid-state theory.

Feynman used Ernst Stueckelberg's interpretation of the positron as if it were an electron moving backward in time.[2] Thus, antiparticles are represented as moving backward along the time axis in Feynman diagrams.

The calculation of probability amplitudes in theoretical particle physics requires the use of rather large and complicated integrals over a large number of variables. These integrals do, however, have a regular structure, and may be represented graphically as Feynman diagrams. A Feynman diagram is a contribution of a particular class of particle paths, which join and split as described by the diagram. More precisely, and technically, a Feynman diagram is a graphical representation of a perturbative contribution to the transition amplitude or correlation function of a quantum mechanical or statistical field theory. Within the canonical formulation of quantum field theory, a Feynman diagram represents a term in the Wick's expansion of the perturbative S-matrix. Alternatively, the path integral formulation of quantum field theory represents the transition amplitude as a weighted sum of all possible histories of the system from the initial to the final state, in terms of either particles or fields. The transition amplitude is then given as the matrix element of the S-matrix between the initial and the final states of the quantum system.

16.1 Motivation and history

When calculating scattering cross-sections in particle physics, the interaction between particles can be described by starting from a free field that describes the incoming and outgoing particles, and including an interaction Hamiltonian to describe how the particles deflect one another. The amplitude for scattering is the sum of each possible interaction history over all possible intermediate particle states. The number of times the interaction Hamiltonian acts is the order of the perturbation expansion, and the time-dependent perturbation theory for fields is known as the Dyson series. When the intermediate states at intermediate times are energy eigenstates (collections of particles with a definite momentum) the series is called old-fashioned perturbation theory.

The Dyson series can be alternatively rewritten as a sum over Feynman diagrams, where at each interaction vertex both the energy and momentum are conserved, but where the length of the energy momentum four vector is not equal to the mass. The Feynman diagrams are much easier to keep track of than old-fashioned terms, because the old-fashioned way treats the particle and antiparticle contributions as separate. Each Feynman diagram is the sum of exponentially many old-fashioned terms, because each internal line can separately represent either a particle or an antiparticle. In a non-relativistic theory, there are no antiparticles and there is no doubling, so each Feynman diagram includes only one term.

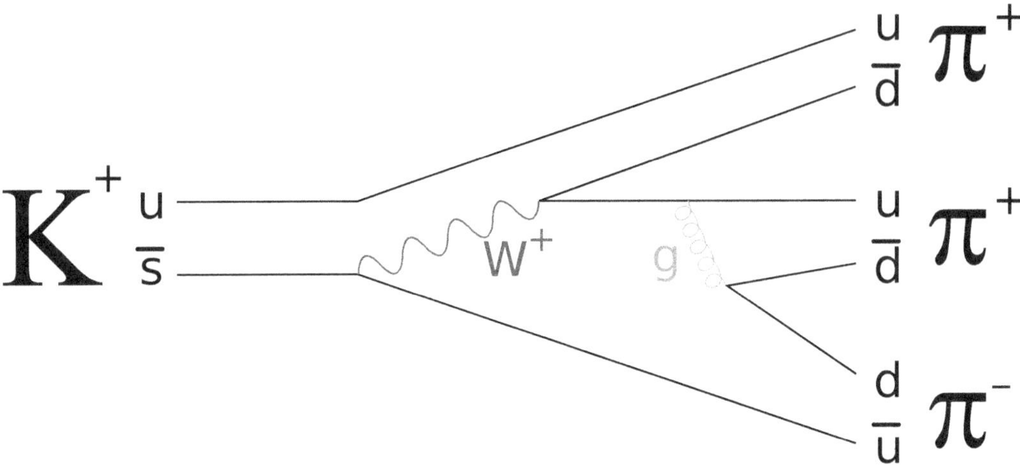

In this diagram, a kaon, made of an up and anti-strange quark, decays both weakly and strongly into three pions, with intermediate steps involving a W boson and a gluon (represented by the green spiral).

Feynman gave a prescription for calculating the amplitude for any given diagram from a field theory Lagrangian—the Feynman rules. Each internal line corresponds to a factor of the corresponding virtual particle's propagator; each vertex where lines meet gives a factor derived from an interaction term in the Lagrangian, and incoming and outgoing lines carry an energy, momentum, and spin.

In addition to their value as a mathematical tool, Feynman diagrams provide deep physical insight into the nature of particle interactions. Particles interact in every way available; in fact, intermediate virtual particles are allowed to propagate faster than light. The probability of each final state is then obtained by summing over all such possibilities. This is closely tied to the functional integral formulation of quantum mechanics, also invented by Feynman–see path integral formulation.

The naïve application of such calculations often produces diagrams whose amplitudes are infinite, because the short-distance particle interactions require a careful limiting procedure, to include particle self-interactions. The technique of renormalization, suggested by Ernst Stueckelberg and Hans Bethe and implemented by Dyson, Feynman, Schwinger, and Tomonaga compensates for this effect and eliminates the troublesome infinities. After renormalization, calculations using Feynman diagrams match experimental results with very high accuracy.

Feynman diagram and path integral methods are also used in statistical mechanics and can even be applied to classical mechanics.[3]

16.1.1 Alternative names

Murray Gell-Mann always referred to Feynman diagrams as **Stueckelberg diagrams**, after a Swiss physicist, Ernst Stueckelberg, who devised a similar notation many years earlier. Stueckelberg was motivated by the need for a manifestly covariant formalism for quantum field theory, but did not provide as automated a way to handle symmetry factors and loops, although he was first to find the correct physical interpretation in terms of forward and backward in time particle paths, all without the path-integral.[4] Historically they were sometimes called **Feynman–Dyson diagrams** or **Dyson graphs**,[5] because when they were introduced the path integral was unfamiliar, and Freeman Dyson's derivation from old-fashioned perturbation theory was easier to follow for physicists trained in earlier methods. However, in 2006 Dyson himself stated that the diagrams should be called *Feynman diagrams* because "he taught us how to use them". This reflects historical fact: Feynman had to lobby hard for the diagrams which confused the establishment physicists trained in equations and graphs.[6]

16.2 Representation of physical reality

In their presentations of fundamental interactions,[7][8] written from the particle physics perspective, Gerard 't Hooft and Martinus Veltman gave good arguments for taking the original, non-regularized Feynman diagrams as the most succinct representation of our present knowledge about the physics of quantum scattering of fundamental particles. Their motivations are consistent with the convictions of James Daniel Bjorken and Sidney Drell:[9] "The Feynman graphs and rules of calculation summarize quantum field theory in a form in close contact with the experimental numbers one wants to understand. Although the statement of the theory in terms of graphs may imply perturbation theory, use of graphical methods in the many-body problem shows that this formalism is flexible enough to deal with phenomena of nonperturbative characters ... Some modification of the Feynman rules of calculation may well outlive the elaborate mathematical structure of local canonical quantum field theory ..." So far there are no opposing opinions. In quantum field theories the Feynman diagrams are obtained from Lagrangian by Feynman rules.

16.3 Particle-path interpretation

A Feynman diagram is a representation of quantum field theory processes in terms of particle paths. The particle trajectories are represented by the lines of the diagram, which can be squiggly or straight, with an arrow or without, depending on the type of particle. A point where lines connect to other lines is an interaction vertex, and this is where the particles meet and interact: by emitting or absorbing new particles, deflecting one another, or changing type.

There are three different types of lines: *internal lines* connect two vertices, *incoming lines* extend from "the past" to a vertex and represent an initial state, and *outgoing lines* extend from a vertex to "the future" and represent the final state. Sometimes, the bottom of the diagram is the past and the top the future; other times, the past is to the left and the future to the right. When calculating correlation functions instead of scattering amplitudes, there is no past and future and all the lines are internal. The particles then begin and end on little x's, which represent the positions of the operators whose correlation is being calculated.

Feynman diagrams are a pictorial representation of a contribution to the total amplitude for a process that can happen in several different ways. When a group of incoming particles are to scatter off each other, the process can be thought of as one where the particles travel over all possible paths, including paths that go backward in time.

Feynman diagrams are often confused with spacetime diagrams and bubble chamber images because they all describe particle scattering. Feynman diagrams are graphs that represent the trajectories of particles in intermediate stages of a scattering process. Unlike a bubble chamber picture, only the sum of all the Feynman diagrams represent any given particle interaction; particles do not choose a particular diagram each time they interact. The law of summation is in accord with the principle of superposition—every diagram contributes to the total amplitude for the process.

16.4 Description

A Feynman diagram represents a perturbative contribution to the amplitude of a quantum transition from some initial quantum state to some final quantum state.

For example, in the process of electron-positron annihilation the initial state is one electron and one positron, the final state: two photons.

The initial state is often assumed to be at the left of the diagram and the final state at the right (although other conventions are also used quite often).

A Feynman diagram consists of points, called vertices, and lines attached to the vertices.

The particles in the initial state are depicted by lines sticking out in the direction of the initial state (e.g., to the left), the particles in the final state are represented by lines sticking out in the direction of the final state (e.g., to the right).

In QED there are two types of particles: electrons/positrons (called fermions) and photons (called gauge bosons). They are represented in Feynman diagrams as follows:

1. Electron in the initial state is represented by a solid line with an arrow pointing toward the vertex (→•).
2. Electron in the final state is represented by a line with an arrow pointing away from the vertex: (•→).
3. Positron in the initial state is represented by a solid line with an arrow pointing away from the vertex: (←•).
4. Positron in the final state is represented by a line with an arrow pointing toward the vertex: (•←).
5. Photon in the initial and the final state is represented by a wavy line (~• and •~).

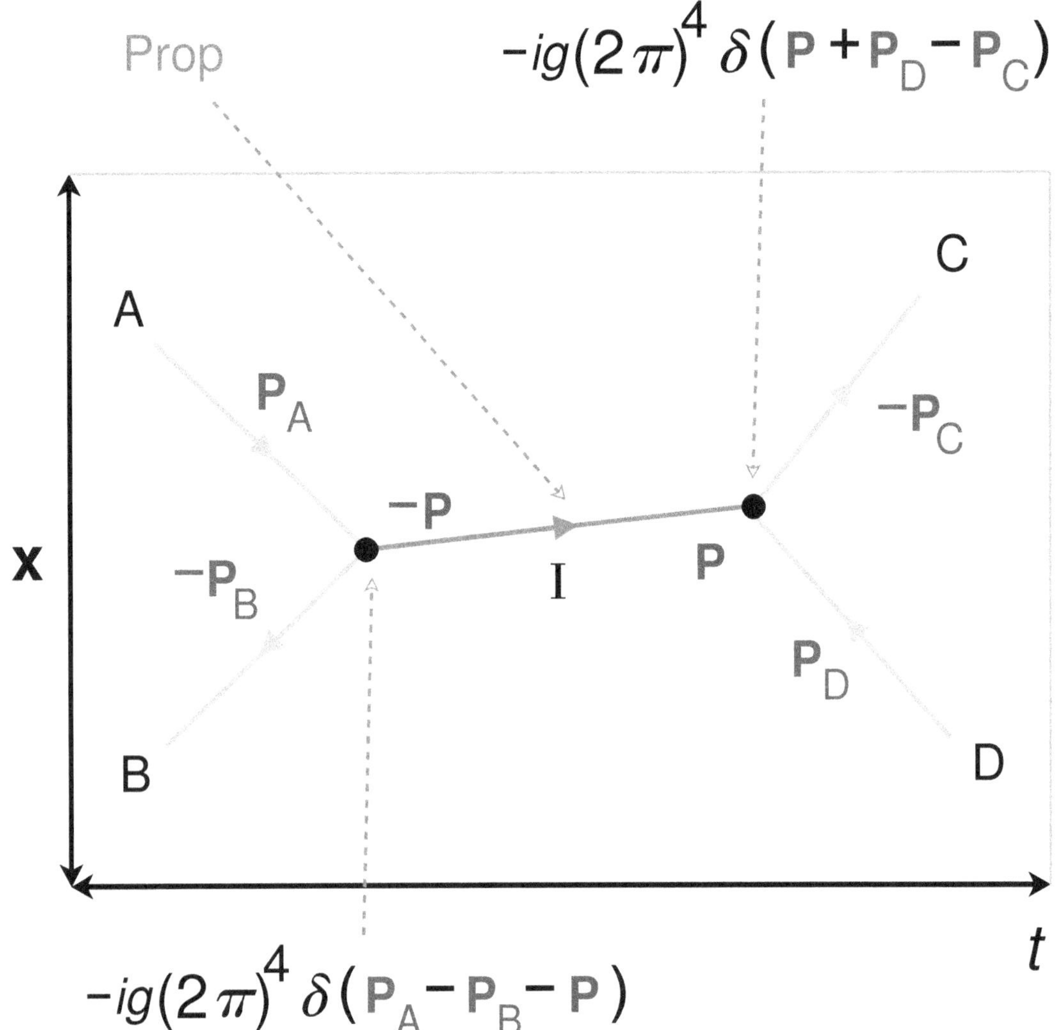

General features of the scattering process $A + B \to C + D$:
*• internal lines **(red)** for intermediate particles and processes, which has a propagator factor ("prop"), external lines **(orange)** for incoming/outgoing particles to/from vertices **(black)**,*
• at each vertex there is 4-momentum conservation using delta functions, 4-momenta entering the vertex are positive while those leaving are negative, the factors at each vertex and internal line are multiplied in the amplitude integral,
*• space **x** and time t axes are not always shown, directions of external lines correspond to passage of time.*

In QED a vertex always has three lines attached to it: one bosonic line, one fermionic line with arrow toward the vertex, and one fermionic line with arrow away from the vertex.

The vertices might be connected by a bosonic or fermionic propagator. A bosonic propagator is represented by a wavy line connecting two vertices (•~•). A fermionic propagator is represented by a solid line (with an arrow in one or another direction) connecting two vertices, (•←•).

The number of vertices gives the order of the term in the perturbation series expansion of the transition amplitude.

16.4.1 Electron-positron annihilation example

The electron-positron annihilation interaction:

$e^+e^- \to 2\gamma$

has a contribution from the second order Feynman diagram shown adjacent:

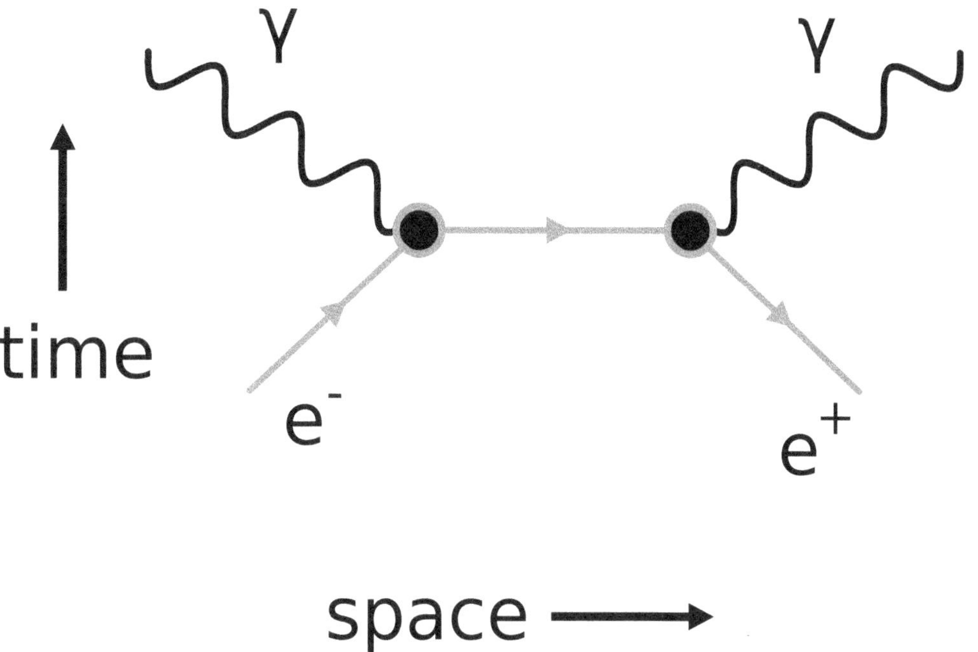

Feynman diagram of electron-positron annihilation

In the initial state (at the bottom; early time) there is one electron (e^-) and one positron (e^+) and in the final state (at the top; late time) there are two photons (γ).

16.5 Canonical quantization formulation

The probability amplitude for a transition of a quantum system from the initial state $|i\rangle$ to the final state $|f\rangle$ is given by the matrix element

$$S_{fi} = \langle f|S|i\rangle \ ,$$

where S is the S-matrix.

In the canonical quantum field theory the S-matrix is represented within the interaction picture by the perturbation series in the powers of the interaction Lagrangian,

$$S = \sum_{n=0}^{\infty} \frac{i^n}{n!} \int \prod_{j=1}^{n} d^4x_j T \prod_{j=1}^{n} L_v(x_j) \equiv \sum_{n=0}^{\infty} S^{(n)} \ ,$$

where L_v is the interaction Lagrangian and T signifies the time-ordered product of operators.

A Feynman diagram is a graphical representation of a term in the Wick's expansion of the time-ordered product in the n -th order term $S^{(n)}$ of the S-matrix,

$$T \prod_{j=1}^{n} L_v(x_j) = \sum_{\text{all possible contractions}} (\pm) N \prod_{j=1}^{n} L_v(x_j) \ ,$$

where N signifies the normal-product of the operators and $(\pm)$ takes care of the possible sign change when commuting the fermionic operators to bring them together for a contraction (a propagator).

16.5.1 Feynman rules

The diagrams are drawn according to the Feynman rules, which depend upon the interaction Lagrangian. For the QED interaction Lagrangian, $L_v = -g\bar{\psi}\gamma^\mu\psi A_\mu$, describing the interaction of a fermionic field ψ with a bosonic gauge field A_μ , the Feynman rules can be formulated in coordinate space as follows:

1. Each integration coordinate x_j is represented by a point (sometimes called a vertex);
2. A bosonic propagator is represented by a wiggly line connecting two points;
3. A fermionic propagator is represented by a solid line connecting two points;
4. A bosonic field $A_\mu(x_i)$ is represented by a wiggly line attached to the point x_i ;
5. A fermionic field $\psi(x_i)$ is represented by a solid line attached to the point x_i with an arrow toward the point;
6. A fermionic field $\bar{\psi}(x_i)$ is represented by a solid line attached to the point x_i with an arrow from the point;

16.5.2 Example: second order processes in QED

The second order perturbation term in the S-matrix is

$$S^{(2)} = \frac{(ie)^2}{2!} \int d^4x\, d^4x'\, T\bar{\psi}(x)\,\gamma^\mu\,\psi(x)\,A_\mu(x)\,\bar{\psi}(x')\,\gamma^\nu\,\psi(x')\,A_\nu(x').$$

Scattering of fermions

The Wick's expansion of the integrand gives (among others) the following term
$N\bar{\psi}(x)\gamma^\mu\psi(x)\bar{\psi}(x')\gamma^\nu\psi(x')\underline{A_\mu(x)A_\nu(x')}$,
where
$\underline{A_\mu(x)A_\nu(x')} = \int \frac{d^4k}{(2\pi)^4}\frac{-ig_{\mu\nu}}{k^2+i0}e^{-ik(x-x')}$
is the electromagnetic contraction (propagator) in the Feynman gauge. This term is represented by the Feynman diagram at the right. This diagram gives contributions to the following processes:

1. e^-e^- scattering (initial state at the right, final state at the left of the diagram);
2. e^+e^+ scattering (initial state at the left, final state at the right of the diagram);
3. e^-e^+ scattering (initial state at the bottom/top, final state at the top/bottom of the diagram).

Compton scattering and annihilation/generation of e^-e^+ pairs

Another interesting term in the expansion is

$$N\bar{\psi}(x)\,\gamma^\mu\,\underline{\psi(x)\,\bar{\psi}(x')}\,\gamma^\nu\,\psi(x')\,A_\mu(x)\,A_\nu(x')\,,$$

where

$$\underline{\psi(x)\bar{\psi}(x')} = \int \frac{d^4p}{(2\pi)^4}\frac{i}{\gamma p - m + i0}e^{-ip(x-x')}$$

is the fermionic contraction (propagator).

16.6 Path integral formulation

In a path-integral, the field Lagrangian, integrated over all possible field histories, defines the probability amplitude to go from one field configuration to another. In order to make sense, the field theory should have a well-defined ground state, and the integral should be performed a little bit rotated into imaginary time, i.e. a Wick Rotation.

16.6.1 Scalar field Lagrangian

A simple example is the free relativistic scalar field in d-dimensions, whose action integral is:

$$S = \int \frac{1}{2}\partial_\mu \phi \partial^\mu \phi d^d x \,.$$

The probability amplitude for a process is:

$$\int_A^B e^{iS} D\phi \,,$$

where A and B are space-like hypersurfaces that define the boundary conditions. The collection of all the $\phi(A)$ on the starting hypersurface give the initial value of the field, analogous to the starting position for a point particle, and the field values $\phi(B)$ at each point of the final hypersurface defines the final field value, which is allowed to vary, giving a different amplitude to end up at different values. This is the field-to-field transition amplitude.

The path integral gives the expectation value of operators between the initial and final state:

$$\int_A^B e^{iS} \phi(x_1)...\phi(x_n) D\phi = \langle A|\phi(x_1)...\phi(x_n)|B\rangle \,,$$

and in the limit that A and B recede to the infinite past and the infinite future, the only contribution that matters is from the ground state (this is only rigorously true if the path-integral is defined slightly rotated into imaginary time). The path integral should be thought of as analogous to a probability distribution, and it is convenient to define it so that multiplying by a constant doesn't change anything:

$$\frac{\int e^{iS}\phi(x_1)...\phi(x_n)D\phi}{\int e^{iS} D\phi} = \langle 0|\phi(x_1)....\phi(x_n)|0\rangle \,.$$

The normalization factor on the bottom is called the *partition function* for the field, and it coincides with the statistical mechanical partition function at zero temperature when rotated into imaginary time.

The initial-to-final amplitudes are ill-defined if one thinks of the continuum limit right from the beginning, because the fluctuations in the field can become unbounded. So the path-integral should be thought of as on a discrete square lattice, with lattice spacing a and the limit $a \to 0$ should be taken carefully. If the final results do not depend on the shape of the lattice or the value of a, then the continuum limit exists.

16.6.2 On a lattice

On a lattice, (i), the field can be expanded in Fourier modes:

$$\phi(x) = \int \frac{dk}{(2\pi)^d} \phi(k) e^{ik\cdot x} = \int_k \phi(k) e^{ikx} \,.$$

Here the integration domain is over k restricted to a cube of side length $2\pi/a$, so that large values of k are not allowed. It is important to note that the k-measure contains the factors of 2π from Fourier transforms, this is the best standard convention for k-integrals in QFT. The lattice means that fluctuations at large k are not allowed to contribute right away, they only start to contribute in the limit $a \to 0$. Sometimes, instead of a lattice, the field modes are just cut off at high values of k instead.

It is also convenient from time to time to consider the space-time volume to be finite, so that the k modes are also a lattice. This is not strictly as necessary as the space-lattice limit, because interactions in k are not localized, but it is convenient for keeping track of the factors in front of the k-integrals and the momentum-conserving delta functions that will arise.

On a lattice, (ii), the action needs to be discretized:

$$S = \sum_{<x,y>} \frac{1}{2}(\phi(x) - \phi(y))^2 ,$$

where $< x, y >$ is a pair of nearest lattice neighbors x and y . The discretization should be thought of as defining what the derivative $\partial_\mu \phi$ means.

In terms of the lattice Fourier modes, the action can be written:

$$S = \int_k ((1 - \cos(k_1)) + (1 - \cos(k_2)) + ... + (1 - \cos(k_d)))\phi_k^* \phi^k .$$

For k near zero this is:

$$S = \int_k \frac{1}{2} k^2 |\phi(k)|^2 .$$

Now we have the continuum Fourier transform of the original action. In finite volume, the quantity $d^d k$ is not infinitesimal, but becomes the volume of a box made by neighboring Fourier modes, or $(2\pi/V)^d$.

The field ϕ is real-valued, so the Fourier transform obeys:

$$\phi(k)^* = \phi(-k) .$$

In terms of real and imaginary parts, the real part of $\phi(k)$ is an even function of k, while the imaginary part is odd. The Fourier transform avoids double-counting, so that it can be written:

$$S = \int_k \frac{1}{2} k^2 \phi(k)\phi(-k)$$

over an integration domain that integrates over each pair (k,–k) exactly once.

For a complex scalar field with action

$$S = \int \frac{1}{2} \partial_\mu \phi^* \partial^\mu \phi d^d x$$

the Fourier transform is unconstrained:

$$S = \int_k \frac{1}{2} k^2 |\phi(k)|^2$$

and the integral is over all k.

Integrating over all different values of $\phi(x)$ is equivalent to integrating over all Fourier modes, because taking a Fourier transform is a unitary linear transformation of field coordinates. When you change coordinates in a multidimensional integral by a linear transformation, the value of the new integral is given by the determinant of the transformation matrix. If

$$y_i = A_{ij} x_j \, ,$$

then

$$\det(A) \int dx_1 dx_2 ... dx_n = \int dy_1 dy_2 ... dy_n \, .$$

If A is a rotation, then

$$A^T A = I$$

so that det $A = \pm 1$, and the sign depends on whether the rotation includes a reflection or not.

The matrix that changes coordinates from $\phi(x)$ to $\phi(k)$ can be read off from the definition of a Fourier transform.

$$A_{kx} = e^{ikx}$$

and the Fourier inversion theorem tells you the inverse:

$$A_{kx}^{-1} = e^{-ikx}$$

which is the complex conjugate-transpose, up to factors of 2π . On a finite volume lattice, the determinant is nonzero and independent of the field values.

$$\det A = 1$$

and the path integral is a separate factor at each value of k.

$$\int \exp\left(\frac{i}{2} \sum_k k^2 \phi^*(k) \phi(k)\right) D\phi = \prod_k \int_{\phi_k} e^{\frac{i}{2} k^2 |\phi_k|^2 d^d k}$$

The factor $d^d k$ is the infinitesimal volume of a discrete cell in k-space, in a square lattice box $d^d k = 1/L^d$, where L is the side-length of the box. Each separate factor is an oscillatory Gaussian, and the width of the Gaussian diverges as the volume goes to infinity.

In imaginary time, the *Euclidean action* becomes positive definite, and can be interpreted as a probability distribution. The probability of a field having values ϕ_k is

$$e^{\int_k -\frac{1}{2}k^2\phi_k^*\phi_k} = \prod_k e^{-k^2|\phi_k|^2 d^d k}$$

The expectation value of the field is the statistical expectation value of the field when chosen according to the probability distribution:

$$\langle \phi(x_1)...\phi(x_n) \rangle = \frac{\int e^{-S}\phi(x_1)...\phi(x_n) D\phi}{\int e^{-S} D\phi}$$

Since the probability of ϕ_k is a product, the value of $\phi(k)$ at each separate value of k is independently Gaussian distributed. The variance of the Gaussian is 1/ (k^2 d^dk), which is formally infinite, but that just means that the fluctuations are unbounded in infinite volume. In any finite volume, the integral is replaced by a discrete sum, and the variance of the integral is V/k^2 .

16.6.3 Monte Carlo

The path integral defines a probabilistic algorithm to generate a Euclidean scalar field configuration. Randomly pick the real and imaginary parts of each Fourier mode at wavenumber k to be a gaussian random variable with variance $1/k^2$. This generates a configuration $\phi_C(k)$ at random, and the Fourier transform gives $\phi_C(x)$. For real scalar fields, the algorithm must generate only one of each pair $\phi(k), \phi(-k)$, and make the second the complex conjugate of the first.

To find any correlation function, generate a field again and again by this procedure, and find the statistical average:

$$\langle \phi(x_1)...\phi(x_n) \rangle = \lim_{|C|\to\infty} \frac{\sum_C \phi_C(x_1)...\phi_C(x_n)}{|C|}$$

where $|C|$ is the number of configurations, and the sum is of the product of the field values on each configuration. The Euclidean correlation function is just the same as the correlation function in statistics or statistical mechanics. The quantum mechanical correlation functions are an analytic continuation of the Euclidean correlation functions.

For free fields with a quadratic action, the probability distribution is a high-dimensional Gaussian, and the statistical average is given by an explicit formula. But the Monte Carlo method also works well for bosonic interacting field theories where there is no closed form for the correlation functions.

16.6.4 Scalar propagator

Each mode is independently Gaussian distributed. The expectation of field modes is easy to calculate:

$$\langle \phi_k \phi_{k'} \rangle = 0$$

for $k \neq k'$, since then the two Gaussian random variables are independent and both have zero mean.

$$\langle \phi_k \phi_k \rangle = \frac{V}{k^2}$$

in finite volume V, when the two k-values coincide, since this is the variance of the Gaussian. In the infinite volume limit,

$$\langle\phi(k)\phi(k')\rangle = \delta(k-k')\frac{1}{k^2}$$

Strictly speaking, this is an approximation: the lattice propagator is:

$$\langle\phi(k)\phi(k')\rangle = \delta(k-k')\frac{1}{2(d-\cos(k_1)+\cos(k_2)...+\cos(k_d))}$$

But near k=0, for field fluctuations long compared to the lattice spacing, the two forms coincide.

It is important to emphasize that the delta functions contain factors of 2π , so that they cancel out the 2π factors in the measure for k integrals.

$$\delta(k) = (2\pi)^d\delta_D(k_1)\delta_D(k_2)...\delta_D(k_d)$$

where $\delta_D(k)$ is the ordinary one-dimensional Dirac delta function. This convention for delta-functions is not universal—some authors keep the factors of 2π in the delta functions (and in the k-integration) explicit.

16.6.5 Equation of motion

The form of the propagator can be more easily found by using the equation of motion for the field. From the Lagrangian, the equation of motion is:

$$\partial_\mu\partial^\mu\phi = 0$$

and in an expectation value, this says:

$$\partial_\mu\partial^\mu\langle\phi(x)\phi(y)\rangle = 0$$

Where the derivatives act on x, and the identity is true everywhere except when x and y coincide, and the operator order matters. The form of the singularity can be understood from the canonical commutation relations to be a delta-function. Defining the (euclidean) *Feynman propagator* Δ as the Fourier transform of the time-ordered two-point function (the one that comes from the path-integral):

$$\partial^2\Delta(x) = i\delta(x)$$

So that:

$$\Delta(k) = \frac{i}{k^2}$$

If the equations of motion are linear, the propagator will always be the reciprocal of the quadratic-form matrix that defines the free Lagrangian, since this gives the equations of motion. This is also easy to see directly from the Path integral. The factor of i disappears in the Euclidean theory.

Wick theorem

Main article: Wick's Theorem

Because each field mode is an independent Gaussian, the expectation values for the product of many field modes obeys *Wick's theorem*:

$$\langle \phi(k_1)\phi(k_2)...\phi(k_n)\rangle$$

is zero unless the field modes coincide in pairs. This means that it is zero for an odd number of ϕ 's, and for an even number of ϕ 's, it is equal to a contribution from each pair separately, with a delta function.

$$\langle \phi(k_1)...\phi(k_{2n})\rangle = \sum \prod_{i,j} \frac{\delta(k_i - k_j)}{k_i^2}$$

where the sum is over each partition of the field modes into pairs, and the product is over the pairs. For example,

$$\langle \phi(k_1)\phi(k_2)\phi(k_3)\phi(k_4)\rangle = \frac{\delta(k_1 - k_2)}{k_1^2}\frac{\delta(k_3 - k_4)}{k_3^2} + \frac{\delta(k_1 - k_3)}{k_3^2}\frac{\delta(k_2 - k_4)}{k_2^2} + \frac{\delta(k_1 - k_4)}{k_1^2}\frac{\delta(k_2 - k_3)}{k_2^2}$$

An interpretation of Wick's theorem is that each field insertion can be thought of as a dangling line, and the expectation value is calculated by linking up the lines in pairs, putting a delta function factor that ensures that the momentum of each partner in the pair is equal, and dividing by the propagator.

Higher Gaussian moments—completing Wick's theorem

There is a subtle point left before Wick's theorem is proved—what if more than two of the phis have the same momentum? If it's an odd number, the integral is zero; negative values cancel with the positive values. But if the number is even, the integral is positive. The previous demonstration assumed that the phis would only match up in pairs.

But the theorem is correct even when arbitrarily many of the phis are equal, and this is a notable property of Gaussian integration:

$$I = \int e^{-ax^2/2}dx = \sqrt{\frac{2\pi}{a}}$$

$$\frac{\partial^n}{\partial a^n} I = \int \frac{x^{2n}}{2^n} e^{-ax^2/2}dx = \frac{1 \cdot 3 \cdot 5 ... \cdot (2n-1)}{2 \cdot 2 \cdot 2 ... \quad \cdot 2} \sqrt{2\pi} a^{-\frac{2n+1}{2}}$$

Dividing by I,

$$\langle x^{2n}\rangle = \frac{\int x^{2n} e^{-ax^2/2}}{\int e^{-ax^2/2}} = 1 \cdot 3 \cdot 5 ... \cdot (2n-1)\frac{1}{a^n}$$

$$\langle x^2\rangle = \frac{1}{a}$$

If Wick's theorem were correct, the higher moments would be given by all possible pairings of a list of 2n x's:

$$\langle x_1 x_2 x_3 ... x_{2n} \rangle$$

where the x's are all the same variable, the index is just to keep track of the number of ways to pair them. The first x can be paired with 2n–1 others, leaving 2n–2. The next unpaired x can be paired with 2n-3 different x's leaving 2n–4, and so on. This means that Wick's theorem, uncorrected, says that the expectation value of x^{2n} should be:

$$\langle x^{2n} \rangle = (2n-1) \cdot (2n-3) .... \cdot 5 \cdot 3 \cdot 1 (\langle x^2 \rangle)^n$$

and this is in fact the correct answer. So Wick's theorem holds no matter how many of the momenta of the internal variables coincide.

Interaction

Interactions are represented by higher order contributions, since quadratic contributions are always Gaussian. The simplest interaction is the quartic self-interaction, with an action:

$$S = \int \partial^\mu \phi \partial_\mu \phi + \frac{\lambda}{4!} \phi^4 .$$

The reason for the combinatorial factor 4! will be clear soon. Writing the action in terms of the lattice (or continuum) Fourier modes:

$$S = \int_k k^2 |\phi(k)|^2 + \int_{k_1 k_2 k_3 k_4} \phi(k_1)\phi(k_2)\phi(k_3)\phi(k_4)\delta(k_1 + k_2 + k_3 + k_4) = S_F + X .$$

Where S_F is the free action, whose correlation functions are given by Wick's theorem. The exponential of S in the path integral can be expanded in powers of λ , giving a series of corrections to the free action.

$$e^{-S} = e^{-S_F}(1 + X + \frac{1}{2!} XX + \frac{1}{3!} XXX + ...)$$

The path integral for the interacting action is then a power series of corrections to the free action. The term represented by X should be thought of as four half-lines, one for each factor of $\phi(k)$. The half-lines meet at a vertex, which contributes a delta-function that ensures that the sum of the momenta are all equal.

To compute a correlation function in the interacting theory, there is a contribution from the X terms now. For example, the path-integral for the four-field correlator:

$$\langle \phi(k_1)\phi(k_2)\phi(k_3)\phi(k_4) \rangle = \frac{\int e^{-S} \phi(k_1)\phi(k_2)\phi(k_3)\phi(k_4) D\phi}{Z}$$

which in the free field was only nonzero when the momenta k were equal in pairs, is now nonzero for all values of the k. The momenta of the insertions $\phi(k_i)$ can now match up with the momenta of the X's in the expansion. The insertions should also be thought of as half-lines, four in this case, which carry a momentum k, but one that is not integrated.

The lowest order contribution comes from the first nontrivial term $e^{-S_F} X$ in the Taylor expansion of the action. Wick's theorem requires that the momenta in the X half-lines, the $\phi(k)$ factors in X, should match up with the momenta of the external half-lines in pairs. The new contribution is equal to:

$$\lambda \frac{1}{k_1^2} \frac{1}{k_2^2} \frac{1}{k_3^2} \frac{1}{k_4^2}.$$

The 4! inside X is canceled because there are exactly 4! ways to match the half-lines in X to the external half-lines. Each of these different ways of matching the half-lines together in pairs contributes exactly once, regardless of the values of the k's, by Wick's theorem.

Feynman diagrams

The expansion of the action in powers of X gives a series of terms with progressively higher number of X's. The contribution from the term with exactly n X's are called n-th order.

The n-th order terms has:

1. 4n internal half-lines, which are the factors of $\phi(k)$ from the X's. These all end on a vertex, and are integrated over all possible k.
2. external half-lines, which are the come from the $\phi(k)$ insertions in the integral.

By Wick's theorem, each pair of half-lines must be paired together to make a *line*, and this line gives a factor of

$$\frac{\delta(k_1 + k_2)}{k_1^2}$$

which multiplies the contribution. This means that the two half-lines that make a line are forced to have equal and opposite momentum. The line itself should be labelled by an arrow, drawn parallel to the line, and labeled by the momentum in the line k. The half-line at the tail end of the arrow carries momentum k, while the half-line at the head-end carries momentum –k. If one of the two half-lines is external, this kills the integral over the internal k, since it forces the internal k to be equal to the external k. If both are internal, the integral over k remains.

The diagrams that are formed by linking the half-lines in the X's with the external half-lines, representing insertions, are the Feynman diagrams of this theory. Each line carries a factor of $\frac{1}{k^2}$, the propagator, and either goes from vertex to vertex, or ends at an insertion. If it is internal, it is integrated over. At each vertex, the total incoming k is equal to the total outgoing k.

The number of ways of making a diagram by joining half-lines into lines almost completely cancels the factorial factors coming from the Taylor series of the exponential and the 4! at each vertex.

Loop order

A forest diagram is one where all the internal lines have momentum that is completely determined by the external lines and the condition that the incoming and outgoing momentum are equal at each vertex. The contribution of these diagrams is a product of propagators, without any integration. A tree diagram is a connected forest diagram.

An example of a tree diagram is the one where each of four external lines end on an X. Another is when three external lines end on an X, and the remaining half-line joins up with another X, and the remaining half-lines of this X run off to external lines. These are all also forest diagrams (as every tree is a forest); an example of a forest that is not a tree is when eight external lines end on two X's.

It is easy to verify that in all these cases, the momenta on all the internal lines is determined by the external momenta and the condition of momentum conservation in each vertex.

A diagram that is not a forest diagram is called a *loop* diagram, and an example is one where two lines of an X are joined to external lines, while the remaining two lines are joined to each other. The two lines joined to each other can have any momentum at all, since they both enter and leave the same vertex. A more complicated example is one where two X's are joined to each other by matching the legs one to the other. This diagram has no external lines at all.

The reason loop diagrams are called loop diagrams is because the number of k-integrals that are left undetermined by momentum conservation is equal to the number of independent closed loops in the diagram, where independent loops are counted as in homology theory. The homology is real-valued (actually R^d valued), the value associated with each line is the momentum. The boundary operator takes each line to the sum of the end-vertices with a positive sign at the head and a negative sign at the tail. The condition that the momentum is conserved is exactly the condition that the boundary of the k-valued weighted graph is zero.

A set of k-values can be relabeled whenever there is a closed loop going from vertex to vertex, never revisiting the same vertex. Such a cycle can be thought of as the boundary of a 2-cell. The k-labelings of a graph that conserves momentum (which has zero boundary) up to redefinitions of k (up to boundaries of 2-cells) define the first homology of a graph. The number of independent momenta that are not determined is then equal to the number of independent homology loops. For many graphs, this is equal to the number of loops as counted in the most intuitive way.

Symmetry factors

The number of ways to form a given Feynman diagram by joining together half-lines is large, and by Wick's theorem, each way of pairing up the half-lines contributes equally. Often, this completely cancels the factorials in the denominator of each term, but the cancellation is sometimes incomplete.

The uncancelled denominator is called the *symmetry factor* of the diagram. The contribution of each diagram to the correlation function must be divided by its symmetry factor.

For example, consider the Feynman diagram formed from two external lines joined to one X, and the remaining two half-lines in the X joined to each other. There are 4×3 ways to join the external half-lines to the X, and then there is only one way to join the two remaining lines to each other. The X comes divided by 4!=4×3×2, but the number of ways to link up the X half lines to make the diagram is only 4×3, so the contribution of this diagram is divided by two.

For another example, consider the diagram formed by joining all the half-lines of one X to all the half-lines of another X. This diagram is called a *vacuum bubble*, because it does not link up to any external lines. There are 4! ways to form this diagram, but the denominator includes a 2! (from the expansion of the exponential, there are two X's) and two factors of 4!. The contribution is multiplied by 4!/(2×4!×4!) = 1/48.

Another example is the Feynman diagram formed from two X's where each X links up to two external lines, and the remaining two half-lines of each X are joined to each other. The number of ways to link an X to two external lines is 4×3, and either X could link up to either pair, giving an additional factor of 2. The remaining two half-lines in the two X's can be linked to each other in two ways, so that the total number of ways to form the diagram is 4×3×4×3×2×2, while the denominator is 4!×4!×2!. The total symmetry factor is 2, and the contribution of this diagram is divided by 2.

The symmetry factor theorem gives the symmetry factor for a general diagram: the contribution of each Feynman diagram must be divided by the order of its group of automorphisms, the number of symmetries that it has.

An automorphism of a Feynman graph is a permutation M of the lines and a permutation N of the vertices with the following properties:

1. If a line l goes from vertex v to vertex v', then M(l) goes from N(v) to N(v'). If the line is undirected, as it is for a real scalar field, then M(l) can go from N(v') to N(v) too.
2. If a line l ends on an external line, M(l) ends on the same external line.
3. If there are different types of lines, M(l) should preserve the type.

This theorem has an interpretation in terms of particle-paths: when identical particles are present, the integral over all intermediate particles must not double-count states that differ only by interchanging identical particles.

Proof: To prove this theorem, label all the internal and external lines of a diagram with a unique name. Then form the diagram by linking the a half-line to a name and then to the other half line.

Now count the number of ways to form the named diagram. Each permutation of the X's gives a different pattern of linking names to half-lines, and this is a factor of n!. Each permutation of the half-lines in a single X gives a factor of 4!. So a named diagram can be formed in exactly as many ways as the denominator of the Feynman expansion.

But the number of unnamed diagrams is smaller than the number of named diagram by the order of the automorphism group of the graph.

Connected diagrams: *linked-cluster theorem*

Roughly speaking, a Feynman diagram is called *connected* if all vertices and propagator lines are linked by a sequence of vertices and propagators of the diagram itself. If one views it as a (undirected) graph it is connected. The remarkable relevance of such diagrams in QFTs is due to the fact that they are sufficient to determine the quantum partition function $Z[J]$. More precisely, connected Feynman diagrams determine

$$iW[J] \equiv \ln Z[J].$$

To see this, one should recall that

$$Z[J] \propto \sum_k D_k$$

with D_k constructed from some (arbitrary) Feynman diagram that can be thought to consist of several connected components C_i . If one encounters n_i (identical) copies of a component C_i within the Feynman diagram D_k one has to include a *symmetry factor* $n_i!$. However, in the end each contribution of a Feynman diagram D_k to the partition function has the generic form

$$\prod_i \frac{C_i^{n_i}}{n_i!}$$

where i labels the (infinite) many connected Feynman diagrams possible.

A scheme to successively create such contributions from the D_k to $Z[J]$ is obtained by

$$\left(\frac{1}{0!} + \frac{C_1}{1!} + \frac{C_1^2}{2!} + \dots\right)\left(1 + C_2 + \frac{1}{2}C_2^2 + \dots\right)\dots$$

and therefore yields

$$Z[J] \propto \prod_i \sum_{n_i=0}^{\infty} \frac{C_i^{n_i}}{n_i!} = \exp \sum_i C_i \propto \exp W[J].$$

To establish the *normalization* $Z_0 = \exp W[0] = 1$ one simply calculates all connected *vacuum diagrams*, i.e., the diagrams without any *sources* J (sometimes referred to as *external legs* of a Feynman diagram).

Vacuum bubbles

An immediate consequence of the linked-cluster theorem is that all vacuum bubbles, diagrams without external lines, cancel when calculating correlation functions. A correlation function is given by a ratio of path-integrals:

$$\langle \phi_1(x_1)...\phi_n(x_n)\rangle = \frac{\int e^{-S}\phi_1(x_1)...\phi_n(x_n)D\phi}{\int e^{-S}D\phi}.$$

The top is the sum over all Feynman diagrams, including disconnected diagrams that do not link up to external lines at all. In terms of the connected diagrams, the numerator includes the same contributions of vacuum bubbles as the denominator:

$$\int e^{-S}\phi_1(x_1)...\phi_n(x_n)D\phi = (\sum E_i)(\exp(\sum_i C_i)).$$

Where the sum over E diagrams includes only those diagrams each of whose connected components end on at least one external line. The vacuum bubbles are the same whatever the external lines, and give an overall multiplicative factor. The denominator is the sum over all vacuum bubbles, and dividing gets rid of the second factor.

The vacuum bubbles then are only useful for determining Z itself, which from the definition of the path integral is equal to:

$$Z = \int e^{-S}D\phi = e^{-HT} = e^{-\rho V}$$

where ρ is the energy density in the vacuum. Each vacuum bubble contains a factor of $\delta(k)$ zeroing the total k at each vertex, and when there are no external lines, this contains a factor of $\delta(0)$, because the momentum conservation is over-enforced. In finite volume, this factor can be identified as the total volume of space time. Dividing by the volume, the remaining integral for the vacuum bubble has an interpretation: it is a contribution to the energy density of the vacuum.

Sources

Correlation functions are the sum of the connected Feynman diagrams, but the formalism treats the connected and disconnected diagrams differently. Internal lines end on vertices, while external lines go off to insertions. Introducing *sources* unifies the formalism, by making new vertices where one line can end.

Sources are external fields, fields that contribute to the action, but are not dynamical variables. A scalar field source is another scalar field h that contributes a term to the (Lorentz) Lagrangian:

$$\int h(x)\phi(x)d^dx = \int h(k)\phi(k)d^dk$$

In the Feynman expansion, this contributes H terms with one half-line ending on a vertex. Lines in a Feynman diagram can now end either on an X vertex, or on an H-vertex, and only one line enters an H vertex. The Feynman rule for an H-vertex is that a line from an H with momentum k gets a factor of h(k).

The sum of the connected diagrams in the presence of sources includes a term for each connected diagram in the absence of sources, except now the diagrams can end on the source. Traditionally, a source is represented by a little "x" with one line extending out, exactly as an insertion.

$$\log(Z[h]) = \sum_{n,C} h(k_1)h(k_2)...h(k_n)C(k_1,...,k_n)$$

where $C(k_1,....,k_n)$ is the connected diagram with n external lines carrying momentum as indicated. The sum is over all connected diagrams, as before.

The field h is not dynamical, which means that there is no path integral over h: h is just a parameter in the Lagrangian, which varies from point to point. The path integral for the field is:

$$Z[h] = \int e^{iS+i\int h\phi} D\phi$$

and it is a function of the values of h at every point. One way to interpret this expression is that it is taking the Fourier transform in field space. If there is a probability density on R^n, the Fourier transform of the probability density is:

$$\int \rho(y) e^{iky} d^n y = \langle e^{iky} \rangle = \langle \prod_{i=1}^{n} e^{ih_i y_i} \rangle$$

The Fourier transform is the expectation of an oscillatory exponential. The path integral in the presence of a source h(x) is:

$$Z[h] = \int e^{iS} e^{i\int_x h(x)\phi(x)} D\phi = \langle e^{ih\phi} \rangle$$

which, on a lattice, is the product of an oscillatory exponential for each field value:

$$\langle \prod_x e^{ih_x \phi_x} \rangle$$

The fourier transform of a delta-function is a constant, which gives a formal expression for a delta function:

$$\delta(x-y) = \int e^{ik(x-y)} dk$$

This tells you what a field delta function looks like in a path-integral. For two scalar fields ϕ and η ,

$$\delta(\phi - \eta) = \int e^{ih(x)(\phi(x)-\eta(x)d^d x} Dh$$

Which integrates over the Fourier transform coordinate, over h. This expression is useful for formally changing field coordinates in the path integral, much as a delta function is used to change coordinates in an ordinary multi-dimensional integral.

The partition function is now a function of the field h, and the physical partition function is the value when h is the zero function:

The correlation functions are derivatives of the path integral with respect to the source:

$$\langle \phi(x) \rangle = \frac{1}{Z} \frac{\partial}{\partial h(x)} Z[h] = \frac{\partial}{\partial h(x)} \log(Z[h]).$$

In Euclidean space, source contributions to the action can still appear with a factor of “i”, so that they still do a Fourier transform.

16.6.6 Spin 1/2; "photons" and "ghosts"

Spin 1/2: Grassmann integrals

The field path-integral can be extended to the Fermi case, but only if the notion of integration is expanded. A Grassmann integral of a free Fermi field is a high-dimensional determinant or Pfaffian, which defines the new type of Gaussian integration appropriate for Fermi fields.

The two fundamental formulas of Grassmann integration are:

$$\int e^{M_{ij}\bar{\psi}^i\psi^j} D\bar{\psi}D\psi = \text{Det}(M)$$

where M is an arbitrary matrix and $\psi,\bar{\psi}$ are independent Grassmann variables for each index i, and

$$\int e^{\frac{1}{2}A_{ij}\psi^i\psi^j} D\psi = \text{Pfaff}(A)$$

Where A is an antisymmetric matrix, ψ is a collection of Grassmann variables, and the 1/2 is to prevent double-counting (since $\psi^i\psi^j = -\psi^j\psi^i$). In matrix notation, where $\bar{\psi}$ and $\bar{\eta}$ are Grassmann valued row vectors, η and ψ are Grassmann valued column vectors, and M is a real valued matrix:

$$Z = \int e^{\bar{\psi}M\psi + \bar{\eta}\psi + \bar{\psi}\eta} D\bar{\psi}D\psi = \int e^{(\bar{\psi}+\bar{\eta}M^{-1})M(\psi+M^{-1}\eta)-\bar{\eta}M^{-1}\eta} D\bar{\psi}D\psi = \text{Det}(M)e^{-\bar{\eta}M^{-1}\eta}$$

Where the last equality is a consequence of the translation invariance of the Grassmann integral. The Grassmann variables η are external sources for ψ , and differentiating with respect to η pulls down factors of $\bar{\psi}$.

$$\langle\bar{\psi}\psi\rangle = \frac{1}{Z}\frac{\partial}{\partial\eta}\frac{\partial}{\partial\bar{\eta}}Z|_{\eta=\bar{\eta}=0} = M^{-1}$$

again, in a schematic matrix notation. The meaning of the formula above is that the derivative with respect to the appropriate component of η and $\bar{\eta}$ gives the matrix element of M^{-1} . This is exactly analogous to the Bosonic path integration formula for a Gaussian integral of a complex Bosonic field:

$$\int e^{\phi^* M\phi + h^*\phi + \phi^* h} D\phi^* D\phi = \frac{e^{h^* M^{-1} h}}{\text{Det}(M)}$$

$$\langle\phi^*\phi\rangle = \frac{1}{Z}\frac{\partial}{\partial h}\frac{\partial}{\partial h^*}Z|_{h=h^*=0} = M^{-1}$$

So that the propagator is the inverse of the matrix in the quadratic part of the action in both the Bose and Fermi case.

For real Grassmann fields, for Majorana fermions, the path integral a Pfaffian times a source quadratic form, and the formulas give the square root of the determinant, just as they do for real Bosonic fields. The propagator is still the inverse of the quadratic part.

The free Dirac Lagrangian:

$$\int \bar{\psi}(\gamma^\mu \partial_\mu - m)\psi$$

formally gives the equations of motion and the anticommutation relations of the Dirac field, just as the Klein Gordon Lagrangian in an ordinary path integral gives the equations of motion and commutation relations of the scalar field. By using the spatial Fourier transform of the Dirac field as a new basis for the Grassmann algebra, the quadratic part of the Dirac action becomes simple to invert:

$$S = \int_k \bar{\psi}(i\gamma^\mu k_\mu - m)\psi.$$

The propagator is the inverse of the matrix M linking $\psi(k)$ and $\bar{\psi}(k)$, since different values of k do not mix together.

$$\langle \bar{\psi}(k')\psi(k)\rangle = \delta(k+k')\frac{1}{\gamma\cdot k - m} = \delta(k+k')\frac{\gamma\cdot k + m}{k^2 - m^2}$$

The analog of Wick's theorem matches psi and psi-bars in pairs:

$$\langle \bar{\psi}(k_1)\bar{\psi}(k_2)...\bar{\psi}(k_n)\psi(k_1')...\psi(k_n)\rangle = \sum_{\text{pairings}} (-1)^S \prod_{\text{pairs } i,j} \delta(k_i - k_j)\frac{1}{\gamma\cdot k_i - m}$$

where S is the sign of the permutation that reorders the sequence of psi-bars and psis to put the ones that are paired up to make the delta-functions next to each other, with the psi-bar coming right before the psi. Since a psi-psi-bar pair is a commuting element of the Grassmann algebra, it doesn't matter what order the pairs are in. If more than one psi/psi-bar pair have the same k, the integral is zero, and it is easy to check that the sum over pairings gives zero in this case (there are always an even number of them). This is the Grassmann analog of the higher Gaussian moments that completed the Bosonic Wick's theorem earlier.

The rules for spin-1/2 Dirac particles are as follows: The propagator is the inverse of the Dirac operator, the lines have arrows just as for a complex scalar field, and the diagram acquires an overall factor of −1 for each closed Fermi loop. If there are an odd number of Fermi loops, the diagram changes sign. Historically, the −1 rule was very difficult for Feynman to discover. He discovered it after a long process of trial and error, since he lacked a proper theory of Grassmann integration.

The rule follows from the observation that the number of Fermi lines at a vertex is always even. Each term in the Lagrangian must always be Bosonic. A Fermi loop is counted by following Fermionic lines until one comes back to the starting point, then removing those lines from the diagram. Repeating this process eventually erases all the Fermionic lines: this is the Euler algorithm to 2-color a graph, which works whenever each vertex has even degree. Note that the number of steps in the Euler algorithm is only equal to the number of independent Fermionic homology cycles in the common special case that all terms in the Lagrangian are exactly quadratic in the Fermi fields, so that each vertex has exactly two Fermionic lines. When there are four-Fermi interactions (like in the Fermi effective theory of the Weak interactions) there are more k-integrals than Fermi loops. In this case, the counting rule should apply the Euler algorithm by pairing up the Fermi lines at each vertex into pairs that together form a bosonic factor of the term in the Lagrangian, and when entering a vertex by one line, the algorithm should always leave with the partner line.

To clarify and prove the rule, consider a Feynman diagram formed from vertices, terms in the Lagrangian, with Fermion fields. The full term is Bosonic, it is a commuting element of the Grassmann algebra, so the order in which the vertices appear is not important. The Fermi lines are linked into loops, and when traversing the loop, one can reorder the vertex terms one after the other as one goes around without any sign cost. The exception is when you return to the starting point, and the final half-line must be joined with the unlinked first half-line. This requires one permutation to move the last psi-bar to go in front of the first psi, and this gives the sign.

This rule is the only visible effect of the exclusion principle in internal lines. When there are external lines, the amplitudes are antisymmetric when two Fermi insertions for identical particles are interchanged. This is automatic in the source formalism, because the sources for Fermi fields are themselves Grassmann valued.

Spin 1: photons

The naive propagator for photons is infinite, since the Lagrangian for the A-field is:

$$S = \int \frac{1}{4} F^{\mu\nu} F_{\mu\nu} = \int -\frac{1}{2}(\partial^{\mu} A_{\nu} \partial_{\mu} A^{\nu} - \partial^{\mu} A_{\mu} \partial_{\nu} A^{\nu}).$$

The quadratic form defining the propagator is non-invertible. The reason is the gauge invariance of the field, adding a gradient to A does not change the physics.

To fix this problem, one needs to fix a gauge. The most convenient way is to demand that the divergence of A is some function f, whose value is random from point to point. It does no harm to integrate over the values of f, since it only determines the choice of gauge. This procedure inserts the following factor into the path integral for A:

$$\int \delta(\partial_{\mu} A^{\mu} - f) e^{-\frac{f^2}{2}} Df.$$

The first factor, the delta function, fixes the gauge. The second factor sums over different values of f that are inequivalent gauge fixings. This is simply

$$e^{-\frac{(\partial_{\mu} A_{\mu})^2}{2}}.$$

The additional contribution from gauge-fixing cancels the second half of the free Lagrangian, giving the Feynman Lagrangian:

$$S = \int \partial^{\mu} A^{\nu} \partial_{\mu} A_{\nu}$$

which is just like four independent free scalar fields, one for each component of A. The Feynman propagator is:

$$\langle A_{\mu}(k) A_{\nu}(k') \rangle = \delta(k + k') \frac{g_{\mu\nu}}{k^2}.$$

The one difference is that the sign of one propagator is wrong in the Lorentz case: the timelike component has an opposite sign propagator. This means that these particle states have negative norm—they are not physical states. In the case of photons, it is easy to show by diagram methods that these states are not physical—their contribution cancels with longitudinal photons to only leave two physical photon polarization contributions for any value of k.

If the averaging over f is done with a coefficient different from 1/2, the two terms don't cancel completely. This gives a covariant Lagrangian with a coefficient λ , which does not affect anything:

$$S = \int \frac{1}{2}(\partial^{\mu} A^{\nu} \partial_{\mu} A_{\nu} - \lambda (\partial_{\mu} A^{\mu})^2)$$

and the covariant propagator for QED is:

$$\langle A_{\mu}(k) A_{\nu}(k') \rangle = \delta(k + k') \frac{g_{\mu\nu} - \lambda \frac{k_{\mu} k_{\nu}}{k^2}}{k^2}.$$

Spin 1: nonabelian ghosts

To find the Feynman rules for nonabelian Gauge fields, the procedure that performs the Gauge fixing must be carefully corrected to account for a change of variables in the path-integral.

The gauge fixing factor has an extra determinant from popping the delta function:

$$\delta(\partial_\mu A_\mu - f)e^{-\frac{f^2}{2}}\mathrm{Det}M$$

To find the form of the determinant, consider first a simple two-dimensional integral of a function f that depends only on r, not on the angle θ . Inserting an integral over theta:

$$\int f(r)dxdy = \int f(r)\int d\theta\delta(y)|\frac{dy}{d\theta}|dxdy$$

The derivative-factor ensures that popping the delta function in θ removes the integral. Exchanging the order of integration,

$$\int f(r)dxdy = \int d\theta\int f(r)\delta(y)|\frac{dy}{d\theta}|dxdy$$

but now the delta-function can be popped in y,

$$\int f(r)dxdy = \int d\theta_0\int f(x)|\frac{dy}{d\theta}|dx\,.$$

The integral over θ just gives an overall factor of 2π , while the rate of change of y with a change in θ is just x, so this exercise reproduces the standard formula for polar integration of a radial function:

$$\int f(r)dxdy = 2\pi\int f(x)xdx$$

In the path-integral for a nonabelian gauge field, the analogous manipulation is:

$$\int DA\int \delta(F(A))\mathrm{Det}(\frac{\partial F}{\partial G})DGe^{iS} = \int DG\int \delta(F(A))\mathrm{Det}(\frac{\partial F}{\partial G})e^{iS}$$

The factor in front is the volume of the gauge group, and it contributes a constant, which can be discarded. The remaining integral is over the gauge fixed action.

$$\int \mathrm{Det}(\frac{\partial F}{\partial G})e^{iS_{GF}}DA$$

To get a covariant gauge, the gauge fixing condition is the same as in the Abelian case:

$$\partial_\mu A^\mu = f$$

Whose variation under an infinitesimal gauge transformation is given by:

$$\partial_\mu D_\mu \alpha$$

where α is the adjoint valued element of the Lie algebra at every point that performs the infinitesimal gauge transformation. This adds the Faddeev Popov determinant to the action:

$$Det(\partial_\mu D_\mu)$$

which can be rewritten as a Grassman integral by introducing ghost fields:

$$\int e^{\bar{\eta}\partial_\mu D^\mu \eta} D\bar{\eta} D\eta$$

The determinant is independent of f, so the path-integral over f can give the Feynman propagator (or a covariant propagator) by choosing the measure for f as in the abelian case. The full gauge fixed action is then the Yang Mills action in Feynman gauge with an additional ghost action:

$$S = \int Tr\partial_\mu A_\nu \partial^\mu A^\nu + f^i_{jk} \partial^\nu A^\mu_i A^j_\mu A^k_\nu + f^i_{jr} f^r_{kl} A_i A_j A^k A^l + Tr\partial_\mu \bar{\eta} \partial^\mu \eta + \bar{\eta} A_j \eta$$

The diagrams are derived from this action. The propagator for the spin-1 fields has the usual Feynman form. There are vertices of degree 3 with momentum factors whose couplings are the structure constants, and vertices of degree 4 whose couplings are products of structure constants. There are additional ghost loops, which cancel out timelike and longitudinal states in A loops.

In the Abelian case, the determinant for covariant gauges does not depend on A, so the ghosts do not contribute to the connected diagrams.

16.7 Particle-path representation

Feynman diagrams were originally discovered by Feynman, by trial and error, as a way to represent the contribution to the S-matrix from different classes of particle trajectories.

16.7.1 Schwinger representation

The Euclidean scalar propagator has a suggestive representation:

$$\frac{1}{p^2 + m^2} = \int_0^\infty e^{-\tau(p^2+m^2)} d\tau$$

The meaning of this identity (which is an elementary integration) is made clearer by Fourier transforming to real space.

$$\Delta(x) = \int_0^\infty d\tau e^{-m^2\tau} \frac{1}{(4\pi\tau)^{d/2}} e^{\frac{-x^2}{4\tau}}$$

The contribution at any one value of τ to the propagator is a Gaussian of width $\sqrt{\tau}$. The total propagation function from 0 to x is a weighted sum over all proper times τ of a normalized Gaussian, the probability of ending up at x after a random walk of time τ .

The path-integral representation for the propagator is then:

$$\Delta(x) = \int_0^\infty d\tau \int DX e^{-\int_0^\tau (\dot{x}^2/2+m^2)d\tau'}$$

which is a path-integral rewrite of the Schwinger representation.

The Schwinger representation is both useful for making manifest the particle aspect of the propagator, and for symmetrizing denominators of loop diagrams.

16.7.2 Combining denominators

The Schwinger representation has an immediate practical application to loop diagrams. For example, For the diagram in the phi-4 theory formed by joining two x's together in two half-lines, and making the remaining lines external, the integral over the internal propagators in the loop is:

$$\int_k \frac{1}{(k^2+m^2)} \frac{1}{((k+p)^2+m^2)} .$$

Here one line carries momentum k and the other k+p. The asymmetry can be fixed by putting everything in the Schwinger representation.

$$\int_{t,t'} e^{-t(k^2+m^2)-t'((k+p)^2+m^2)} dt dt' .$$

Now the exponent mostly depends on t+t',

$$\int_{t,t'} e^{-(t+t')(k^2+m^2)-t'2p\cdot k-t'p^2} ,$$

except for the asymmetrical little bit. Defining the variable u=(t+t') and v = t'/u, the variable u goes from 0 to infinity, while v goes from 0 to 1. The variable u is the total proper time for the loop, while v parametrizes the fraction of the proper time on the top of the loop vs. the bottom.

The Jacobian for this transformation of variables is easy to work out from the identities:

$$d(uv) = dt' \quad du = dt + dt' ,$$

and "wedging" gives

$$udu \wedge dv = dt \wedge dt'$$

This allows the u integral to be evaluated explicitly:

$$\int_{u,v} ue^{-u(k^2+m^2+v2p\cdot k+vp^2)} = \int \frac{1}{(k^2+m^2+v2p\cdot k-vp^2)^2} dv$$

leaving only the v -integral. This method, invented by Schwinger but usually attributed to Feynman, is called *combining denominator*. Abstractly, it is the elementary identity:

$$\frac{1}{AB} = \int_0^1 \frac{1}{(vA+(1-v)B)^2} dv$$

But this form does not provide the physical motivation for introducing v — v is the proportion of proper time on one of the legs of the loop.

Once the denominators are combined, a shift in k to $k' = k + vp$ symmetrizes everything:

$$\int_0^1 \int \frac{1}{(k^2+m^2+v2p\cdot k+vp^2)^2} dk dv = \int_0^1 \int \frac{1}{(k'^2+m^2+v(1-v)p^2)^2} dk' dv$$

This form shows that the moment that p^2 is more negative than 4 times the mass of the particle in the loop, which happens in a physical region of Lorentz space, the integral has a cut. This is exactly when the external momentum can create physical particles.

When the loop has more vertices, there are more denominators to combine:

$$\int dk \frac{1}{(k^2+m^2)} \frac{1}{((k+p_1)^2+m^2)} \cdots \frac{1}{((k+p_n)^2+m^2)}$$

The general rule follows from the Schwinger prescription for n+1 denominators:

$$\frac{1}{D_0 D_1 ... D_n} = \int_0^\infty \cdots \int_0^\infty e^{-u_0 D_0 ... -u_n D_n} du_0 ... du_n \,.$$

The integral over the Schwinger parameters u_i can be split up as before into an integral over the total proper time $u=u_0+u_1...+u_n$ and an integral over the fraction of the proper time in all but the first segment of the loop $v_i=u_i/u$ for $i\in\{1,2,...,n\}$. The v's are positive and add up to less than 1, so that the v integral is over an n dimensional simplex.

The Jacobian for the coordinate transformation can be worked out as before:

$$du = du_0 + du_1 ... + du_n$$

$$d(uv_i) = du_i \,.$$

"Wedging" all these equation together, one obtains

$$u^n du \wedge dv_1 \wedge dv_2 ... \wedge dv_n = du_0 \wedge du_1 ... \wedge du_n \,.$$

This gives the integral:

$$\int_0^\infty \int_{\text{simplex}} u^n e^{-u(v_0 D_0 + v_1 D_1 + v_2 D_2 ... + v_n D_n)} dv_1 ... dv_n du \,,$$

where the simplex is the region defined by the conditions $v_i > 0$ and $\sum_{i=1}^n v_i < 1$ as well as $v_0 = 1 - \sum_{i=1}^n v_i$. Performing the u integral gives the general prescription for combining denominators:

$$\frac{1}{D_0 ... D_n} = n! \int_{\text{simplex}} \frac{1}{(v_0 D_0 + v_1 D_1 ... + v_n D_n)^{n+1}} dv_1 dv_2 ... dv_n$$

Since the numerator of the integrand is not involved, the same prescription works for any loop, no matter what the spins are carried by the legs. The interpretation of the parameters v_i is that they are the fraction of the total proper time spent on each leg.

16.7.3 Scattering

The correlation functions of a quantum field theory describe the scattering of particles. The definition of "particle" in relativistic field theory is not self-evident, because if you try to determine the position so that the uncertainty is less than the compton wavelength, the uncertainty in energy is large enough to produce more particles and antiparticles of the same type from the vacuum. This means that the notion of a single-particle state is to some extent incompatible with the notion of an object localized in space.

In the 1930s, Wigner gave a mathematical definition for single-particle states: they are a collection of states that form an irreducible representation of the Poincaré group. Single particle states describe an object with a finite mass, a well defined momentum, and a spin. This definition is fine for protons and neutrons, electrons and photons, but it excludes quarks, which are permanently confined, so the modern point of view is more accommodating: a particle is anything whose interaction can be described in terms of Feynman diagrams, which have an interpretation as a sum over particle trajectories.

A field operator can act to produce a one-particle state from the vacuum, which means that the field operator $\phi(x)$ produces a superposition of Wigner particle states. In the free field theory, the field produces one particle states only. But when there are interactions, the field operator can also produce 3-particle, 5-particle (if there is no +/– symmetry also 2, 4, 6 particle) states too. To compute the scattering amplitude for single particle states only requires a careful limit, sending the fields to infinity and integrating over space to get rid of the higher-order corrections.

The relation between scattering and correlation functions is the LSZ-theorem: The scattering amplitude for n particles to go to m-particles in a scattering event is the given by the sum of the Feynman diagrams that go into the correlation function for n+m field insertions, leaving out the propagators for the external legs.

For example, for the $\lambda\phi^4$ interaction of the previous section, the order λ contribution to the (Lorentz) correlation function is:

$$\langle \phi(k_1)\phi(k_2)\phi(k_3)\phi(k_4) \rangle = \frac{i}{k_1^2} \frac{i}{k_2^2} \frac{i}{k_3^2} \frac{i}{k_4^2} i\lambda$$

Stripping off the external propagators, that is, removing the factors of i/k^2 , gives the invariant scattering amplitude M:

$$M = i\lambda$$

which is a constant, independent of the incoming and outgoing momentum. The interpretation of the scattering amplitude is that the sum of $|M|^2$ over all possible final states is the probability for the scattering event. The normalization of the single-particle states must be chosen carefully, however, to ensure that M is a relativistic invariant.

Non-relativistic single particle states are labeled by the momentum k, and they are chosen to have the same norm at every value of k. This is because the nonrelativistic unit operator on single particle states is:

$$\int dk|k\rangle\langle k|$$

In relativity, the integral over the k-states for a particle of mass m integrates over a hyperbola in E,k space defined by the energy–momentum relation:

$$E^2 - k^2 = m^2$$

If the integral weighs each k point equally, the measure is not Lorentz invariant. The invariant measure integrates over all values of k and E, restricting to the hyperbola with a Lorentz invariant delta function:

$$\int \delta(E^2 - k^2 - m^2)|E,k\rangle\langle E,k|dEdk = \int \frac{dk}{2E}|k\rangle\langle k|$$

So the normalized k-states are different from the relativistically normalized k-states by a factor of $\sqrt{E} = (k^2 - m^2)^{\frac{1}{4}}$

The invariant amplitude M is then the probability amplitude for relativistically normalized incoming states to become relativistically normalized outgoing states.

For nonrelativistic values of k, the relativistic normalization is the same as the nonrelativistic normalization (up to a constant factor $\sqrt{m}$). In this limit, the ϕ^4 invariant scattering amplitude is still constant. The particles created by the field phi scatter in all directions with equal amplitude.

The nonrelativistic potential, which scatters in all directions with an equal amplitude (in the Born approximation), is one whose Fourier transform is constant—a delta-function potential. The lowest order scattering of the theory reveals the non-relativistic interpretation of this theory—it describes a collection of particles with a delta-function repulsion. Two such particles have an aversion to occupying the same point at the same time.

16.8 Nonperturbative effects

Thinking of Feynman diagrams as a perturbation series, nonperturbative effects like tunneling do not show up, because any effect that goes to zero faster than any polynomial does not affect the Taylor series. Even bound states are absent, since at any finite order particles are only exchanged a finite number of times, and to make a bound state, the binding force must last forever.

But this point of view is misleading, because the diagrams not only describe scattering, but they also are a representation of the short-distance field theory correlations. They encode not only asymptotic processes like particle scattering, they also describe the multiplication rules for fields, the operator product expansion. Nonperturbative tunneling processes involve field configurations that on average get big when the coupling constant gets small, but each configuration is a coherent superposition of particles whose local interactions are described by Feynman diagrams. When the coupling is small, these become collective processes that involve large numbers of particles, but where the interactions between each of the particles is simple.

This means that nonperturbative effects show up asymptotically in resummations of infinite classes of diagrams, and these diagrams can be locally simple. The graphs determine the local equations of motion, while the allowed large-scale configurations describe non-perturbative physics. But because Feynman propagators are nonlocal in time, translating a field process to a coherent particle language is not completely intuitive, and has only been explicitly worked out in certain special cases. In the case of nonrelativistic bound states, the Bethe–Salpeter equation describes the class of diagrams to include to describe a relativistic atom. For quantum chromodynamics, the Shifman Vainshtein Zakharov sum rules describe non-perturbatively excited long-wavelength field modes in particle language, but only in a phenomenological way.

The number of Feynman diagrams at high orders of perturbation theory is very large, because there are as many diagrams as there are graphs with a given number of nodes. Nonperturbative effects leave a signature on the way in which the number of diagrams and resummations diverge at high order. It is only because non-perturbative effects appear in hidden form in diagrams that it was possible to analyze nonperturbative effects in string theory, where in many cases a Feynman description is the only one available.

16.9 In popular culture

- The use of the above diagram of the virtual particle producing a quark–antiquark pair was featured in the television sit-com *The Big Bang Theory*, in the episode *"The Bat Jar Conjecture"*.
- *PhD Comics* of January 11, 2012, shows Feynman diagrams that *visualize and describe quantum academic interactions*, i.e. the paths followed by Ph.D. students when interacting with their advisors[10]

16.10 See also

- Julian Schwinger#Schwinger and Feynman
- Stueckelberg–Feynman interpretation
- Invariance mechanics
- Penguin diagram
- Path integral formulation
- Propagator
- List of Feynman diagrams
- Angular momentum diagrams (quantum mechanics)

16.11 Notes

[1] "Physics and Feynman's Diagrams" by David Kaiser, *American Scientist*, Volume 93, p. 156

[2] Feynman, Richard (1949). "The Theory of Positrons". *Physical Review* **76** (76): 749. Bibcode:1949PhRv...76..749F. doi:10.1103/PhysRev.76.749. In this solution, the "negative energy states" appear in a form which may be pictured (as by Stückelberg) in space-time as waves traveling away from the external potential backwards in time. Experimentally, such a wave corresponds to a positron approaching the potential and annihilating the electron.

[3] R. Penco, D. Mauro (2006). "Perturbation theory via Feynman diagrams in classical mechanics". v2. arXiv:hep-th/0605061.

[4] George Johnson (July 2000). "The Jaguar and the Fox". *The Atlantic*. Retrieved February 26, 2013.

[5] Gribbin, John and Mary. *Richard Feynman: A Life in Science*, Penguin-Putnam, 1997 Ch 5.

[6] Leonard Mlodinow. *Feynman's Rainbow*. Vintage, 2011. p. 29

[7] Gerardus 't Hooft, Martinus Veltman, *Diagrammar*, CERN Yellow Report 1973, reprinted in G. 't Hooft, *Under the Spell of Gauge Principle* (World Scientific, Singapore, 1994), Introduction online

[8] Martinus Veltman, *Diagrammatica: The Path to Feynman Diagrams*, Cambridge Lecture Notes in Physics, ISBN 0-521-45692-4

[9] Bjorken, J. D.; Drell, S. D. (1965). "Relativistic Quantum Fields". New York: McGraw-Hill. p. viii.

[10] Jorge Cham, Academic Interaction - Feynman Diagrams, January 11, 2012.

16.12 References

- Gerardus 't Hooft, Martinus Veltman, *Diagrammar*, CERN Yellow Report 1973, online
- David Kaiser, *Drawing Theories Apart: The Dispersion of Feynman Diagrams in Postwar Physics*, Chicago: University of Chicago Press, 2005. ISBN 0-226-42266-6
- Martinus Veltman, *Diagrammatica: The Path to Feynman Diagrams*, Cambridge Lecture Notes in Physics, ISBN 0-521-45692-4 (expanded, updated version of above)
- Mark Srednicki, *Quantum Field Theory*, online Script (2006)

16.13 External links

- AMS article: "What's New in Mathematics: Finite-dimensional Feynman Diagrams"
- Draw Feynman diagrams explained by Flip Tanedo at Quantumdiaries.com
- Drawing Feynman diagrams with FeynDiagram C++ library that produces PostScript output.
- Feynman Diagram Examples using Thorsten Ohl's Feynmf LaTeX package.
- Online Diagram Tool A graphical application for creating publication ready diagrams.
- JaxoDraw A Java program for drawing Feynman diagrams.
- SCaViS – a Java program that can be used for drawing Feynman diagrams using Python scripts
- Bowley, Roger; Copeland, Ed (2010). "Feynman Diagrams". *Sixty Symbols*. Brady Haran for the University of Nottingham.

Chapter 17

Bosonic string theory

Bosonic string theory is the original version of string theory, developed in the late 1960s. It is so called because it only contains bosons in the spectrum.

In the 1980s, supersymmetry was discovered in the context of string theory, and a new version of string theory called superstring theory (supersymmetric string theory) became the real focus. Nevertheless, bosonic string theory remains a very useful model to understand many general features of perturbative string theory, and many theoretical difficulties of superstrings can actually already be found in the context of bosonic strings.

17.1 Problems

Although bosonic string theory has many attractive features, it falls short as a viable physical model in two significant areas.

First, it predicts only the existence of bosons whereas many physical particles are fermions.

Second, it predicts the existence of a mode of the string with imaginary mass, implying that the theory has an instability to a process known as "tachyon condensation".

In addition, bosonic string theory in a general spacetime dimension displays inconsistencies due to the conformal anomaly. But, as was first noticed by Claud Lovelace, in a spacetime of 26 dimensions (25 dimensions of space and one of time), the critical dimension for the theory, the anomaly cancels. This high dimensionality is not necessarily a problem for string theory, because it can be formulated in such a way that along the 22 excess dimensions spacetime is folded up to form a small torus or other compact manifold. This would leave only the familiar four dimensions of spacetime visible to low energy experiments. The existence of a critical dimension where the anomaly cancels is a general feature of all string theories.

17.2 Types of Bosonic strings

There are four possible bosonic string theories, depending on whether open strings are allowed and whether strings have a specified orientation. Recall that a theory of open strings also must include closed strings; open strings can be thought as having their endpoints fixed on a D25-brane that fills all of spacetime. A specific orientation of the string means that only interaction corresponding to an orientable worldsheet are allowed (e.g., two strings can only merge with equal orientation). A sketch of the spectra of the four possible theories is as follows:

Note that all four theories have a negative energy tachyon ($M^2 = -\frac{1}{\alpha'}$) and a massless graviton.

The rest of this article applies to the closed, oriented theory, corresponding to borderless, orientable worldsheets.

17.3 Mathematics

17.3.1 Path integral perturbation theory

Bosonic string theory can be said[1] to be defined by the path integral quantization of the Polyakov action:

$$I_0[g,X]=\frac{T}{8\pi}\int_M d^2\xi\sqrt{g}g^{mn}\partial_m x^\mu\partial_n x^\nu G_{\mu\nu}(x)$$

$X^\mu(\xi)$ is the field on the worldsheet describing the embedding of the string in 25+1 spacetime; in the Polyakov formulation, g is not to be understood as the induced metric from the embedding, but as an independent dynamical field. G is the metric on the target spacetime, which is usually taken to be the Minkowski metric in the perturbative theory. Under a Wick rotation, this is brought to a Euclidean metric $G_{\mu\nu}=\delta_{\mu\nu}$. M is the worldsheet as a topological manifold parametrized by the ξ coordinates. T is the string tension and related to the Regge slope as $T=\frac{1}{2\pi\alpha'}$.

I_0 has diffeomorphism and Weyl invariance. Weyl symmetry is broken upon quantization (Conformal anomaly) and therefore this action has to be supplemented with a counterterm, along with an hypothetical purely topological term, proportional to the Euler characteristic:

$$I=I_0+\lambda\chi(M)+\mu_0^2\int_M d^2\xi\sqrt{g}$$

The explicit breaking of Weyl invariance by the counterterm can be cancelled away in the critical dimension 26.

Physical quantities are then constructed from the (Euclidean) partition function and N-point function:

$$Z=\sum_{h=0}^{\infty}\int\frac{\mathcal{D}g_{mn}\mathcal{D}X^\mu}{\mathcal{N}}\exp(-I[g,X])$$

$$\langle V_{i_1}(k_1^\mu)\cdots V_{i_p}(k_p^\mu)\rangle=\sum_{h=0}^{\infty}\int\frac{\mathcal{D}g_{mn}\mathcal{D}X^\mu}{\mathcal{N}}\exp(-I[g,X])V_{i_1}(k_1^\mu)\cdots V_{i_p}(k_p^\mu)$$

The discrete sum is a sum over possible topologies, which for euclidean bosonic orientable closed strings are com-

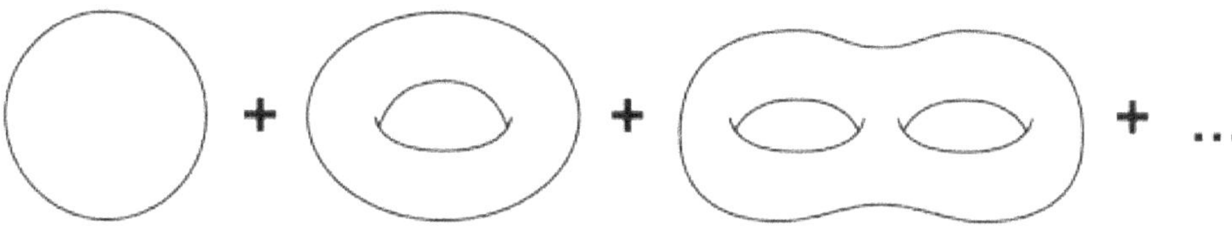

The perturbative series is expressed as a sum over topologies, indexed by the genus.

pact orientable Riemannian surfaces and are thus identified by a genus h . A normalization factor $\mathcal{N}$ is introduced to compensate overcounting from symmetries. While the computation of the partition function correspond to the cosmological constant, the N-point function, including p vertex operators, describes the scattering amplitude of strings.

The symmetry group of the action actually reduces drastically the integration space to a finite dimensional manifold. The g path-integral in the partition function is *de facto* a sum over possible Riemannian structures; however, quotienting with respect to Weyl transformations allows us to only consider conformal structures, that is, equivalence classes of metrics under the identifications of metrics related by

$$g'(\xi)=e^{\sigma(\xi)}g(\xi)$$

Since the world-sheet is two dimensional, there is a 1-1 correspondence between conformal structures and complex structures. One still has to quotient away diffeomorphisms. This leaves us with an integration over the space of all possible complex structures modulo diffeomorphisms, which is simply the moduli space of the given topological surface, and is in fact a finite-dimensional complex manifold. The fundamental problem of perturbative bosonic strings therefore becomes the parametrization of Moduli space, which is non-trivial for genus $h\geq 4$.

h = 0

At tree-level, corresponding to genus 0, the cosmological constant vanishes: $Z_0 = 0$.

The four-point function for the scattering of four tachyons is the Shapiro-Virasoro amplitude:

$$A_4 \propto (2\pi)^{26}\delta^{26}(k)\frac{\Gamma(-1-s/2)\Gamma(-1-t/2)\Gamma(-1-u/2)}{\Gamma(2+s/2)\Gamma(2+t/2)\Gamma(2+u/2)}$$

Where k is the total momentum and s , t , u are the Mandelstam variables.

h = 1

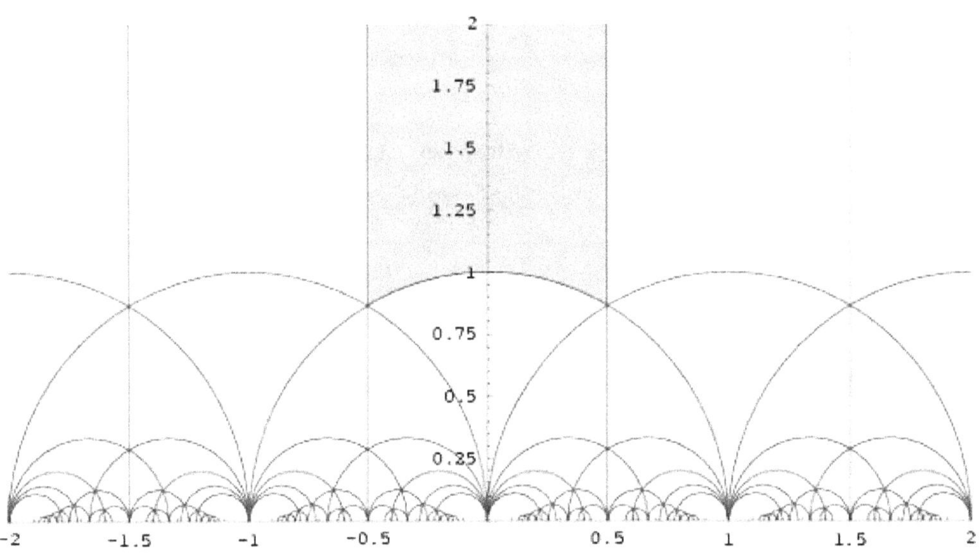

The shaded region is a possible fundamental domain for the modular group.

Genus 1 is the torus, and corresponds to the one-loop level. The partition function amounts to:

$$Z_1 = \int_{\mathcal{M}_1} \frac{d^2\tau}{8\pi^2\tau_2^2}\frac{1}{(4\pi^2\tau_2)^{12}}\left|\eta(\tau)\right|^{-48}$$

τ is a complex number with positive imaginary part τ_2 ; $\mathcal{M}_1$, holomorphic to the moduli space of the torus, is any fundamental domain for the modular group $PSL(2,\mathbb{Z})$ acting on the upper half-plane, for example $\{\tau_2 > 0, |\tau|^2 > 1, -\frac{1}{2} < \tau_1 < \frac{1}{2}\}$. $\eta(\tau)$ is the Dedekind eta function. The integrand is of course invariant under the modular group: the measure $\frac{d^2\tau}{\tau_2^2}$ is simply the Poincaré metric which has PSL(2,R) as isometry group; the rest of the integrand is also invariant by virtue of $\tau_2 \rightarrow |c\tau + d|^2\tau_2$ and the fact that $\eta(\tau)$ is a modular form of weight 1/2.

This integral diverges. This is due to the presence of the tachyon and is related to the instability of the perturbative vacuum.

17.4 See also

- Nambu–Goto action
- Polyakov action

17.5 Notes

[1] D'Hoker, Phong

17.6 References

D'Hoker, Eric and Phong, D. H. (Oct 1988). "The geometry of string perturbation theory,". *Rev. Mod. Phys.* (American Physical Society,) **60** (4): 917–1065. Bibcode:1988RvMP...60..917D. doi:10.1103/RevModPhys.60.917.

Belavin, A.A. and Knizhnik, V.G. (Feb 1986). "Complex geometry and the theory of quantum strings,". *ZhETF* **91** (2): 364–390.

17.7 External links

- How many string theories are there?
- PIRSA:C09001 - Introduction to the Bosonic String

Chapter 18

Superstring theory

"Superstring" redirects here. For the converse relation of "substring", see Superstring (formal languages). For the bundle of firecrackers, see Superstring (fireworks).

Superstring theory is an attempt to explain all of the particles and fundamental forces of nature in one theory by modelling them as vibrations of tiny supersymmetric strings.

'Superstring theory' is a shorthand for **supersymmetric string theory** because unlike bosonic string theory, it is the version of string theory that incorporates fermions and supersymmetry.

Since the second superstring revolution the five superstring theories are regarded as different limits of a single theory tentatively called M-theory, or simply string theory.

18.1 Background

The deepest problem in theoretical physics is harmonizing the theory of general relativity, which describes gravitation and applies to large-scale structures (stars, galaxies, super clusters), with quantum mechanics, which describes the other three fundamental forces acting on the atomic scale.

The development of a quantum field theory of a force invariably results in infinite (and therefore useless) probabilities. Physicists have developed mathematical techniques (renormalization) to eliminate these infinities which work for three of the four fundamental forces—electromagnetic, strong nuclear and weak nuclear forces—but not for gravity. The development of a quantum theory of gravity must therefore come about by different means than those used for the other forces.[1]

According to the theory, the fundamental constituents of reality are strings of the Planck length (about 10^{-33} cm) which vibrate at resonant frequencies. Every string, in theory, has a unique resonance, or harmonic. Different harmonics determine different fundamental particles. The tension in a string is on the order of the Planck force (10^{44} newtons). The graviton (the proposed messenger particle of the gravitational force), for example, is predicted by the theory to be a string with wave amplitude zero.

18.1.1 Lack of experimental evidence

Superstring theory is based on supersymmetry. No supersymmetric particles have been discovered and recent research at LHC and Tevatron has excluded some of the ranges.[2][3][4][5] For instance, the mass constraint of the Minimal Supersymmetric Standard Model squarks has been up to 1.1 TeV, and gluinos up to 500 GeV.[6] No report on suggesting large extra dimensions has been delivered from LHC. There have been no principles so far to limit the number of vacua in the concept of a landscape of vacua.[7]

Some particle physicists became disappointed[8] by the lack of experimental verification of supersymmetry, and some have already discarded it; Jon Butterworth at the University College London said that we had no sign of supersymmetry, even in higher energy region, excluding the superpartners of the top quark up to a few TeV. Ben Allanach at the University of Cambridge states that if we do not discover any new particles in the next trial at the LHC, then we can say it is unlikely to discover supersymmetry at CERN in the foreseeable future.[8]

18.2 Extra dimensions

See also: Why does consistency require 10 dimensions?

Our physical space is observed to have only three large dimensions and—taken together with duration as the fourth dimension—a physical theory must take this into account. However, nothing prevents a theory from including more than 4 dimensions. In the case of string theory, consistency requires spacetime to have 10 (3+1+6) dimensions. The fact that we see only 3 dimensions of space can be explained by one of two mechanisms: either the extra dimensions are compactified on a very small scale, or else our world may live on a 3-dimensional submanifold corresponding to a brane, on which all known particles besides gravity would be restricted.

If the extra dimensions are compactified, then the extra six dimensions must be in the form of a Calabi–Yau manifold. Within the more complete framework of M-theory, they would have to take form of a G2 manifold. Calabi-Yaus are interesting mathematical spaces in their own right. A particular exact symmetry of string/M-theory called T-duality (which exchanges momentum modes for winding number and sends compact dimensions of radius R to radius 1/R),[9] has led to the discovery of equivalences between different Calabi-Yaus called Mirror Symmetry.

Superstring theory is not the first theory to propose extra spatial dimensions. It can be seen as building upon the Kaluza–Klein theory which proposed a 4+1-dimensional theory of gravity. When compactified on a circle, the gravity in the extra dimension precisely describes electromagnetism from the perspective of the 3 remaining large space dimensions. Thus the original Kaluza–Klein theory is a prototype for the unification of gauge and gravity interactions, at least at the classical level, however it is known to be insufficient to describe nature for a variety of reasons (missing weak and strong forces, lack of parity violation, etc.) A more complex compact geometry is needed to reproduce the known gauge forces. This is not all: In order to obtain a consistent, fundamental, quantum theory the upgrade to string theory is also necessary, not just the extra dimensions.

18.3 Number of superstring theories

Theoretical physicists were troubled by the existence of five separate string theories. A possible solution for this dilemma was suggested at the beginning of what is called the second superstring revolution in the 1990s, which suggests that the five string theories might be different limits of a single underlying theory, called M-theory. This remains a conjecture.[10]

The five consistent superstring theories are:

- The type I string has one supersymmetry in the ten-dimensional sense (16 supercharges). This theory is special in the sense that it is based on unoriented open and closed strings, while the rest are based on oriented closed strings.
- The type II string theories have two supersymmetries in the ten-dimensional sense (32 supercharges). There are actually two kinds of type II strings called type IIA and type IIB. They differ mainly in the fact that the IIA theory is non-chiral (parity conserving) while the IIB theory is chiral (parity violating).
- The heterotic string theories are based on a peculiar hybrid of a type I superstring and a bosonic string. There are two kinds of heterotic strings differing in their ten-dimensional gauge groups: the heterotic $E_8 \times E_8$ string and the heterotic SO(32) string. (The name heterotic SO(32) is slightly inaccurate since among the SO(32) Lie groups, string theory singles out a quotient Spin(32)/Z_2 that is not equivalent to SO(32).)

Chiral gauge theories can be inconsistent due to anomalies. This happens when certain one-loop Feynman diagrams cause a quantum mechanical breakdown of the gauge symmetry. The anomalies were canceled out via the Green–Schwarz mechanism.

Even though there are only five superstring theories, in order to make detailed predictions for real experiments, information is needed about exactly what physical configuration the theory is in. This considerably complicates efforts to test string theory because there is an astronomically high number – 10^{500} or more – of configurations that meet some of the basic requirements to be consistent with our world. Along with the extreme remoteness of the Planck scale, this is the other major reason it is hard to test superstring theory.

Another approach to the number of superstring theories refers to the mathematical structure called composition algebra. In the findings of abstract algebra there are just seven composition algebras over the field of real numbers. In 1990 physicists R. Foot and G.C. Joshi in Australia stated that "the seven classical superstring theories are in one-to-one correspondence to the seven composition algebras."[11]

18.4 Integrating general relativity and quantum mechanics

General relativity typically deals with situations involving large mass objects in fairly large regions of spacetime whereas quantum mechanics is generally reserved for scenarios at the atomic scale (small spacetime regions). The two are very rarely used together, and the most common case in which they are combined is in the study of black holes. Having "peak density", or the maximum amount of matter possible in a space, and very small area, the two must be used in synchrony in order to predict conditions in such places; yet, when used together, the equations fall apart, spitting out impossible answers, such as imaginary distances and less than one dimension.

The major problem with their congruence is that, at Planck scale (a fundamental small unit of length) lengths, general relativity predicts a smooth, flowing surface, while quantum mechanics predicts a random, warped surface, neither of which are anywhere near compatible. Superstring theory resolves this issue, replacing the classical idea of point particles with loops. These loops have an average diameter of the Planck length, with extremely small variances, which completely ignores the quantum mechanical predictions of Planck-scale length dimensional warping.

Singularities are avoided because the observed consequences of "Big Crunches" never reach zero size. In fact, should the universe begin a "big crunch" sort of process, string theory dictates that the universe could never be smaller than the size of a string, at which point it would actually begin expanding.

18.5 Mathematics

18.5.1 D-branes

D-branes are membrane-like objects in 10D string theory. They can be thought of as occurring as a result of a Kaluza–Klein compactification of 11D M-theory which contains membranes. Because compactification of a geometric theory produces extra vector fields the D-branes can be included in the action by adding an extra U(1) vector field to the string action.

$$\partial_z \to \partial_z + iA_z(z, \overline{z})$$

In **type I** open string theory, the ends of open strings are always attached to D-brane surfaces. A string theory with more gauge fields such as SU(2) gauge fields would then correspond to the compactification of some higher-dimensional theory above 11 dimensions which is not thought to be possible to date. Furthemore, the tachyons attached to the D-branes, show, the instability of those d-branes with respect to the annihilation.We will consider that tachyon total energy is (or reflects) the total energy of the D-branes.

18.5.2 Why five superstring theories?

For a 10 dimensional supersymmetric theory we are allowed a 32-component Majorana spinor. This can be decomposed into a pair of 16-component Majorana-Weyl (chiral) spinors. There are then various ways to construct an invariant depending on whether these two spinors have the same or opposite chiralities:

The heterotic superstrings come in two types SO(32) and $E_8 \times E_8$ as indicated above and the type I superstrings include open strings.

18.6 Beyond superstring theory

It is conceivable that the five superstring theories are approximated to a theory in higher dimensions possibly involving membranes. Because the action for this involves quartic terms and higher so is not Gaussian, the functional integrals are very difficult to solve and so this has confounded the top theoretical physicists. Edward Witten has popularised the concept of a theory in 11 dimensions M-theory involving membranes interpolating from the known symmetries of superstring theory. It may turn out that there exist membrane models or other non-membrane models in higher dimensions which may become acceptable when new unknown symmetries of nature are found, such as noncommutative geometry for example. It is thought, however, that 16 is probably the maximum since O(16) is a maximal subgroup of E8 the largest exceptional lie group and also is more than large enough to contain the Standard Model. Quartic integrals of the non-functional kind are easier to solve so there is hope for the future. This is the series solution which is always convergent when a is non-zero and negative:

$$\int_{-\infty}^{\infty} \exp(ax^4 + bx^3 + cx^2 + dx + f)\, dx = e^f \sum_{n,m,p=0}^{\infty} \frac{b^{4n}}{(4n)!} \frac{c^{2m}}{(2m)!} \frac{d^{4p}}{(4p)!} \frac{\Gamma(3n + m + p + \frac{1}{4})}{a^{3n+m+p+\frac{1}{4}}}$$

In the case of membranes the series would correspond to sums of various membrane interactions that are not seen in string theory.

18.6.1 Compactification

Investigating theories of higher dimensions often involves looking at the 10 dimensional superstring theory and interpreting some of the more obscure results in terms of compactified dimensions. For example D-branes are seen as compactified membranes from 11D M-theory. Theories of higher dimensions such as 12D F-theory and beyond will produce other effects such as gauge terms higher than $U(1)$. The components of the extra vector fields (A) in the D-brane actions can be thought of as extra coordinates (X) in disguise. However, the *known* symmetries including supersymmetry currently restrict the spinors to have 32-components which limits the number of dimensions to 11 (or 12 if you include two time dimensions.) Some commentators (e.g. John Baez et al.) have speculated that the exceptional lie groups E_6, E_7 and E_8 having maximum orthogonal subgroups O(10), O(12) and O(16) may be related to theories in 10, 12 and 16 dimensions; 10 dimensions corresponding to string theory and the 12 and 16 dimensional theories being yet undiscovered but would be theories based on 3-branes and 7-branes respectively. However this is a minority view within the string community. Since E_7 is in some sense F_4 quaternified and E_8 is F_4 octonified, then the 12 and 16 dimensional theories, if they did exist, may involve the noncommutative geometry based on the quaternions and octonions respectively. From the above discussion, it can be seen that physicists have many ideas for extending superstring theory beyond the current 10 dimensional theory, but so far none have been successful.

18.6.2 Kac–Moody algebras

Since strings can have an infinite number of modes, the symmetry used to describe string theory is based on infinite dimensional Lie algebras. Some Kac–Moody algebras that have been considered as symmetries for M-theory have been E_{10} and E_{11} and their supersymmetric extensions.

18.7 See also

- AdS/CFT
- dS/CFT correspondence
- Grand unification theory
- Large Hadron Collider
- List of string theory topics
- Quantum gravity
- String field theory

18.8 Notes

[1] Polchinski, Joseph. *String Theory: Volume I.* Cambridge University Press, p. 4.

[2] Woit, Peter (February 22, 2011). "Implications of Initial LHC Searches for Supersymmetry".

[3] Cassel, S.; Ghilencea, D. M.; Kraml, S.; Lessa, A.; Ross, G. G. (2011). "Fine-tuning implications for complementary dark matter and LHC SUSY searches". *Journal of High Energy Physics* **2011** (5): 120. arXiv:1101.4664. Bibcode:2011JHEP...05..120C. doi:10.1007/JHEP05(2011)120.

[4] Falkowski, Adam (Jester) (February 16, 2011). "What LHC tells about SUSY". *resonaances.blogspot.com*. Archived from the original on March 22, 2014. Retrieved March 22, 2014.

[5] Tapper, Alex (24 March 2010). "Early SUSY searches at the LHC" (PDF). Imperial College London.

[6] CMS Collaboration (2011). "Search for Supersymmetry at the LHC in Events with Jets and Missing Transverse Energy". *Physical Review Letters* **107** (22): 221804. arXiv:1109.2352. Bibcode:2011PhRvL.107v1804C. doi:10.1103/PhysRevLett.107.221804. PMID 22182023.

[7] Shifman, M. (2012). "Frontiers Beyond the Standard Model: Reflections and Impressionistic Portrait of the Conference". *Modern Physics Letters A* **27** (40): 1230043. Bibcode:2012MPLA...2730043S. doi:10.1142/S0217732312300431.

[8] Jha, Alok (August 6, 2013). "One year on from the Higgs boson find, has physics hit the buffers?". *The Guardian*. photograph: Harold Cunningham/Getty Images (London: GMG). ISSN 0261-3077. OCLC 60623878. Archived from the original on March 22, 2014. Retrieved March 22, 2014.

[9] Polchinski, Joseph. *String Theory: Volume I.* Cambridge University Press, p. 247.

[10] Polchinski, Joseph. *String Theory: Volume II.* Cambridge University Press, p. 198.

[11] Foot, R.; Joshi, G. C. (1990). "Nonstandard signature of spacetime, superstrings, and the split composition algebras". *Letters in Mathematical Physics* **19**: 65–71. Bibcode:1990LMaPh..19...65F. doi:10.1007/BF00402262.

18.9 References

- Kaku, Michio (1999). *Introduction to Superstring and M-Theory* (2nd ed.). New York, USA: Springer-Verlag.
- Shen, Sinyan (1982). *Introduction to Superfluidity* (2nd ed.). Beijing, China: Science Press.
- Greene, Brian (2000). *The Elegant Universe: Superstrings, Hidden Dimensions, and the Quest for the Ultimate Theory*. Random House Inc.

18.10 External links

- Wellcome Collection video on superstring theory
- The Official Superstring theory website: http://superstringtheory.com/index.html

Chapter 19

Type I string theory

In theoretical physics, **type I string theory** is one of five consistent supersymmetric string theories in ten dimensions. It is the only one whose strings are unoriented (both orientations of a string are equivalent) and which contains not only closed strings, but also open strings.

The classic 1976 work of Ferdinando Gliozzi, Joel Scherk and David Olive paved the way to a systematic understanding of the rules behind string spectra in cases where only closed strings are present via modular invariance but did not lead to similar progress for models with closed strings, despite the fact that the original discussion was based on the type I string theory.

As first proposed by Augusto Sagnotti in 1987, the type I string theory can be obtained as an orientifold of type IIB string theory, with 32 half-D9-branes added in the vacuum to cancel various anomalies.

At low energies, type I string theory is described by the N=1 supergravity (type I supergravity) in ten dimensions coupled to the SO(32) supersymmetric Yang–Mills theory. The discovery in 1984 by Michael Green and John H. Schwarz that anomalies in type I string theory cancel sparked the first superstring revolution. However, a key property of these models, shown by A. Sagnotti in 1992, is that in general the Green-Schwarz mechanism takes a more general form, and involves several two forms in the cancellation mechanism.

The relation between the type-IIB string theory and the type-I string theory has a large number of surprising consequences, both in ten and in lower dimensions, that were first displayed by the String Theory group at the University of Rome "Tor Vergata" in the early 1990s. It opened the way to the construction of entire new classes of string spectra with or without supersymmetry. Joseph Polchinski's work on D-branes provided a geometrical interpretation for these results in terms of extended objects (D-brane, orientifold).

In the 1990s it was first argued by Edward Witten that type I string theory with the string coupling constant g is equivalent to the SO(32) heterotic string with the coupling $1/g$. This equivalence is known as S-duality.

19.1 References

- F. Gliozzi, J. Scherk and D.I. Olive, "Supersymmetry, Supergravity Theories And The Dual Spinor Model", *Nucl. Phys. B* **122** (1977) 253.
- E. Witten, "String theory dynamics in various dimensions", *Nucl. Phys. B* **443** (1995) 85. arXiv:hep-th/9503124.
- J. Polchinski, S. Chaudhuri and C.V. Johnson, "Notes on D-Branes", arXiv:hep-th/9602052.
- C. Angelantonj and A. Sagnotti, "Open strings", *Phys. Rept.* **1** [(Erratum-ibid.) 339] arXiv:hep-th/0204089.

Chapter 20

Type II string theory

In theoretical physics, **type II string theory** is a unified term that includes both **type IIA strings** and **type IIB strings** theories. Type II string theory accounts for two of the five consistent superstring theories in ten dimensions. Both theories have the maximal amount of supersymmetry — namely 32 supercharges — in ten dimensions. Both theories are based on oriented closed strings. On the worldsheet, they differ only in the choice of GSO projection.

20.1 Type IIA string

At low energies, type IIA string theory is described by type IIA supergravity in ten dimensions which is a non-chiral theory (i.e. left-right symmetric) with (1,1) d=10 supersymmetry; the fact that the anomalies in this theory cancel is therefore trivial.

In the 1990s it was realized by Edward Witten (building on previous insights by Michael Duff, Paul Townsend, and others) that the limit of type IIA string theory in which the string coupling goes to infinity becomes a new 11-dimensional theory called M-theory.

The mathematical treatment of type IIA string theory belongs to symplectic topology and algebraic geometry, particularly Gromov–Witten invariants.

20.2 Type IIB string

At low energies, type IIB string theory is described by type IIB supergravity in ten dimensions which is a chiral theory (left-right asymmetric) with (2,0) d=10 supersymmetry; the fact that the anomalies in this theory cancel is therefore nontrivial.

In the 1990s it was realized that type II string theory with the string coupling constant g is equivalent to the same theory with the coupling $1/g$. This equivalence is known as S-duality.

Orientifold of type IIB string theory leads to type I string theory.

The mathematical treatment of type IIB string theory belongs to algebraic geometry, specifically the deformation theory of complex structures originally studied by Kunihiko Kodaira and Donald C. Spencer.

In 1997 Juan Maldacena gave some arguments indicating that type IIB string theory is equivalent to a Supersymmetric Yang Mills theory with 4 supersymmetries and gauge group $SU(N)$, in the 't Hooft limit; it was the first suggestion concerning the AdS/CFT correspondence.[1]

20.3 Relationship between the type II theories

In the late 1980s, it was realized that type IIA string theory is related to type IIB string theory by T-duality.

20.4 See also

- Superstring theory
- Type I string
- Heterotic string

20.5 References

[1] J. Maldacena, "The Large N Limit of Superconformal Field Theories and Supergravity" arXiv:hep-th/9711200

Chapter 21

Heterotic string theory

This article is about string theory. For heterosis in biology, see Heterosis.

In string theory, a **heterotic string** is a closed string (or loop) which is a hybrid ('heterotic') of a superstring and a bosonic string. There are two kinds of heterotic string, the heterotic SO(32) and the heterotic $E_8 \times E_8$, abbreviated to **HO** and **HE**. Heterotic string theory was first developed in 1985 by David Gross, Jeffrey Harvey, Emil Martinec, and Ryan Rohm (the so-called "Princeton String Quartet"[1]), in one of the key papers that fueled the first superstring revolution.

In string theory, the left-moving and the right-moving excitations are completely decoupled,[2] and it is possible to construct a string theory whose left-moving (counter-clockwise) excitations are treated as a bosonic string propagating in $D = 26$ dimensions, while the right-moving (clock-wise) excitations are treated as a superstring in $D = 10$ dimensions.

The mismatched 16 dimensions must be compactified on an even, self-dual lattice (a discrete subgroup of a linear space). There are two possible even self-dual lattices in 16 dimensions, and it leads to two types of the heterotic string. They differ by the gauge group in 10 dimensions. One gauge group is SO(32) (the HO string) while the other is $E_8 \times E_8$ (the HE string).[3]

These two gauge groups also turned out to be the only two anomaly-free gauge groups that can be coupled to the N = 1 supergravity in 10 dimensions other than $U(1)^{496}$ and $E_8 \times U(1)^{248}$, which is suspected to lie in the swampland.

Every heterotic string must be a closed string, not an open string; it is not possible to define any boundary conditions that would relate the left-moving and the right-moving excitations because they have a different character.

A heterotic string is embedded in the membrane that creates harmonics on the string which translate into mass and energy through mechanisms discussed above.

21.1 String duality

String duality is a class of symmetries in physics that link different string theories. In the 1990s, it was realized that the strong coupling limit of the HO theory is type I string theory — a theory that also contains open strings; this relation is called S-duality. The HO and HE theories are also related by T-duality.

Because the various superstring theories were shown to be related by dualities, it was proposed that that each type of string was a different aspect of a single underlying theory called M-theory.

21.2 References

[1] Dennis Overbye, "String theory, at 20, explains it all (or not)". *NY Times*, 2004-12-07

[2] *String Theory and M-Theory* by Becker, Becker and Schwarz (2006), p. 253

[3] Joseph Polchinski (1998). *String Theory: Volume 2*, p. 45.

Chapter 22

S-duality

This article is about S-duality (strong–weak duality) in physics. For the mathematical S-duality (Spanier–Whitehead duality), see S-duality (homotopy theory).

In theoretical physics, **S-duality** is an equivalence of two physical theories, which may be either quantum field theories or string theories. S-duality is useful for doing calculations in theoretical physics because it relates a theory in which calculations are difficult to a theory in which they are easier.[1]

In quantum field theory, S-duality generalizes a well known fact from classical electrodynamics, namely the invariance of Maxwell's equations under the interchange of electric and magnetic fields. One of the earliest known examples of S-duality in quantum field theory is Montonen–Olive duality which relates two versions of a quantum field theory called N = 4 supersymmetric Yang–Mills theory. Recent work of Anton Kapustin and Edward Witten suggests that Montonen–Olive duality is closely related to a research program in mathematics called the geometric Langlands program. Another realization of S-duality in quantum field theory is Seiberg duality, which relates two versions of a theory called N=1 supersymmetric Yang–Mills theory.

There are also many examples of S-duality in string theory. The existence of these string dualities implies that seemingly different formulations of string theory are actually physically equivalent. This led to the realization, in the mid-1990s, that all of the five consistent superstring theories are just different limiting cases of a single eleven-dimensional theory called M-theory.[2]

22.1 Overview

In quantum field theory and string theory, a coupling constant is a number that controls the strength of interactions in the theory. For example, the strength of gravity is described by a number called Newton's constant, which appears in Newton's law of gravity and also in the equations of Albert Einstein's general theory of relativity. Similarly, the strength of the electromagnetic force is described by a coupling constant, which is related to the charge carried by a single proton.

To compute observable quantities in quantum field theory or string theory, physicists typically apply the methods of perturbation theory. In perturbation theory, quantities called probability amplitudes, which determine the probability for various physical processes to occur, are expressed as sums of infinitely many terms, where each term is proportional to a power of the coupling constant g :

$$A = A_0 + A_1 g + A_2 g^2 + A_3 g^3 + \dots$$

In order for such an expression to make sense, the coupling constant must be less than 1 so that the higher powers of g become negligibly small and the sum is finite. If the coupling constant is not less than 1, then the terms of this sum will grow larger and larger, and the expression gives a meaningless infinite answer. In this case the theory is said to be *strongly coupled*, and one cannot use perturbation theory to make predictions.

For certain theories, S-duality provides a way of doing computations at strong coupling by translating these computations into different computations in a weakly coupled theory. S-duality is a particular example of a general notion of duality in physics. The term *duality* refers to a situation where two seemingly different physical systems turn out to be equivalent in a nontrivial way. If two theories are related by a duality, it means that one theory can be transformed in some way so that it ends up looking just like the other theory. The two theories are then said to be *dual* to one another under the transformation. Put differently, the two theories are mathematically different descriptions of the same phenomena.

S-duality is useful because it relates a theory with coupling constant g to an equivalent theory with coupling constant $1/g$. Thus it relates a strongly coupled theory (where the coupling constant g is much greater than 1) to a weakly coupled theory (where the coupling constant $1/g$ is much less than 1 and computations are possible). For this reason, S-duality is called a **strong-weak duality**.

22.2 S-duality in quantum field theory

22.2.1 A symmetry of Maxwell's equations

In classical physics, the behavior of the electric and magnetic field is described by a system of equations known as Maxwell's equations. Working in the language of vector calculus and assuming that no electric charges or currents are present, these equations can be written[3]

$$\begin{aligned}
\nabla \cdot \mathbf{E} &= 0, \\
\nabla \cdot \mathbf{B} &= 0, \\
\nabla \times \mathbf{E} &= -\frac{\partial \mathbf{B}}{\partial t}, \\
\nabla \times \mathbf{B} &= \frac{1}{c^2}\frac{\partial \mathbf{E}}{\partial t}.
\end{aligned}$$

Here **E** is a vector (or more precisely a *vector field* whose magnitude and direction may vary from point to point in space) representing the electric field, **B** is a vector representing the magnetic field, t is time, and c is the speed of light. The other symbols in these equations refer to the divergence and curl, which are concepts from vector calculus.

An important property of these equations[4] is their invariance under the transformation that simultaneously replaces the electric field **E** by the magnetic field **B** and replaces **B** by $-1/c^2\mathbf{E}$:

$$\begin{aligned}
\mathbf{E} &\to \mathbf{B} \\
\mathbf{B} &\to -\frac{1}{c^2}\mathbf{E}.
\end{aligned}$$

In other words, given a pair of electric and magnetic fields that solve Maxwell's equations, it is possible to describe a new physical setup in which these electric and magnetic fields are essentially interchanged, and the new fields will again give a solution of Maxwell's equations. This situation is the most basic manifestation of S-duality in quantum field theory.

22.2.2 Montonen–Olive duality

Main article: Montonen–Olive duality

In quantum field theory, the electric and magnetic fields are unified into a single entity called the electromagnetic field, and this field is described by a special type of quantum field theory called a gauge theory or Yang–Mills theory. In a gauge theory, the physical fields have a high degree of symmetry which can be understood mathematically using the notion of a Lie group. This Lie group is known as the gauge group. The electromagnetic field is described by a very simple gauge theory corresponding to the abelian gauge group U(1), but there are other gauge theories with more complicated non-abelian gauge groups.[5]

It is natural to ask whether there is an analog in gauge theory of the symmetry interchanging the electric and magnetic fields in Maxwell's equations. The answer was given in the late 1970s by Claus Montonen and David Olive,[6] building on earlier work of Peter Goddard, Jean Nuyts, and Olive.[7] Their work provides an example of S-duality now known as Montonen–Olive duality. Montonen–Olive duality applies to a very special type of gauge theory called N = 4 supersymmetric Yang–Mills theory, and it says that two such theories may be equivalent in a certain precise sense.[1] If one of the theories has a gauge group G , then the dual theory has gauge group ${}^LG$ where ${}^LG$ denotes the Langlands dual group which is in general different from G .[8]

An important quantity in quantum field theory is complexified coupling constant. This is a complex number defined by the formula[9]

$$\tau = \frac{\theta}{2\pi} + \frac{4\pi i}{g^2}$$

where θ is the theta angle, a quantity appearing in the Lagrangian that defines the theory,[9] and g is the coupling constant. For example, in the Yang–Mills theory that describes the electromagnetic field, this number g is simply the elementary charge e carried by a single proton.[1] In addition to exchanging the gauge groups of the two theories, Montonen–Olive duality transforms a theory with complexified coupling coupling constant τ to a theory with complexified constant $-1/\tau$.[9]

22.2.3 Relation to the Langlands program

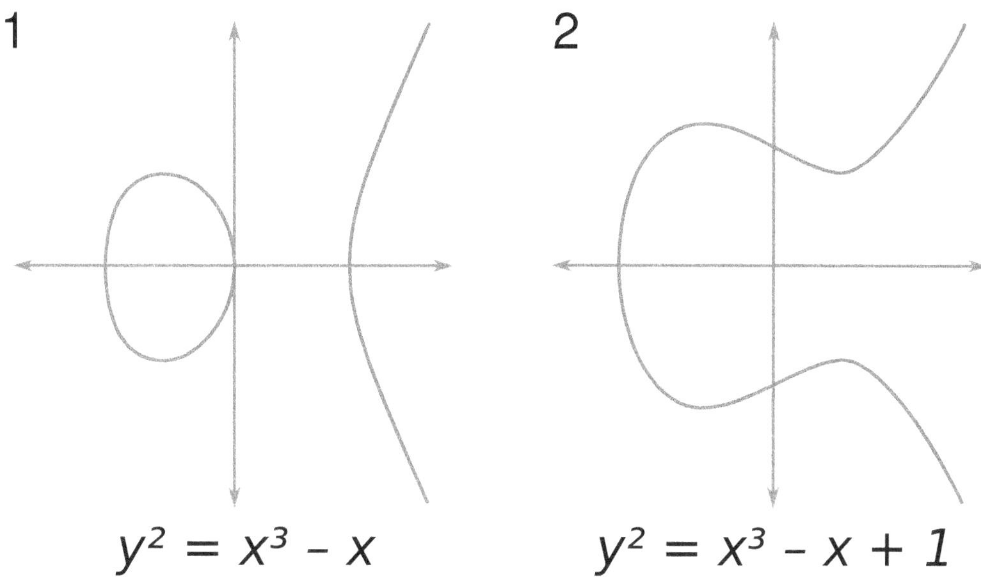

The geometric Langlands correspondence is a relationship between abstract geometric objects associated to an algebraic curve such as the elliptic curves illustrated above.

Main article: Langlands program

In mathematics, the classical Langlands correspondence is a collection of results and conjectures relating number theory to the branch of mathematics known as representation theory.[10] Formulated by Robert Langlands in the late 1960s, the Langlands correspondence is related to important conjectures in number theory such as the Taniyama-Shimura conjecture, which includes Fermat's last theorem as a special case.[10]

In spite of its importance in number theory, establishing the Langlands correspondence in the number theoretic context has proved extremely difficult.[10] As a result, some mathematicians have worked on a related conjecture known as the geometric Langlands correspondence. This is a geometric reformulation of the classical Langlands correspondence which is obtained by replacing the number fields appearing in the original version by function fields and applying techniques from algebraic geometry.[10]

In a paper from 2007, Anton Kapustin and Edward Witten suggested that the geometric Langlands correspondence can be viewed as a mathematical statement of Montonen–Olive duality.[11] Starting with two Yang–Mills theories related by S-duality, Kapustin and Witten showed that one can construct a pair of quantum field theories in two-dimensional spacetime. By analyzing what this dimensional reduction does to certain physical objects called D-branes, they showed that one can recover the mathematical ingredients of the geometric Langlands correspondence.[12] Their work shows that the Langlands correspondence is closely related to S-duality in quantum field theory, with possible applications in both subjects.[10]

22.2.4 Seiberg duality

Main article: Seiberg duality

Another realization of S-duality in quantum field theory is Seiberg duality, first introduced by Nathan Seiberg around 1995.[13] Unlike Montonen–Olive duality, which relates two versions of the maximally supersymmetric gauge theory in four-dimensional spacetime, Seiberg duality relates less symmetric theories called N=1 supersymmetric gauge theories. The two N=1 theories appearing in Seiberg duality are not identical, but they give rise to the same physics at large distances. Like Montonen–Olive duality, Seiberg duality generalizes the symmetry of Maxwell's equations that interchanges electric and magnetic fields.

22.3 S-duality in string theory

Up until the mid 1990s, physicists working on string theory believed there were five distinct versions of the theory: type I, type IIA, type IIB, and the two flavors of heterotic string theory (SO(32) and $E_8 \times E_8$). The different theories allow different types of strings, and the particles that arise at low energies exhibit different symmetries.

In the mid 1990s, physicists noticed that these five string theories are actually related by highly nontrivial dualities. One of these dualities is S-duality. The existence of S-duality in string theory was first proposed by Ashoke Sen in 1994.[14] It was shown that type IIB string theory with the coupling constant g is equivalent via S-duality to the same string theory with the coupling constant $1/g$. Similarly, type I string theory with the coupling g is equivalent to the SO(32) heterotic string theory with the coupling constant $1/g$.

The existence of these dualities showed that the five string theories were in fact not all distinct theories. In 1995, at the string theory conference at University of Southern California, Edward Witten made the surprising suggestion that all five of these theories were just different limits of a single theory now known as M-theory.[15] Witten's proposal was based on the observation that type IIA and $E_8 \times E_8$ heterotic string theories are closely related to a gravitational theory called eleven-dimensional supergravity. His announcement led to a flurry of work now known as the second superstring revolution.

22.4 See also

- T-duality
- Mirror symmetry
- AdS/CFT correspondence

22.5 Notes

[1] Frenkel 2009, p.2

[2] Zwiebach 2009, p.325

[3] Griffiths 1999, p.326

[4] Griffiths 1999, p.327

[5] For an introduction to quantum field theory in general including the basics of gauge theory, see Zee 2010.

A diagram of string theory dualities. Yellow lines indicate S-duality. Blue lines indicate T-duality.

[6] Montonen and Olive 1977

[7] Goddard, Nuyts, and Olive 1977

[8] Frenkel 2009, p.5

[9] Frenkel 2009, p.12

[10] Frenkel 2007

[11] Kapustin and Witten 2007

[12] Aspinwall et al. 2009, p.415

[13] Seiberg 1995

[14] Sen 1994

[15] Witten 1995

22.6 References

- Aspinwall, Paul; Bridgeland, Tom; Craw, Alastair; Douglas, Michael; Gross, Mark; Kapustin, Anton; Moore, Gregory; Segal, Graeme; Szendröi, Balázs; Wilson, P.M.H., eds. (2009). *Dirichlet Branes and Mirror Symmetry*. American Mathematical Society. ISBN 978-0-8218-3848-8.

- Frenkel, Edward (2007). "Lectures on the Langlands program and conformal field theory". *Frontiers in number theory, physics, and geometry II* (Springer): 387–533. arXiv:hep-th/0512172. Bibcode:2005hep.th...12172F.

- Frenkel, Edward (2009). "Gauge theory and Langlands duality". *Seminaire Bourbaki.*

- Goddard, Peter; Nuyts, Jean; Olive, David (1977). "Gauge theories and magnetic charge". *Nuclear Physics B* **125** (1): 1–28. Bibcode:1977NuPhB.125....1G. doi:10.1016/0550-3213(77)90221-8.

- Griffiths, David (1999). *Introduction to Electrodynamics*. New Jersey: Prentice-Hall.

- Kapustin, Anton; Witten, Edward (2007). "Electric-magnetic duality and the geometric Langlands program". *Communications in Number Theory and Physics* **1** (1): 1–236. arXiv:hep-th/0604151. Bibcode:2007CNTP....1....1K. doi:10.4310/cntp.2007.v1.n1.a1.

- Montonen, Claus; Olive, David (1977). "Magnetic monopoles as gauge particles?". *Physics Letters B* **72** (1): 117–120. Bibcode:1977PhLB...72..117M. doi:10.1016/0370-2693(77)90076-4.

- Seiberg, Nathan (1995). "Electric-magnetic duality in supersymmetric non-Abelian gauge theories". *Nuclear Physics B* **435** (1): 129–146. arXiv:hep-th/9411149. Bibcode:1995NuPhB.435..129S. doi:10.1016/0550-3213(94)00023-8.

- Sen, Ashoke (1994). "Strong-weak coupling duality in four-dimensional string theory". *International Journal of Modern Physics A* **9** (21): 3707–3750. arXiv:hep-th/9402002. Bibcode:1994IJMPA...9.3707S. doi:10.1142/S0217751X9400

- Witten, Edward (March 13–18, 1995). "Some problems of strong and weak coupling". *Proceedings of Strings '95: Future Perspectives in String Theory.* World Scientific.

- Witten, Edward (1995). "String theory dynamics in various dimensions". *Nuclear Physics B* **443** (1): 85–126. arXiv:hep-th/9503124. Bibcode:1995NuPhB.443...85W. doi:10.1016/0550-3213(95)00158-O.

- Zee, Anthony (2010). *Quantum Field Theory in a Nutshell* (2nd ed.). Princeton University Press. ISBN 978-0-691-14034-6.

- Zwiebach, Barton (2009). *A First Course in String Theory.* Cambridge University Press. ISBN 978-0-521-88032-9.

Chapter 23

T-duality

In theoretical physics, **T-duality** is an equivalence of two physical theories, which may be either quantum field theories or string theories. In the simplest example of this relationship, one of the theories describes strings propagating in an imaginary spacetime shaped like a circle of some radius R , while the other theory describes strings propagating on a spacetime shaped like a circle of radius $1/R$. The two theories are equivalent in the sense that all observable quantities in one description are identified with quantities in the dual description. For example, momentum in one description takes discrete values and is equal to the number of times the string winds around the circle in the dual description.

The idea of T-duality can be extended to more complicated theories, including superstring theories. The existence of these dualities implies that seemingly different superstring theories are actually physically equivalent. This led to the realization, in the mid-1990s, that all of the five consistent superstring theories are just different limiting cases of a single eleven-dimensional theory called M-theory.

In general, T-duality relates two theories with different spacetime geometries. In this way, T-duality suggests a possible scenario in which the classical notions of geometry break down in a theory of Planck scale physics.[1] The geometric relationships suggested by T-duality are also important in pure mathematics. Indeed, according to the SYZ conjecture of Andrew Strominger, Shing-Tung Yau, and Eric Zaslow, T-duality is closely related to another duality called mirror symmetry, which has important applications in a branch of mathematics called enumerative algebraic geometry.

23.1 Overview

23.1.1 Strings and duality

T-duality is a particular example of a general notion of duality in physics. The term *duality* refers to a situation where two seemingly different physical systems turn out to be equivalent in a nontrivial way. If two theories are related by a duality, it means that one theory can be transformed in some way so that it ends up looking just like the other theory. The two theories are then said to be *dual* to one another under the transformation. Put differently, the two theories are mathematically different descriptions of the same phenomena.

Like many of the dualities studied in theoretical physics, T-duality was discovered in the context of string theory.[2] In string theory, particles are modeled not as zero-dimensional points but as one-dimensional extended objects called strings. The physics of strings can be studied in various numbers of dimensions. In addition to three familiar dimensions from everyday experience (up/down, left/right, forward/backward), string theories may include one or more compact dimensions which are curled up into circles.

A standard analogy for this is to consider multidimensional object such as a garden hose.[3] If the hose is viewed from a sufficient distance, it appears to have only one dimension, its length. However, as one approaches the hose, one discovers that it contains a second dimension, its circumference. Thus, an ant crawling inside it would move in two dimensions. Such extra dimensions are important in T-duality, which relates a theory in which strings propagate on a circle of some radius R to a theory in which strings propagate on a circle of radius $1/R$.

23.1.2 Winding numbers

Main article: Winding number

In mathematics, the winding number of a curve in the plane around a given point is an integer representing the total number of times that curve travels counterclockwise around the point. The notion of winding number is important in the mathematical description of T-duality where it is used to measure the winding of strings around compact extra dimensions.

For example, the image below shows several examples of curves in the plane, illustrated in red. Each curve is assumed to be closed, meaning it has no endpoints, and is allowed to intersect itself. Each curve has an orientation given by the arrows in the picture. In each situation, there is a distinguished point in the plane, illustrated in black. The *winding number* of the curve around this distinguished point is equal to the total number of counterclockwise turns that the curve makes around this point.

When counting the total number of turns, counterclockwise turns count as positive, while clockwise turns counts as negative. For example, if the curve first circles the origin four times counterclockwise, and then circles the origin once clockwise, then the total winding number of the curve is three. According to this scheme, a curve that does not travel around the distinguished point at all has winding number zero, while a curve that travels clockwise around the point has negative winding number. Therefore, the winding number of a curve may be any integer. The pictures above show curves with winding numbers between −2 and 3:

23.1.3 Quantized momenta

The simplest theories in which T-duality arises are two-dimensional sigma models with circular target spaces. These are simple quantum field theories that describe propagation of strings in an imaginary spacetime shaped like a circle. The strings can thus be modeled as curves in the plane that are confined to lie in a circle, say of radius R , about the origin. In what follows, the strings are assumed to be closed (that is, without endpoints).

Denote this circle by S^1_R . One can think of this circle as a copy of the real line with two points identified if they differ by a multiple of the circle's circumference $2\pi R$. It follows that the state of a string at any given time can be represented as a function $\varphi(\theta)$ of a single real parameter θ . Such a function can be expanded in a Fourier series as

$$\varphi(\theta) = mR\theta + x + \sum_{n\neq 0} c_n e^{in\theta}$$

Here m denotes the winding number of the string around the circle, and the constant mode $x = c_0$ of the Fourier series has been singled out. Since this expression represents the configuration of a string at a fixed time, all coefficients (x and the c_n) are also functions of time.

Let $\dot{x}$ denote the time derivative of the constant mode x . This represents a type of momentum in the theory. One can show, using the fact that the strings considered here are closed, that this momentum can only take on discrete values of the form $\dot{x} = n/R$ for some integer n . In more physical language, one says that the momentum spectrum is *quantized*.

23.1.4 An equivalence of theories

In the situation described above, the total energy, or Hamiltonian, of the string is given by the expression

$$H = (mR)^2 + \dot{x}^2 + \sum_n |\dot{c}_n|^2 + n^2|c_n|^2$$

Since the momenta of the theory are quantized, the first two terms in this formula are $(mR)^2 + (n/R)^2$, and this expression is unchanged when one simultaneously replaces the radius R by $1/R$ and exchanges the winding number m and the integer n . The summation in the expression for H is similarly unaffected by these changes, so the total energy is unchanged. In fact, this equivalence of Hamiltonians descends to an equivalence of two quantum mechanical theories: One of these theories describes strings propagating on a circle of radius R , while the other describes string propagating in a circle of radius $1/R$ with momentum and winding numbers interchanged. This equivalence of theories is the simplest manifestation of T-duality.

23.2 Superstrings

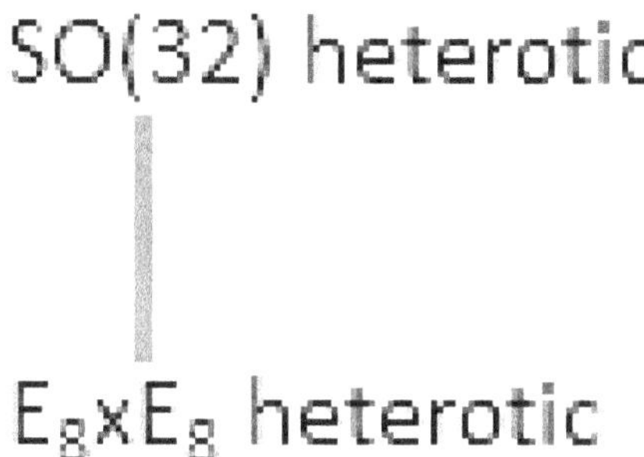

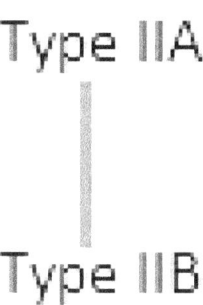

A diagram of string theory dualities. Yellow lines indicate S-duality. Blue lines indicate T-duality.

Up until the mid 1990s, physicists working on string theory believed there were five distinct versions of the theory: type I, type IIA, type IIB, and the two flavors of heterotic string theory (SO(32) and $E_8 \times E_8$). The different theories allow different types of strings, and the particles that arise at low energies exhibit different symmetries.

In the mid 1990s, physicists noticed that these five string theories are actually related by highly nontrivial dualities. One of these dualities is T-duality. For example, it was shown that type IIA string theory is equivalent to type IIB string theory via T-duality and also that the two versions of heterotic string theory are related by T-duality.

The existence of these dualities showed that the five string theories were in fact not all distinct theories. In 1995, at the string theory conference at University of Southern California, Edward Witten made the surprising suggestion that all five of these theories were just different limits of a single theory now known as M-theory.[4] Witten's proposal was based on the observation that different superstring theories are linked by dualities and the fact that type IIA and

$E_8 \times E_8$ heterotic string theories are closely related to a gravitational theory called eleven-dimensional supergravity. His announcement led to a flurry of work now known as the second superstring revolution.

23.3 Mirror symmetry

Main article: Mirror symmetry (string theory)

In string theory and algebraic geometry, the term "mirror symmetry" refers to a phenomenon involving complicated

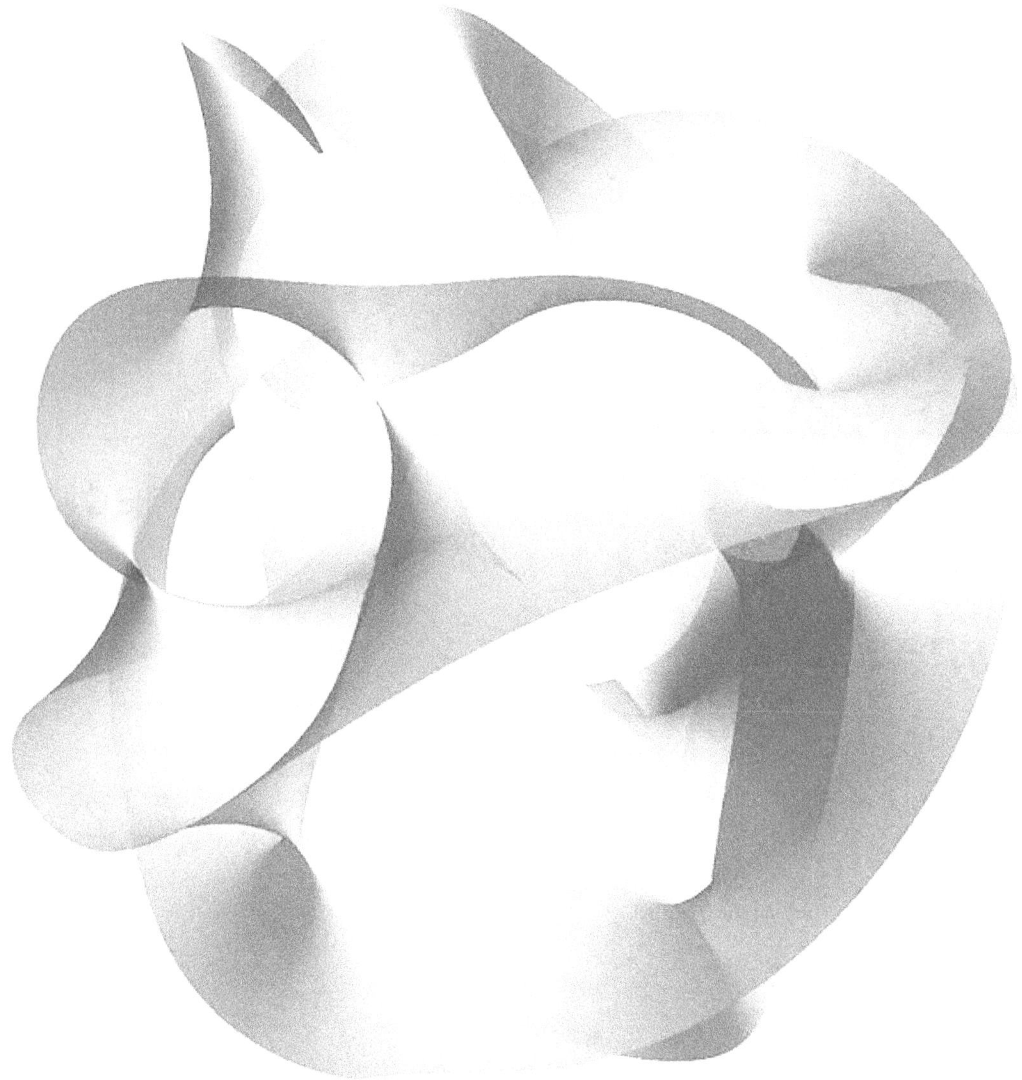

A hypersurface of a six-dimensional Calabi–Yau manifold.

shapes called Calabi-Yau manifolds. These manifolds provide an interesting geometry on which strings can propagate, and the resulting theories may have applications in particle physics.[5] In the late 1980s, it was noticed that such a Calabi-Yau manifold does not uniquely determine the physics of the theory. Instead, one finds that there are *two* Calabi-Yau manifolds that give rise to the same physics.[6] These manifolds are said to be "mirror" to one another. This mirror duality is an important computational tool in string theory, and it has allowed mathematicians to solve difficult problems in enumerative geometry.[7]

One approach to understanding mirror symmetry is the SYZ conjecture, which was suggested by Andrew Strominger, Shing-Tung Yau, and Eric Zaslow in 1996.[8] According to the SYZ conjecture, mirror symmetry can be understood

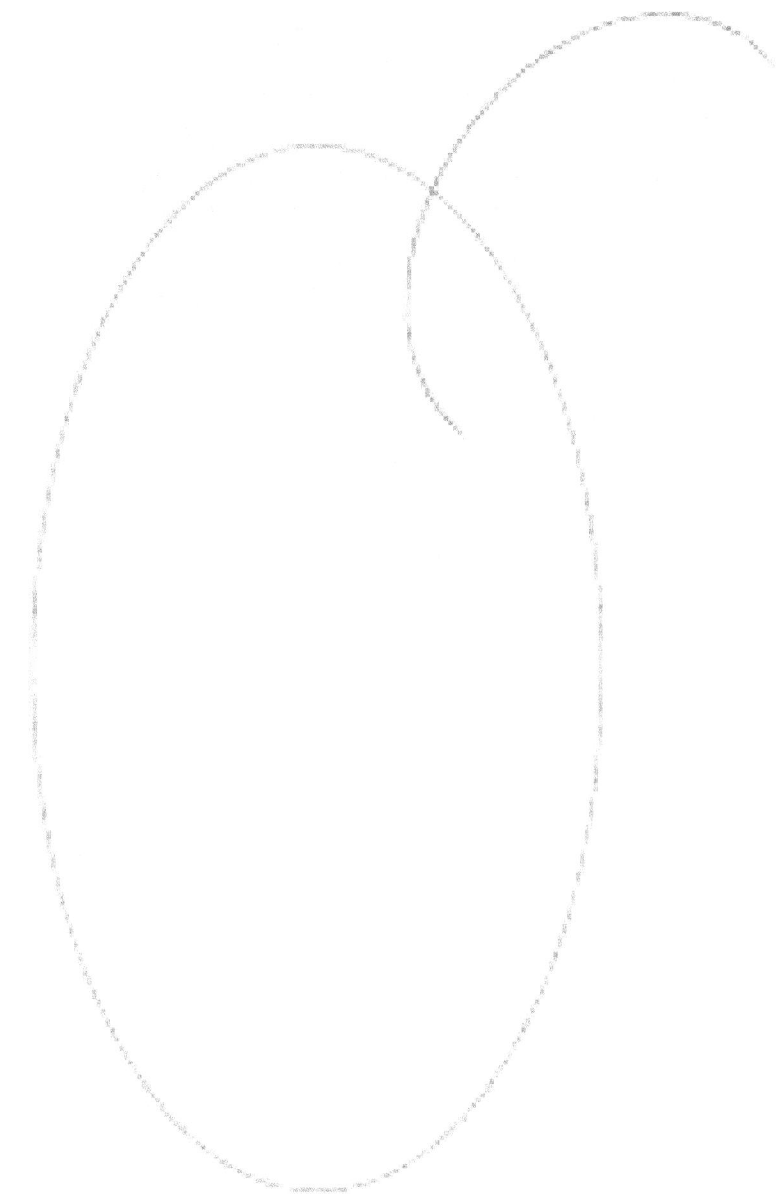

A torus is the cartesian product of two circles.

by dividing a complicated Calabi-Yau manifold into simpler pieces and considering the effects of T-duality on these pieces.[9]

The simplest example of a Calabi-Yau manifold is a torus (a surface shaped like a donut). Such a surface can be viewed as the product of two circles. This means that the torus can be viewed as the union of a collection of longitudinal circles (such as the red circle in the image). There is an auxiliary space which says how these circles are organized, and this space is itself a circle (the pink circle). This space is said to *parametrize* the longitudinal circles on the torus. In this case, mirror symmetry is equivalent to T-duality acting on the longitudinal circles, changing their radii from R to $1/R$.

The SYZ conjecture generalizes this idea to the more complicated case of six-dimensional Calabi-Yau manifolds like the one illustrated above. As in the case of a torus, one can divide a six-dimensional Calabi-Yau manifold into simpler pieces, which in this case are 3-tori (three-dimensional objects which generalize the notion of a torus) parametrized by a 3-sphere (a three-dimensional generalization of a sphere).[10] T-duality can be extended from circles to the three-dimensional tori appearing in this decomposition, and the SYZ conjecture states that mirror symmetry is equivalent to the simultaneous application of T-duality to these three-dimensional tori.[11] In this way, the SYZ conjecture provides a geometric picture of how mirror symmetry acts on a Calabi-Yau manifold.

23.4 See also

- S-duality
- Mirror symmetry
- AdS/CFT correspondence

23.5 Notes

[1] Seiberg 2006

[2] Other dualities that arise in string theory are S-duality, U-duality, mirror symmetry, and the AdS/CFT correspondence.

[3] This analogy is used for example in Greene 2000, p.186

[4] Witten 1995

[5] Candelas et al. 1985

[6] Dixon 1988; Lerche, Vafa, and Warner 1989

[7] Zaslow 2008

[8] Strominger, Yau, and Zaslow 1996

[9] Yau and Nadis 2010, p.174

[10] More precisely, there is a 3-torus associated to every point on the three-sphere except at certain bad points, which correspond to singular tori. See Yau and Nadis 2010, pp.176–7.

[11] Yau and Nadis 2010, p.178

23.6 References

- Candelas, Philip; Horowitz, Gary; Strominger, Andrew; Witten, Edward (1985). “Vacuum configurations for superstrings”. *Nuclear Physics B* **258**: 46–74. Bibcode:1985NuPhB.258...46C. doi:10.1016/0550-3213(85)90602-9.

- Dixon, Lance (1988). “Some world-sheet properties of superstring compactifications, on orbifolds and otherwise”. *ICTP Ser. Theoret. Phys.* **4**: 67–126.

- Greene, Brian (2000). *The Elegant Universe: Superstrings, Hidden Dimensions, and the Quest for the Ultimate Theory*. Random House. ISBN 978-0-9650888-0-0.

- Lerche, Wolfgang; Vafa, Cumrun; Warner, Nicholas (1989). “Chiral rings in $N = 2$ superconformal theories”. *Nuclear Physics B* **324** (2): 427–474. Bibcode:1989NuPhB.324..427L. doi:10.1016/0550-3213(89)90474-4.

- Seiberg, Nathan (2006). "Emergent Spacetime". arXiv:hep-th/0601234.

- Strominger, Andrew; Yau, Shing-Tung; Zaslow, Eric (1996). "Mirror symmetry is T-duality". *Nuclear Physics B* **479** (1): 243–259. arXiv:hep-th/9606040. Bibcode:1996NuPhB.479..243S. doi:10.1016/0550-3213(96)00434-8.

- Witten, Edward (March 13–18, 1995). "Some problems of strong and weak coupling". *Proceedings of Strings '95: Future Perspectives in String Theory*. World Scientific.

- Witten, Edward (1995). "String theory dynamics in various dimensions". *Nuclear Physics B* **443** (1): 85–126. arXiv:hep-th/9503124. Bibcode:1995NuPhB.443...85W. doi:10.1016/0550-3213(95)00158-O.

- Yau, Shing-Tung; Nadis, Steve (2010). *The Shape of Inner Space: String Theory and the Geometry of the Universe's Hidden Dimensions*. Basic Books. ISBN 978-0-465-02023-2.

- Zaslow, Eric (2008). "Mirror Symmetry". In Gowers, Timothy. *The Princeton Companion to Mathematics*. ISBN 978-0-691-11880-2.

Chapter 24

M-theory

For a more accessible and less technical introduction to this topic, see Introduction to M-theory.

M-theory is a theory in physics that unifies all consistent versions of superstring theory. The existence of such a theory was first conjectured by Edward Witten at a string theory conference at the University of Southern California in the spring of 1995. Witten's announcement initiated a flurry of research activity known as the second superstring revolution.

Prior to Witten's announcement, string theorists had identified five versions of superstring theory. Although these theories appeared at first to be very different, work by several physicists showed that the theories were related in intricate and nontrivial ways. In particular, physicists found that apparently distinct theories could be unified by mathematical transformations called S-duality and T-duality. Witten's conjecture was based in part on the existence of these dualities and in part on the relationship of the string theories to a field theory called eleven-dimensional supergravity.

Although a complete formulation of M-theory is not known, the theory should describe two- and five-dimensional objects called branes and should be approximated by eleven-dimensional supergravity at low energies. Modern attempts to formulate M-theory are typically based on matrix theory or the AdS/CFT correspondence. According to Witten, M should stand for "magic", "mystery", or "membrane" according to taste, and the true meaning of the title should be decided when a more fundamental formulation of the theory is known.[1]

Investigations of the mathematical structure of M-theory have spawned important theoretical results in physics and mathematics. More speculatively, M-theory may provide a framework for developing a unified theory of all of the fundamental forces of nature. Attempts to connect M-theory to experiment typically focus on compactifying its extra dimensions to construct candidate models of our four-dimensional world, although so far none have been verified to give rise to physics as observed at, for instance, the Large Hadron Collider.

24.1 Background

24.1.1 Quantum gravity and strings

Main articles: Quantum gravity and String theory

One of the deepest problems in modern physics is the problem of quantum gravity. The current understanding of gravity is based on Albert Einstein's general theory of relativity, which is formulated within the framework of classical physics. However, nongravitational forces are described within the framework of quantum mechanics, a radically different formalism for describing physical phenomena based on probability.[lower-alpha 1] A quantum theory of gravity is needed in order to reconcile general relativity with the principles of quantum mechanics,[lower-alpha 2] but difficulties arise when one attempts to apply the usual prescriptions of quantum theory to the force of gravity.[lower-alpha 3]

String theory is a theoretical framework that attempts to reconcile gravity and quantum mechanics. In string theory, the point-like particles of particle physics are replaced by one-dimensional objects called strings. String theory describes how strings propagate through space and interact with each other. In a given version of string theory, there is only one kind of string, which may look like a small loop or segment of ordinary string, and it can vibrate in different

The fundamental objects of string theory are open and closed strings.

ways. On distance scales larger than the string scale, a string will look just like an ordinary particle, with its mass, charge, and other properties determined by the vibrational state of the string. In this way, all of the different elementary particles may be viewed as vibrating strings. One of the vibrational states of a string gives rise to the graviton, a quantum mechanical particle that carries gravitational force.[lower-alpha 4]

There are several versions of string theory: type I, type IIA, type IIB, and two flavors of heterotic string theory ($SO(32)$ and $E_8 \times E_8$). The different theories allow different types of strings, and the particles that arise at low energies exhibit different symmetries. For example, the type I theory includes both open strings (which are segments with endpoints) and closed strings (which form closed loops), while types IIA and IIB include only closed strings.[2] Each of these five string theories arises as a special limiting case of M-theory. This theory, like its string theory predecessors, is an example of a quantum theory of gravity. It describes a force just like the familiar gravitational force subject to the rules of quantum mechanics.[3]

24.1.2 Number of dimensions

Main article: Compactification (physics)

In everyday life, there are three familiar dimensions of space: height, width and depth. Einstein's general theory of relativity treats time as a dimension on par with the three spatial dimensions; in general relativity, space and time are not modeled as separate entities but are instead unified to a four-dimensional spacetime. In this framework, the phenomenon of gravity is viewed as a consequence of the geometry of spacetime.[4]

In spite of the fact that the universe is well described by four-dimensional spacetime, there are several reasons why physicists consider theories in other dimensions. In some cases, by modeling spacetime in a different number of dimensions, a theory becomes more mathematically tractable, and one can perform calculations and gain general insights more easily.[lower-alpha 5] There are also situations where theories in two or three spacetime dimensions are useful for describing phenomena in condensed matter physics.[5] Finally, there exist scenarios in which there could actually be more than four dimensions of spacetime which have nonetheless managed to escape detection.[6]

One notable feature of string theory and M-theory is that these theories require extra dimensions of spacetime for their mathematical consistency. In string theory, spacetime is ten-dimensional, while in M-theory it is eleven-dimensional. In order to describe real physical phenomena using these theories, one must therefore imagine scenarios in which these extra dimensions would not be observed in experiments.[7]

Compactification is one way of modifying the number of dimensions in a physical theory.[lower-alpha 6] In compactifica-

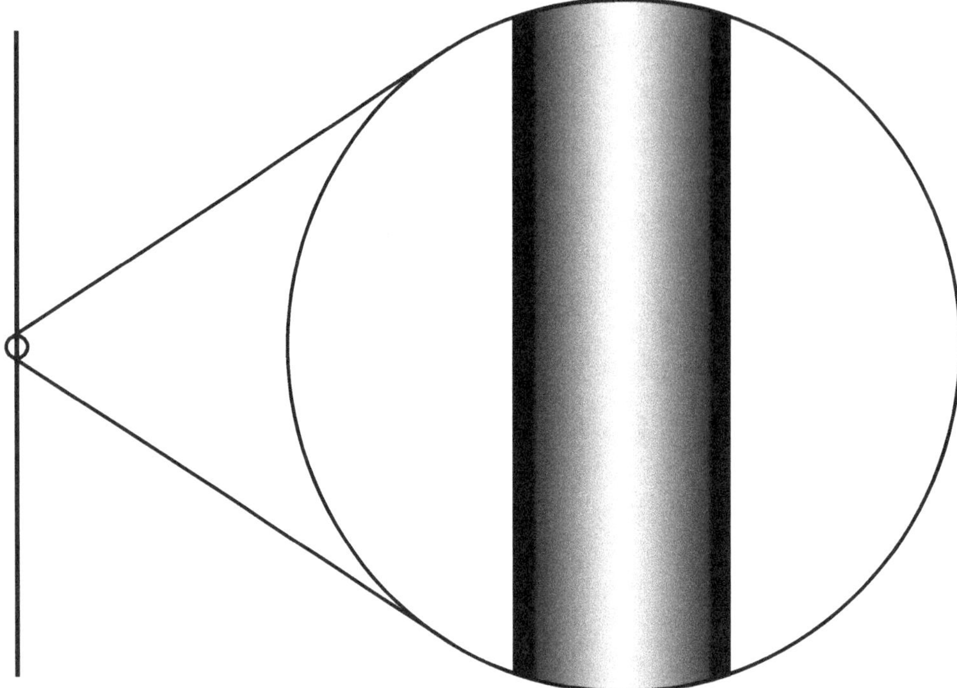

An example of compactification: At large distances, a two dimensional surface with one circular dimension looks one-dimensional.

tion, some of the extra dimensions are assumed to "close up" on themselves to form circles.[8] In the limit where these curled up dimensions become very small, one obtains a theory in which spacetime has effectively a lower number of dimensions. A standard analogy for this is to consider a multidimensional object such as a garden hose. If the hose is viewed from a sufficient distance, it appears to have only one dimension, its length. However, as one approaches the hose, one discovers that it contains a second dimension, its circumference. Thus, an ant crawling on the surface of the hose would move in two dimensions.[lower-alpha 7]

24.1.3 Dualities

Main articles: S-duality and T-duality

Theories that arise as different limits of M-theory turn out to be related in highly nontrivial ways. One of the relationships that can exist between these different physical theories is called S-duality. This is a relationship which says that a collection of strongly interacting particles in one theory can, in some cases, be viewed as a collection of weakly interacting particles in a completely different theory. Roughly speaking, a collection of particles is said to be strongly interacting if they combine and decay often and weakly interacting if they do so infrequently. Type I string theory turns out to be equivalent by S-duality to the $SO(32)$ heterotic string theory. Similarly, type IIB string theory is related to itself in a nontrivial way by S-duality.[10]

Another relationship between different string theories is T-duality. Here one considers strings propagating around a circular extra dimension. T-duality states that a string propagating around a circle of radius R is equivalent to a string propagating around a circle of radius $1/R$ in the sense that all observable quantities in one description are identified with quantities in the dual description. For example, a string has momentum as it propagates around a circle, and it can also wind around the circle one or more times. The number of times the string winds around a circle is called the winding number. If a string has momentum p and winding number n in one description, it will have momentum n and winding number p in the dual description. For example, type IIA string theory is equivalent to type IIB string theory via T-duality, and the two versions of heterotic string theory are also related by T-duality.[10]

In general, the term *duality* refers to a situation where two seemingly different physical systems turn out to be equivalent in a nontrivial way. If two theories are related by a duality, it means that one theory can be transformed in some way so that it ends up looking just like the other theory. The two theories are then said to be *dual* to one another

A diagram of string theory dualities. Yellow arrows indicate S-duality. Blue arrows indicate T-duality. These dualities may be combined to obtain equivalences of any of the five theories with M-theory.[9]

under the transformation. Put differently, the two theories are mathematically different descriptions of the same phenomena.[11]

24.1.4 Supersymmetry

Main article: Supersymmetry

Another important theoretical idea that plays a role in M-theory is supersymmetry. This is a mathematical relation that exists in certain physical theories between a class of particles called bosons and a class of particles called fermions. Roughly speaking, fermions are the constituents of matter, while bosons mediate interactions between particles. In theories with supersymmetry, each boson has a counterpart which is a fermion, and vice versa. When supersymmetry is imposed as a local symmetry, one automatically obtains a quantum mechanical theory that includes gravity. Such a theory is called a supergravity theory.[12]

A theory of strings that incorporates the idea of supersymmetry is called a superstring theory. There are several different versions of superstring theory which are all subsumed within the M-theory framework. At low energies, the superstring theories are approximated by supergravity in ten spacetime dimensions. Similarly, M-theory is approximated at low energies by supergravity in eleven dimensions.[3]

24.1.5 Branes

Main article: Brane

In string theory and related theories such as supergravity theories, a brane is a physical object that generalizes the notion of a point particle to higher dimensions. For example, a point particle can be viewed as a brane of dimension zero, while a string can be viewed as a brane of dimension one. It is also possible to consider higher-dimensional branes. In dimension p, these are called p-branes. Branes are dynamical objects which can propagate through spacetime according to the rules of quantum mechanics. They can have mass and other attributes such as charge. A p-brane sweeps out a $(p+1)$-dimensional volume in spacetime called its *worldvolume*. Physicists often study fields analogous to the electromagnetic field which live on the worldvolume of a brane. The word brane comes from the word "membrane" which refers to a two-dimensional brane.[13]

In string theory, the fundamental objects that give rise to elementary particles are the one-dimensional strings. Although the physical phenomena described by M-theory are still poorly understood, physicists know that the theory describes two- and five-dimensional branes. Much of the current research in M-theory attempts to better understand the properties of these branes.[lower-alpha 8]

24.2 History and development

24.2.1 Kaluza–Klein theory

Main article: Kaluza–Klein theory

In the early 20th century, physicists and mathematicians including Albert Einstein and Hermann Minkowski pioneered the use of four-dimensional geometry for describing the physical world.[14] These efforts culminated in the formulation of Einstein's general theory of relativity, which relates gravity to the geometry of four-dimensional spacetime.[15]

The success of general relativity led to efforts to apply higher dimensional geometry to explain other forces. In 1919, work by Theodor Kaluza showed that by passing to five-dimensional spacetime, one can unify gravity and electromagnetism into a single force.[15] This idea was improved by physicist Oskar Klein, who suggested that the additional dimension proposed by Kaluza could take the form of a circle with radius around 10^{-30} cm.[16]

The Kaluza–Klein theory and subsequent attempts by Einstein to develop unified field theory were never completely successful. In part this was because Kaluza–Klein theory predicted a particle that has never been shown to exist, and in part because it was unable to correctly predict the ratio of an electron's mass to its charge. In addition, these theories were being developed just as other physicists were beginning to discover quantum mechanics, which would ultimately prove successful in describing known forces such as electromagnetism, as well as new nuclear forces that were being discovered throughout the middle part of the century. Thus it would take almost fifty years for the idea of new dimensions to be taken seriously again.[17]

24.2.2 Early work on supergravity

Main article: Supergravity

New concepts and mathematical tools provided fresh insights into general relativity, giving rise to a period in the

In the 1980s, Edward Witten contributed to the understanding of supergravity theories. In 1995, he introduced M-theory, sparking the second superstring revolution.

1960s and 70s now known as the golden age of general relativity.[18] In the mid-1970s, physicists began studying higher-dimensional theories combining general relativity with supersymmetry, the so-called supergravity theories.[19]

General relativity does not place any limits on the possible dimensions of spacetime. Although the theory is typically formulated in four dimensions, one can write down the same equations for the gravitational field in any number of dimensions. Supergravity is more restrictive because it places an upper limit on the number of dimensions.[12] In 1978, work by Werner Nahm showed that the maximum spacetime dimension in which one can formulate a consistent supersymmetric theory is eleven.[20] In the same year, Eugene Cremmer, Bernard Julia, and Joel Scherk of the École Normale Supérieure showed that supergravity not only permits up to eleven dimensions but is in fact most elegant in this maximal number of dimensions.[21][22]

Initially, many physicists hoped that by compactifying eleven-dimensional supergravity, it might be possible to construct realistic models of our four-dimensional world. The hope was that such models would provide a unified description of the four fundamental forces of nature: electromagnetism, the strong and weak nuclear forces, and gravity. Interest in eleven-dimensional supergravity soon waned as various flaws in this scheme were discovered. One of the problems was that the laws of physics appear to distinguish between clockwise and counterclockwise, a phenomenon known as chirality. Edward Witten and others observed this chirality property cannot be readily derived by compactifying from eleven dimensions.[22]

In the first superstring revolution in 1984, many physicists turned to string theory as a unified theory of particle physics and quantum gravity. Unlike supergravity theory, string theory was able to accommodate the chirality of the standard model, and it provided a theory of gravity consistent with quantum effects.[22] Another feature of string theory that many physicists were drawn to in the 1980s and 1990s was its high degree of uniqueness. In ordinary particle theories, one can consider any collection of elementary particles whose classical behavior is described by an arbitrary Lagrangian. In string theory, the possibilities are much more constrained: by the 1990s, physicists had argued that there were only five consistent supersymmetric versions of the theory.[22]

24.2.3 Relationships between string theories

Although there were only a handful of consistent superstring theories, it remained a mystery why there was not just one consistent formulation.[22] However, as physicists began to examine string theory more closely, they realized that these theories are related in intricate and nontrivial ways.[23]

In the late 1970s, Claus Montonen and David Olive had conjectured a special property of certain physical theories.[24] A sharpened version of their conjecture concerns a theory called N=4 supersymmetric Yang–Mills theory, which describes particles similar to the quarks and gluons that make up atomic nuclei. The strength with which the particles of this theory interact is measured by a number called the coupling constant. The result of Montonen and Olive, now known as Montonen–Olive duality, states that N=4 supersymmetric Yang–Mills theory with coupling constant g is equivalent to the same theory with coupling constant $1/g$. In other words, a system of strongly interacting particles (large coupling constant) has an equivalent description as a system of weakly interacting particles (small coupling constant) and vice versa.[25]

In the 1990s, several theorists generalized Montonen–Olive duality to the S-duality relationship, which connects different string theories. Ashoke Sen studied S-duality in the context of heterotic strings in four dimensions.[26][27] Chris Hull and Paul Townsend showed that type IIB string theory with a large coupling constant is equivalent via S-duality to the same theory with small coupling constant.[28] Theorists also found that different string theories may be related by T-duality. This duality implies that strings propagating on completely different spacetime geometries may be physically equivalent.[29]

24.2.4 Membranes and fivebranes

String theory extends ordinary particle physics by promoting zero-dimensional point particles to one-dimensional objects called strings. In the late 1980s, it was natural for theorists to attempt to formulate other extensions in which particles are replaced by two-dimensional supermembranes or by higher-dimensional objects called branes. Such objects had been considered as early as 1962 by Paul Dirac,[30] and they were reconsidered by a small but enthusiastic group of physicists in the 1980s.[22]

Supersymmetry severely restricts the possible number of dimensions of a brane. In 1987, Eric Bergshoeff, Ergin Sezgin, and Paul Townsend showed that eleven-dimensional supergravity includes two-dimensional branes.[31] Intuitively, these objects look like sheets or membranes propagating through the eleven-dimensional spacetime. Shortly after this discovery, Michael Duff, Paul Howe, Takeo Inami, and Kellogg Stelle considered a particular compactification of eleven-dimensional supergravity with one of the dimensions curled up into a circle.[32] In this setting, one can imagine the membrane wrapping around the circular dimension. If the radius of the circle is sufficiently small, then this membrane looks just like a string in ten-dimensional spacetime. In fact, Duff and his collaborators showed that this construction reproduces exactly the strings appearing in type IIA superstring theory.[25]

In 1990, Andrew Strominger published a similar result which suggested that strongly interacting strings in ten dimensions might have an equivalent description in terms of weakly interacting five-dimensional branes.[33] Initially, physicists were unable to prove this relationship for two important reasons. On the one hand, the Montonen–Olive duality was still unproven, and so Strominger's conjecture was even more tenuous. On the other hand, there were many technical issues related to the quantum properties of five-dimensional branes.[34] The first of these problems was solved in 1993 when Ashoke Sen established that certain physical theories require the existence of objects with both electric and magnetic charge which were predicted by the work of Montonen and Olive.[35]

In spite of this progress, the relationship between strings and five-dimensional branes remained conjectural because theorists were unable to quantize the branes. Starting in 1991, a team of researchers including Michael Duff, Ramzi Khuri, Jianxin Lu, and Ruben Minasian considered a special compactification of string theory in which four of the ten dimensions curl up. If one considers a five-dimensional brane wrapped around these extra dimensions, then the brane looks just like a one-dimensional string. In this way, the conjectured relationship between strings and branes was reduced to a relationship between strings and strings, and the latter could be tested using already established theoretical techniques.[29]

24.2.5 Second superstring revolution

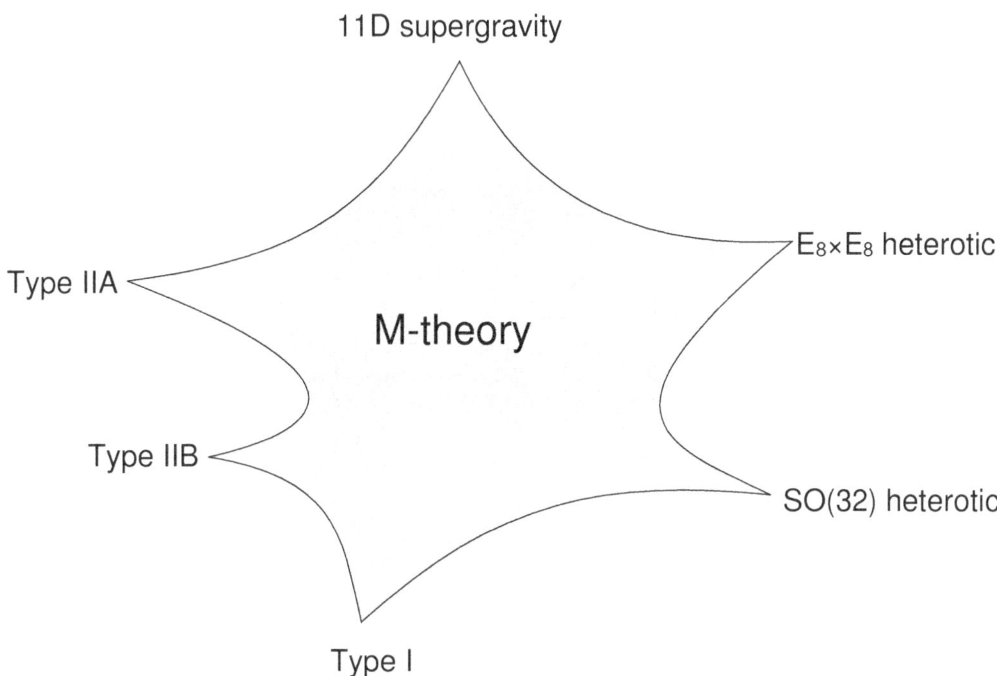

A schematic illustration of the relationship between M-theory, the five superstring theories, and eleven-dimensional supergravity. The shaded region represents a family of different physical scenarios that are possible in M-theory. In certain limiting cases corresponding to the cusps, it is natural to describe the physics using one of the six theories labeled there.

Main article: Second superstring revolution

Speaking at the string theory conference at the University of Southern California in 1995, Edward Witten of the Institute for Advanced Study made the surprising suggestion that all five superstring theories were in fact just different limiting cases of a single theory in eleven spacetime dimensions. Witten's announcement drew together all of the previous results on S- and T-duality and the appearance of two- and five-dimensional branes in string theory.[36] In the months following Witten's announcement, hundreds of new papers appeared on the Internet confirming that the new theory involved membranes in an important way.[37] Today this flurry of work is known as the second superstring revolution.[38]

One of the important developments following Witten's announcement was Witten's work in 1996 with string theorist Petr Hořava.[39][40] Witten and Hořava studied M-theory on a special spacetime geometry with two ten-dimensional boundary components. Their work shed light on the mathematical structure of M-theory and suggested possible ways of connecting M-theory to real world physics.[41]

24.2.6 Origin of the term

Initially, some physicists suggested that the new theory was a fundamental theory of membranes, but Witten was skeptical of the role of membranes in the theory. In a paper from 1996, Hořava and Witten wrote

> As it has been proposed that the eleven-dimensional theory is a supermembrane theory but there are some reasons to doubt that interpretation, we will non-committally call it the M-theory, leaving to the future the relation of M to membranes.[39]

In the absence of an understanding of the true meaning and structure of M-theory, Witten has suggested that the *M* should stand for "magic", "mystery", or "membrane" according to taste, and the true meaning of the title should be decided when a more fundamental formulation of the theory is known.[1]

24.3 Matrix theory

24.3.1 BFSS matrix model

Main article: Matrix theory (physics)

In mathematics, a matrix is a rectangular array of numbers or other data. In physics, a matrix model is a particular kind of physical theory whose mathematical formulation involves the notion of a matrix in an important way. A matrix model describes the behavior of a set of matrices within the framework of quantum mechanics.[42][43]

One important example of a matrix model is the BFSS matrix model proposed by Tom Banks, Willy Fischler, Stephen Shenker, and Leonard Susskind in 1997. This theory describes the behavior of a set of nine large matrices. In their original paper, these authors showed, among other things, that the low energy limit of this matrix model is described by eleven-dimensional supergravity. These calculations led them to propose that the BFSS matrix model is exactly equivalent to M-theory. The BFSS matrix model can therefore be used as a prototype for a correct formulation of M-theory and a tool for investigating the properties of M-theory in a relatively simple setting.[42]

24.3.2 Noncommutative geometry

Main articles: Noncommutative geometry and Noncommutative quantum field theory

In geometry, it is often useful to introduce coordinates. For example, in order to study the geometry of the Euclidean plane, one defines the coordinates x and y as the distances between any point in the plane and a pair of axes. In ordinary geometry, the coordinates of a point are numbers, so they can be multiplied, and the product of two coordinates does not depend on the order of multiplication. That is, $xy = yx$. This property of multiplication is known as the commutative law, and this relationship between geometry and the commutative algebra of coordinates is the starting point for much of modern geometry.[44]

Noncommutative geometry is a branch of mathematics that attempts to generalize this situation. Rather than working with ordinary numbers, one considers some similar objects, such as matrices, whose multiplication does not satisfy the commutative law (that is, objects for which xy is not necessarily equal to yx). One imagines that these noncommuting objects are coordinates on some more general notion of "space" and proves theorems about these generalized spaces by exploiting the analogy with ordinary geometry.[45]

In a paper from 1998, Alain Connes, Michael R. Douglas, and Albert Schwarz showed that some aspects of matrix models and M-theory are described by a noncommutative quantum field theory, a special kind of physical theory in which the coordinates on spacetime do not satisfy the commutativity property.[43] This established a link between matrix models and M-theory on the one hand, and noncommutative geometry on the other hand. It quickly led to the discovery of other important links between noncommutative geometry and various physical theories.[46][47]

24.4 AdS/CFT correspondence

24.4.1 Overview

Main article: AdS/CFT correspondence

The application of quantum mechanics to physical objects such as the electromagnetic field, which are extended in space and time, is known as quantum field theory.[lower-alpha 9] In particle physics, quantum field theories form the basis for our understanding of elementary particles, which are modeled as excitations in the fundamental fields. Quantum field theories are also used throughout condensed matter physics to model particle-like objects called quasiparticles.[lower-alpha 10]

One approach to formulating M-theory and studying its properties is provided by the anti-de Sitter/conformal field theory (AdS/CFT) correspondence. Proposed by Juan Maldacena in late 1997, the AdS/CFT correspondence is a theoretical result which implies that M-theory is in some cases equivalent to a quantum field theory.[48] In addition to providing insights into the mathematical structure of string and M-theory, the AdS/CFT correspondence has shed light on many aspects of quantum field theory in regimes where traditional calculational techniques are ineffective.[49]

In the AdS/CFT correspondence, the geometry of spacetime is described in terms of a certain vacuum solution of Einstein's equation called anti-de Sitter space.[50] In very elementary terms, anti-de Sitter space is a mathematical model of spacetime in which the notion of distance between points (the metric) is different from the notion of distance in ordinary Euclidean geometry. It is closely related to hyperbolic space, which can be viewed as a disk as illustrated on the left.[51] This image shows a tessellation of a disk by triangles and squares. One can define the distance between points of this disk in such a way that all the triangles and squares are the same size and the circular outer boundary is infinitely far from any point in the interior.[52]

Now imagine a stack of hyperbolic disks where each disk represents the state of the universe at a given time. The resulting geometric object is three-dimensional anti-de Sitter space.[51] It looks like a solid cylinder in which any cross section is a copy of the hyperbolic disk. Time runs along the vertical direction in this picture. The surface of this cylinder plays an important role in the AdS/CFT correspondence. As with the hyperbolic plane, anti-de Sitter space is curved in such a way that any point in the interior is actually infinitely far from this boundary surface.[52]

This construction describes a hypothetical universe with only two space dimensions and one time dimension, but it can be generalized to any number of dimensions. Indeed, hyperbolic space can have more than two dimensions and one can "stack up" copies of hyperbolic space to get higher-dimensional models of anti-de Sitter space.[51]

An important feature of anti-de Sitter space is its boundary (which looks like a cylinder in the case of three-dimensional anti-de Sitter space). One property of this boundary is that, within a small region on the surface around any given point, it looks just like Minkowski space, the model of spacetime used in nongravitational physics.[53] One can therefore consider an auxiliary theory in which "spacetime" is given by the boundary of anti-de Sitter space. This observation is the starting point for AdS/CFT correspondence, which states that the boundary of anti-de Sitter space can be regarded as the "spacetime" for a quantum field theory. The claim is that this quantum field theory is equivalent to the gravitational theory on the bulk anti-de Sitter space in the sense that there is a "dictionary" for translating entities and calculations in one theory into their counterparts in the other theory. For example, a single particle in the gravitational theory might correspond to some collection of particles in the boundary theory. In addition, the predictions in the two theories are quantitatively identical so that if two particles have a 40 percent chance of colliding in the gravitational theory, then the corresponding collections in the boundary theory would also have a 40 percent chance of colliding.[54]

24.4.2 6D (2,0) superconformal field theory

Main article: 6D (2,0) superconformal field theory

One particular realization of the AdS/CFT correspondence states that M-theory on the product space $AdS_7 \times S^4$ is equivalent to the so-called (2,0)-theory on the six-dimensional boundary.[48] Here "(2,0)" refers to the particular type of supersymmetry that appears in the theory. In this example, the spacetime of the gravitational theory is effectively seven-dimensional (hence the notation AdS_7), and there are four additional "compact" dimensions (encoded by the S^4 factor). In the real world, spacetime is four-dimensional, at least macroscopically, so this version of the correspondence does not provide a realistic model of gravity. Likewise, the dual theory is not a viable model of any real-world system since it describes a world with six spacetime dimensions.[lower-alpha 11]

Nevertheless, the (2,0)-theory has proven to be important for studying the general properties of quantum field theories. Indeed, this theory subsumes many mathematically interesting effective quantum field theories and points to new

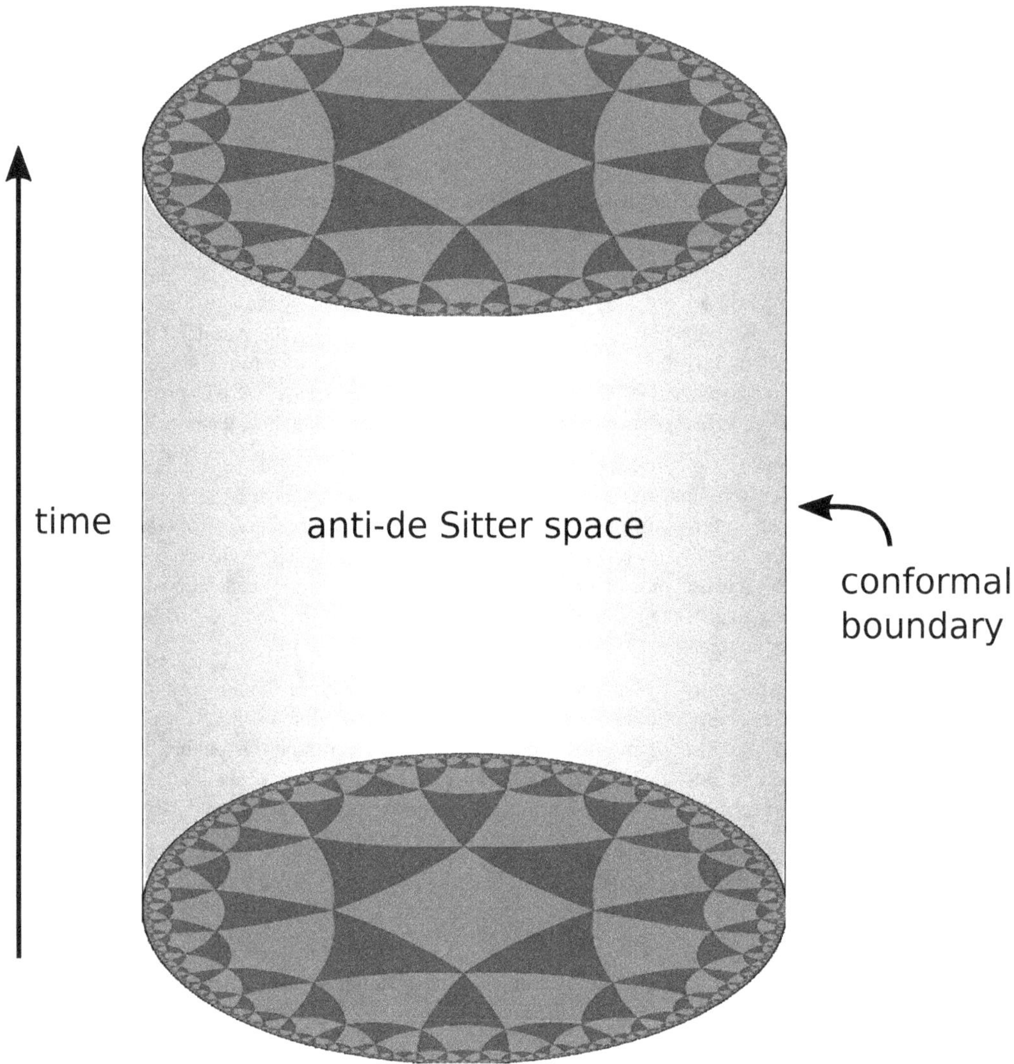

Three-dimensional anti-de Sitter space is like a stack of hyperbolic disks, each one representing the state of the universe at a given time. One can study theories of quantum gravity such as M-theory in the resulting spacetime.

dualities relating these theories. For example, Luis Alday, Davide Gaiotto, and Yuji Tachikawa showed that by compactifying this theory on a surface, one obtains a four-dimensional quantum field theory, and there is a duality known as the AGT correspondence which relates the physics of this theory to certain physical concepts associated with the surface itself.[55] More recently, theorists have extended these ideas to study the theories obtained by compactifying down to three dimensions.[56]

In addition to its applications in quantum field theory, the (2,0)-theory has spawned important results in pure mathematics. For example, the existence of the (2,0)-theory was used by Witten to give a "physical" explanation for a conjectural relationship in mathematics called the geometric Langlands correspondence.[57] In subsequent work, Witten showed that the (2,0)-theory could be used to understand a concept in mathematics called Khovanov homology.[58] Developed by Mikhail Khovanov around 2000, Khovanov homology provides a tool in knot theory, the branch of mathematics that studies and classifies the different shapes of knots.[59] Another application of the (2,0)-theory in mathematics is the work of Davide Gaiotto, Greg Moore, and Andrew Neitzke, which used physical ideas to derive new results in hyperkähler geometry.[60]

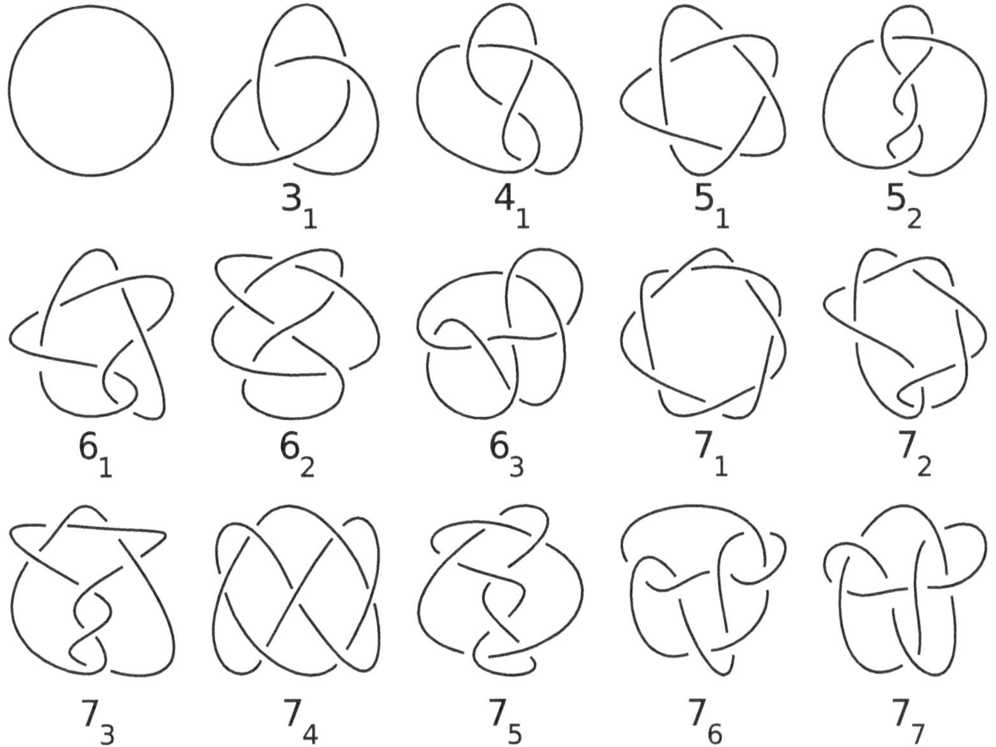

The six-dimensional (2,0)-theory has been used to understand results from the mathematical theory of knots.

24.4.3 ABJM superconformal field theory

Main article: ABJM superconformal field theory

Another realization of the AdS/CFT correspondence states that M-theory on $AdS_4 \times S^7$ is equivalent to a quantum field theory called the ABJM theory in three dimensions. In this version of the correspondence, seven of the dimensions of M-theory are curled up, leaving four non-compact dimensions. Since the spacetime of our universe is four-dimensional, this version of the correspondence provides a somewhat more realistic description of gravity.[61]

The ABJM theory appearing in this version of the correspondence is also interesting for a variety of reasons. Introduced by Aharony, Bergman, Jafferis, and Maldacena, it is closely related to another quantum field theory called Chern–Simons theory. The latter theory was popularized by Witten in the late 1980s because of its applications to knot theory.[62] In addition, the ABJM theory serves as a semi-realistic simplified model for solving problems that arise in condensed matter physics.[61]

24.5 Phenomenology

24.5.1 Overview

Main article: String phenomenology

In addition to being an idea of considerable theoretical interest, M-theory provides a framework for constructing models of real world physics that combine general relativity with the standard model of particle physics. Phenomenology is the branch of theoretical physics in which physicists construct realistic models of nature from more abstract theoretical ideas. String phenomenology is the part of string theory that attempts to construct realistic models of particle physics based on string and M-theory.[63]

Typically, such models are based on the idea of compactification.[lower-alpha 12] Starting with the ten- or eleven-dimensional

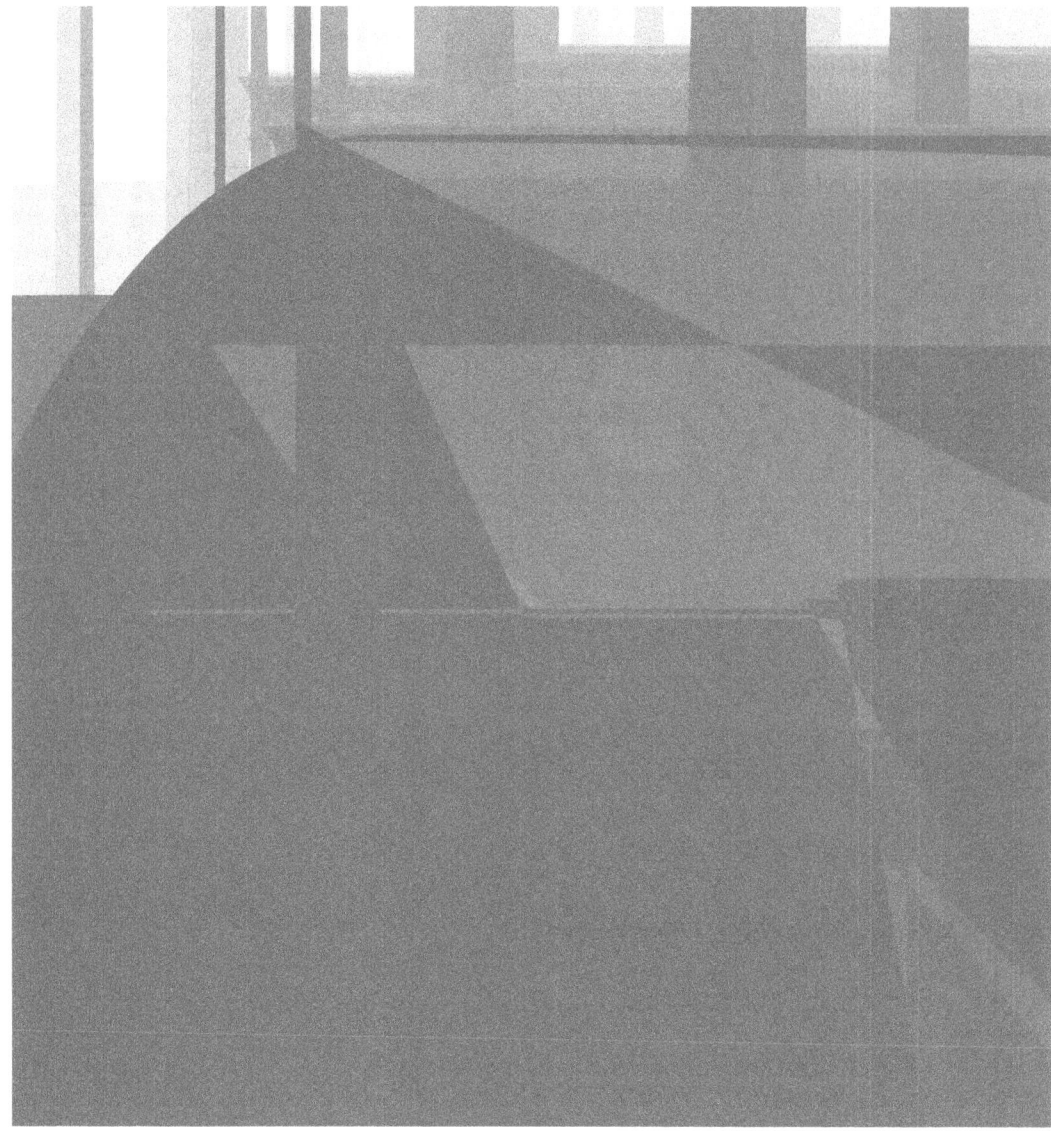

A cross section of a Calabi–Yau manifold

spacetime of string or M-theory, physicists postulate a shape for the extra dimensions. By choosing this shape appropriately, they can construct models roughly similar to the standard model of particle physics, together with additional undiscovered particles,[64] usually supersymmetric partners to analogues of known particles. One popular way of deriving realistic physics from string theory is to start with the heterotic theory in ten dimensions and assume that the six extra dimensions of spacetime are shaped like a six-dimensional Calabi–Yau manifold. This is a special kind of geometric object named after mathematicians Eugenio Calabi and Shing-Tung Yau.[65] Calabi–Yau manifolds offer many ways of extracting realistic physics from string theory. Other similar methods can be used to construct models with physics resembling to some extent that of our four-dimensional world based on M-theory.[66]

Partly because of theoretical and mathematical difficulties and partly because of the extremely high energies (beyond what is technologically possible for the foreseeable future) needed to test these theories experimentally, there is so far no experimental evidence that would unambiguously point to any of these models being a correct fundamental description of nature. This has led some in the community to criticize these approaches to unification and question the value of continued research on these problems.[67]

24.5.2 Compactification on G_2 manifolds

In one approach to M-theory phenomenology, theorists assume that the seven extra dimensions of M-theory are shaped like a G_2 manifold. This is a special kind of seven-dimensional shape constructed by mathematician Dominic Joyce of the University of Oxford.[68] These G_2 manifolds are still poorly understood mathematically, and this fact has made it difficult for physicists to fully develop this approach to phenomenology.[69]

For example, physicists and mathematicians often assume that space has a mathematical property called smoothness, but this property cannot be assumed in the case of a G_2 manifold if one wishes to recover the physics of our four-dimensional world. Another problem is that G_2 manifolds are not complex manifolds, so theorists are unable to use tools from the branch of mathematics known as complex analysis. Finally, there are many open questions about the existence, uniqueness, and other mathematical properties of G_2 manifolds, and mathematicians lack a systematic way of searching for these manifolds.[69]

24.5.3 Heterotic M-theory

Because of the difficulties with G_2 manifolds, most attempts to construct realistic theories of physics based on M-theory have taken a more indirect approach to compactifying eleven-dimensional spacetime. One approach, pioneered by Witten, Hořava, Burt Ovrut, and others, is known as heterotic M-theory. In this approach, one imagines that one of the eleven dimensions of M-theory is shaped like a circle. If this circle is very small, then the spacetime becomes effectively ten-dimensional. One then assumes that six of the ten dimensions form a Calabi–Yau manifold. If this Calabi–Yau manifold is also taken to be small, one is left with a theory in four-dimensions.[69]

Heterotic M-theory has been used to construct models of brane cosmology in which the observable universe is thought to exist on a brane in a higher dimensional ambient space. It has also spawned alternative theories of the early universe that do not rely on the theory of cosmic inflation.[69]

24.6 References

24.6.1 Notes

[1] For a standard introduction to quantum mechanics, see Griffiths 2004.

[2] The necessity of a quantum mechanical description of gravity follows from the fact that one cannot consistently couple a classical system to a quantum one. See Wald 1984, p. 382.

[3] From a technical point of view, the problem is that the theory one gets in this way is not renormalizable and therefore cannot be used to make meaningful physical predictions. See Zee 2010, p. 72 for a discussion of this issue.

[4] For an accessible introduction to string theory, see Greene 2000.

[5] For example, in the context of the AdS/CFT correspondence, theorists often formulate and study theories of gravity in unphysical numbers of spacetime dimensions.

[6] Dimensional reduction is another way of modifying the number of dimensions.

[7] This analogy is used for example in Greene 2000, p. 186.

[8] For example, see the subsections on the 6D (2,0) superconformal field theory and ABJM superconformal field theory.

[9] A standard text is Peskin and Schroeder 1995.

[10] For an introduction to the applications of quantum field theory to condensed matter physics, see Zee 2010.

[11] For a review of the (2,0)-theory, see Moore 2012.

[12] Brane world scenarios provide an alternative way of recovering real world physics from string theory. See Randall and Sundrum 1999.

24.6.2 Citations

[1] Duff 1996, sec. 1

[2] Zwiebach 2009, p. 324

[3] Becker, Becker, and Schwarz 2007, p. 12

[4] Wald 1984, p. 4

[5] Zee 2010, Parts V and VI

[6] Zwiebach 2009, p. 9

[7] Zwiebach 2009, p. 8

[8] Yau and Nadis 2010, Ch. 6

[9] Becker, Becker, and Schwarz 2007, pp. 339–347

[10] Becker, Becker, and Schwarz 2007

[11] Zwiebach 2009, p. 376

[12] Duff 1998, p. 64

[13] Moore 2005

[14] Yau and Nadis 2010, p. 9

[15] Yau and Nadis 2010, p. 10

[16] Yau and Nadis 2010, p. 12

[17] Yau and Nadis 2010, p. 13

[18] Wald 1984, p. 3

[19] van Nieuwenhuizen 1981

[20] Nahm 1978

[21] Cremmer, Julia, and Scherk 1978

[22] Duff 1998, p. 65

[23] Duff 1998

[24] Montonen and Olive 1977

[25] Duff 1998, p. 66

[26] Sen 1994a

[27] Sen 1994b

[28] Hull and Townsend 1995

[29] Duff 1998, p. 67

[30] Dirac 1962

[31] Bergshoeff, Sezgin, and Townsend 1987

[32] Duff et al. 1987

[33] Strominger 1990

[34] Duff 1998, pp 66–67

[35] Sen 1993

[36] Witten 1995

[37] Duff 1998, pp. 67–68

[38] Becker, Becker, and Schwarz 2007, p. 296

[39] Hořava and Witten 1996a

[40] Hořava and Witten 1996b

[41] Duff 1998, p. 68

[42] Banks et al. 1997

[43] Connes, Douglas, and Schwarz 1998

[44] Connes 1994, p. 1

[45] Connes 1994

[46] Nekrasov and Schwarz 1998

[47] Seiberg and Witten 1999

[48] Maldacena 1998

[49] Klebanov and Maldacena 2009

[50] Klebanov and Maldacena 2009, p. 28

[51] Maldacena 2005, p. 60

[52] Maldacena 2005, p. 61

[53] Zwiebach 2009, p. 552

[54] Maldacena 2005, pp. 61–62

[55] Alday, Gaiotto, and Tachikawa 2010

[56] Dimofte, Gaiotto, and Gukov 2010

[57] Witten 2009

[58] Witten 2012

[59] Khovanov 2000

[60] Gaiotto, Moore, and Neitzke 2013

[61] Aharony et al. 2008

[62] Witten 1989

[63] Dine 2000

[64] Candelas et al. 1985

[65] Yau and Nadis 2010, p. ix

[66] Yau and Nadis 2010, pp. 147–150

[67] Woit 2006

[68] Yau and Nadis 2010, p. 149

[69] Yau and Nadis 2010, p. 150

24.6.3 Bibliography

- Aharony, Ofer; Bergman, Oren; Jafferis, Daniel Louis; Maldacena, Juan (2008). "*N*=6 superconformal Chern-Simons-matter theories, M2-branes and their gravity duals". *Journal of High Energy Physics* **2008** (10): 091. arXiv:0806.1218. Bibcode:2008JHEP...10..091A. doi:10.1088/1126-6708/2008/10/091.

- Alday, Luis; Gaiotto, Davide; Tachikawa, Yuji (2010). "Liouville correlation functions from four-dimensional gauge theories". *Letters in Mathematical Physics* **91** (2): 167–197. arXiv:0906.3219. Bibcode:2010LMaPh..91..167A. doi:10.1007/s11005-010-0369-5.

- Banks, Tom; Fischler, Willy; Schenker, Stephen; Susskind, Leonard (1997). "M theory as a matrix model: A conjecture". *Physical Review D* **55** (8): 5112. arXiv:hep-th/9610043. Bibcode:1997PhRvD..55.5112B. doi:10.1103/physrevd.55.5112.

- Becker, Katrin; Becker, Melanie; Schwarz, John (2007). *String theory and M-theory: A modern introduction.* Cambridge University Press. ISBN 978-0-521-86069-7.

- Bergshoeff, Eric; Sezgin, Ergin; Townsend, Paul (1987). "Supermembranes and eleven-dimensional supergravity". *Physics Letters B* **189** (1): 75–78. Bibcode:1987PhLB..189...75B. doi:10.1016/0370-2693(87)91272-X.

- Candelas, Philip; Horowitz, Gary; Strominger, Andrew; Witten, Edward (1985). "Vacuum configurations for superstrings". *Nuclear Physics B* **258**: 46–74. Bibcode:1985NuPhB.258...46C. doi:10.1016/0550-3213(85)90602-9.

- Connes, Alain (1994). *Noncommutative Geometry*. Academic Press. ISBN 978-0-12-185860-5.

- Connes, Alain; Douglas, Michael; Schwarz, Albert (1998). "Noncommutative geometry and matrix theory". *Journal of High Energy Physics*. 19981 (2): 003. arXiv:hep-th/9711162. Bibcode:1998JHEP...02..003C. doi:10.1088/1126-6708/1998/02/003.

- Cremmer, Eugene; Julia, Bernard; Scherk, Joel (1978). "Supergravity theory in eleven dimensions". *Physics Letters B* **76** (4): 409–412. Bibcode:1978PhLB...76..409C. doi:10.1016/0370-2693(78)90894-8.

- Dimofte, Tudor; Gaiotto, Davide; Gukov, Sergei (2010). "Gauge theories labelled by three-manifolds". *Communications in Mathematical Physics* **325** (2): 367–419. Bibcode:2014CMaPh.325..367D. doi:10.1007/s00220-013-1863-2.

- Dine, Michael (2000). "TASI Lectures on M Theory Phenomenology". arXiv:hep-th/0003175.

- Dirac, Paul (1962). "An extensible model of the electron". *Proceedings of the Royal Society of London.* A. Mathematical and Physical Sciences **268** (1332): 57–67. Bibcode:1962RSPSA.268...57D. doi:10.1098/rspa.1962.0124.

- Duff, Michael (1996). "M-theory (the theory formerly known as strings)". *International Journal of Modern Physics A* **11** (32): 6523–41. arXiv:hep-th/9608117. Bibcode:1996IJMPA..11.5623D. doi:10.1142/S0217751X96002583.

- Duff, Michael (1998). "The theory formerly known as strings". *Scientific American* **278** (2): 64–9. doi:10.1038/scientificamerican 64.

- Duff, Michael; Howe, Paul; Inami, Takeo; Stelle, Kellogg (1987). "Superstrings in *D*=10 from supermembranes in *D*=11". *Nuclear Physics B* **191** (1): 70–74. Bibcode:1987PhLB..191...70D. doi:10.1016/0370-2693(87)91323-2.

- Gaiotto, Davide; Moore, Gregory; Neitzke, Andrew (2013). "Wall-crossing, Hitchin systems, and the WKB approximation". *Advances in Mathematics* **2341**: 239–403. arXiv:0907.3987. doi:10.1016/j.aim.2012.09.027.

- Greene, Brian (2000). *The Elegant Universe: Superstrings, Hidden Dimensions, and the Quest for the Ultimate Theory*. Random House. ISBN 978-0-9650888-0-0.

- Griffiths, David (2004). *Introduction to Quantum Mechanics*. Pearson Prentice Hall. ISBN 978-0-13-111892-8.

- Hořava, Petr; Witten, Edward (1996a). "Heterotic and Type I string dynamics from eleven dimensions". *Nuclear Physics B* **460** (3): 506–524. arXiv:hep-th/9510209. Bibcode:1996NuPhB.460..506H. doi:10.1016/0550-3213(95)00621-4.

- Hořava, Petr; Witten, Edward (1996b). "Eleven dimensional supergravity on a manifold with boundary". *Nuclear Physics B* **475** (1): 94–114. arXiv:hep-th/9603142. Bibcode:1996NuPhB.475...94H. doi:10.1016/0550-3213(96)00308-2.

- Hull, Chris; Townsend, Paul (1995). "Unity of superstring dualities". *Nuclear Physics B* **4381** (1): 109–137. arXiv:hep-th/9410167. Bibcode:1995NuPhB.438..109H. doi:10.1016/0550-3213(94)00559-W.

- Khovanov, Mikhail (2000). "A categorification of the Jones polynomial". *Duke Mathematical Journal* **1011** (3): 359–426. doi:10.1215/S0012-7094-00-10131-7.

- Klebanov, Igor; Maldacena, Juan (2009). "Solving Quantum Field Theories via Curved Spacetimes" (PDF). *Physics Today* **62**: 28. Bibcode:2009PhT....62a..28K. doi:10.1063/1.3074260. Retrieved May 2013.

- Maldacena, Juan (1998). "The Large N limit of superconformal field theories and supergravity". *Advances in Theoretical and Mathematical Physics* **2**: 231–252. arXiv:hep-th/9711200. Bibcode:1998AdTMP...2..231M. doi:10.1063/1.59653.

- Maldacena, Juan (2005). "The Illusion of Gravity" (PDF). *Scientific American* **293** (5): 56–63. Bibcode:2005SciAm.293e..56M. doi:10.1038/scientificamerican1105-56. PMID 16318027. Retrieved July 2013.

- Montonen, Claus; Olive, David (1977). "Magnetic monopoles as gauge particles?". *Physics Letters B* **72** (1): 117–120. Bibcode:1977PhLB...72..117M. doi:10.1016/0370-2693(77)90076-4.

- Moore, Gregory (2005). "What is ... a Brane?" (PDF). *Notices of the AMS* **52**: 214. Retrieved June 2013.

- Moore, Gregory (2012). "Lecture Notes for Felix Klein Lectures" (PDF). Retrieved 14 August 2013.

- Nahm, Walter (1978). "Supersymmetries and their representations". *Nuclear Physics B* **135** (1): 149–166. Bibcode:1978NuPhB.135..149N. doi:10.1016/0550-3213(78)90218-3.

- Nekrasov, Nikita; Schwarz, Albert (1998). "Instantons on noncommutative $\mathbf{R}^4$ and (2,0) superconformal six dimensional theory". *Communications in Mathematical Physics* **198** (3): 689–703. arXiv:hep-th/9802068. Bibcode:1998CMaPh.198..689N. doi:10.1007/s002200050490.

- Peskin, Michael; Schroeder, Daniel (1995). *An Introduction to Quantum Field Theory*. Westview Press. ISBN 978-0-201-50397-5.

- Randall, Lisa; Sundrum, Raman (1999). "An alternative to compactification". *Physical Review Letters* **83** (23): 4690. arXiv:hep-th/9906064. Bibcode:1999PhRvL..83.4690R. doi:10.1103/PhysRevLett.83.4690.

- Seiberg, Nathan; Witten, Edward (1999). "String Theory and Noncommutative Geometry". *Journal of High Energy Physics* **1999** (9): 032. arXiv:hep-th/9908142. Bibcode:1999JHEP...09..032S. doi:10.1088/1126-6708/1999/09/032.

- Sen, Ashoke (1993). "Electric-magnetic duality in string theory". *Nuclear Physics B* **404** (1): 109–126. arXiv:hep-th/9207053. Bibcode:1993NuPhB.404..109S. doi:10.1016/0550-3213(93)90475-5.

- Sen, Ashoke (1994a). "Strong-weak coupling duality in four-dimensional string theory". *International Journal of Modern Physics A* **9** (21): 3707–3750. arXiv:hep-th/9402002. Bibcode:1994IJMPA...9.3707S. doi:10.1142/S0217751X94C

- Sen, Ashoke (1994b). "Dyon-monopole bound states, self-dual harmonic forms on the multi-monopole moduli space, and *SL*(2,**Z**) invariance in string theory". *Physics Letters B* **329** (2): 217–221. arXiv:hep-th/9402032. Bibcode:1994PhLB..329..217S. doi:10.1016/0370-2693(94)90763-3.

- Strominger, Andrew (1990). "Heterotic solitons". *Nuclear Physics B* **343** (1): 167–184. Bibcode:1990NuPhB.343..167S. doi:10.1016/0550-3213(90)90599-9.

- van Nieuwenhuizen, Peter (1981). "Supergravity". *Physics Reports* **68** (4): 189–398. Bibcode:1981PhR....68..189V. doi:10.1016/0370-1573(81)90157-5.

- Wald, Robert (1984). *General Relativity*. University of Chicago Press. ISBN 978-0-226-87033-5.

- Witten, Edward (1989). "Quantum Field Theory and the Jones Polynomial". *Communications in Mathematical Physics* **121** (3): 351–399. Bibcode:1989CMaPh.121..351W. doi:10.1007/BF01217730. MR 0990772.

- Witten, Edward (1995). "String theory dynamics in various dimensions". *Nuclear Physics B* **443** (1): 85–126. arXiv:hep-th/9503124. Bibcode:1995NuPhB.443...85W. doi:10.1016/0550-3213(95)00158-O.

- Witten, Edward (2009). "Geometric Langlands from six dimensions". arXiv:0905.2720 [hep-th].

- Witten, Edward (2012). "Fivebranes and knots". *Quantum Topology* **3** (1): 1–137. doi:10.4171/QT/26.

- Woit, Peter (2006). *Not Even Wrong: The Failure of String Theory and the Search for Unity in Physical Law*. Basic Books. p. 105. ISBN 0-465-09275-6.

- Yau, Shing-Tung; Nadis, Steve (2010). *The Shape of Inner Space: String Theory and the Geometry of the Universe's Hidden Dimensions*. Basic Books. ISBN 978-0-465-02023-2.

- Zee, Anthony (2010). *Quantum Field Theory in a Nutshell* (2nd ed.). Princeton University Press. ISBN 978-0-691-14034-6.

- Zwiebach, Barton (2009). *A First Course in String Theory*. Cambridge University Press. ISBN 978-0-521-88032-9.

24.7 External links

- The Elegant Universe—A three-hour miniseries with Brian Greene on the series *Nova* (original PBS broadcast dates: October 28, 8–10 p.m. and November 4, 8–9 p.m., 2003). Various images, texts, videos and animations explaining string theory and M-theory.

- Superstringtheory.com—The "Official String Theory Web Site", created by Patricia Schwarz. References on string theory and M-theory for the layperson and expert.

- Not Even Wrong—Peter Woit's blog on physics in general, and string theory in particular.

Chapter 25

AdS/CFT correspondence

In theoretical physics, the **anti-de Sitter/conformal field theory correspondence**, sometimes called **Maldacena duality** or **gauge/gravity duality**, is a conjectured relationship between two kinds of physical theories. On one side of the correspondence are conformal field theories (CFT) which are quantum field theories, including theories similar to the Yang–Mills theories that describe elementary particles. On the other side are anti-de Sitter spaces (AdS) which are used in theories of quantum gravity, formulated in terms of string theory or M-theory.

The duality represents a major advance in our understanding of string theory and quantum gravity.[1] This is because it provides a non-perturbative formulation of string theory with certain boundary conditions and because it is the most successful realization of the holographic principle, an idea in quantum gravity originally proposed by Gerard 't Hooft and promoted by Leonard Susskind.

It also provides a powerful toolkit for studying strongly coupled quantum field theories.[2] Much of the usefulness of the duality results from the fact that it is a strong-weak duality: when the fields of the quantum field theory are strongly interacting, the ones in the gravitational theory are weakly interacting and thus more mathematically tractable. This fact has been used to study many aspects of nuclear and condensed matter physics by translating problems in those subjects into more mathematically tractable problems in string theory.

The AdS/CFT correspondence was first proposed by Juan Maldacena in late 1997. Important aspects of the correspondence were elaborated in articles by Steven Gubser, Igor Klebanov, and Alexander Markovich Polyakov, and by Edward Witten. By 2010, Maldacena's article had over 7000 citations, becoming the most highly cited article in the field of high energy physics.[3]

25.1 Background

25.1.1 Quantum gravity and strings

Main articles: Quantum gravity and String theory

Our current understanding of gravity is based on Albert Einstein's general theory of relativity.[4] Formulated in 1915, general relativity explains gravity in terms of the geometry of space and time, or spacetime. It is formulated in the language of classical physics[5] developed by physicists such as Isaac Newton and James Clerk Maxwell. The other nongravitational forces are explained in the framework of quantum mechanics. Developed in the first half of the twentieth century by a number of different physicists, quantum mechanics provides a radically different way of describing physical phenomena based on probability.[6]

Quantum gravity is the branch of physics that seeks to describe gravity using the principles of quantum mechanics. Currently, the most popular approach to quantum gravity is string theory,[7] which models elementary particles not as zero-dimensional points but as one-dimensional objects called strings. In the AdS/CFT correspondence, one typically considers theories of quantum gravity derived from string theory or its modern extension, M-theory.[8]

In everyday life, there are three familiar dimensions of space (up/down, left/right, and forward/backward), and there is one dimension of time. Thus, in the language of modern physics, one says that spacetime is four-dimensional.[9] One peculiar feature of string theory and M-theory is that these theories require extra dimensions of spacetime for their mathematical consistency: in string theory spacetime is ten-dimensional, while in M-theory it is eleven-dimensional.[10] The quantum gravity theories appearing in the AdS/CFT correspondence are typically obtained from string and M-theory by a process known as compactification. This produces a theory in which spacetime has effectively a lower number of dimensions and the extra dimensions are "curled up" into circles.[11]

A standard analogy for compactification is to consider a multidimensional object such as a garden hose. If the hose is viewed from a sufficient distance, it appears to have only one dimension, its length, but as one approaches the hose, one discovers that it contains a second dimension, its circumference. Thus, an ant crawling inside it would move in two dimensions.[12]

25.1.2 Quantum field theory

Main articles: Quantum field theory and Conformal field theory

The application of quantum mechanics to physical objects such as the electromagnetic field, which are extended in space and time, is known as quantum field theory.[13] In particle physics, quantum field theories form the basis for our understanding of elementary particles, which are modeled as excitations in the fundamental fields. Quantum field theories are also used throughout condensed matter physics to model particle-like objects called quasiparticles.[14]

In the AdS/CFT correspondence, one considers, in addition to a theory of quantum gravity, a certain kind of quantum field theory called a conformal field theory. This is a particularly symmetric and mathematically well behaved type of quantum field theory.[15] Such theories are often studied in the context of string theory, where they are associated with the surface swept out by a string propagating through spacetime, and in statistical mechanics, where they model systems at a thermodynamic critical point.[16]

25.2 Overview of the correspondence

25.2.1 The geometry of anti-de Sitter space

For more details on the mathematics described here, see Anti-de Sitter space.

In the AdS/CFT correspondence, one considers string theory or M-theory on an anti-de Sitter background. This means that the geometry of spacetime is described in terms of a certain vacuum solution of Einstein's equation called anti-de Sitter space.[17]

In very elementary terms, anti-de Sitter space is a mathematical model of spacetime in which the notion of distance between points (the metric) is different from the notion of distance in ordinary Euclidean geometry. It is closely related to hyperbolic space, which can be viewed as a disk as illustrated on the right.[18] This image shows a tessellation of a disk by triangles and squares. One can define the distance between points of this disk in such a way that all the triangles and squares are the same size and the circular outer boundary is infinitely far from any point in the interior.[19]

Now imagine a stack of hyperbolic disks where each disk represents the state of the universe at a given time. The resulting geometric object is three-dimensional anti-de Sitter space.[18] It looks like a solid cylinder in which any cross section is a copy of the hyperbolic disk. Time runs along the vertical direction in this picture. The surface of this cylinder plays an important role in the AdS/CFT correspondence. As with the hyperbolic plane, anti-de Sitter space is curved in such a way that any point in the interior is actually infinitely far from this boundary surface.[20]

This construction describes a hypothetical universe with only two space and one time dimension, but it can be generalized to any number of dimensions. Indeed, hyperbolic space can have more than two dimensions and one can "stack up" copies of hyperbolic space to get higher-dimensional models of anti-de Sitter space.[18]

25.2.2 The idea of AdS/CFT

An important feature of anti-de Sitter space is its boundary (which looks like a cylinder in the case of three-dimensional anti-de Sitter space). One property of this boundary is that, locally around any point, it looks just like Minkowski space, the model of spacetime used in nongravitational physics.[21]

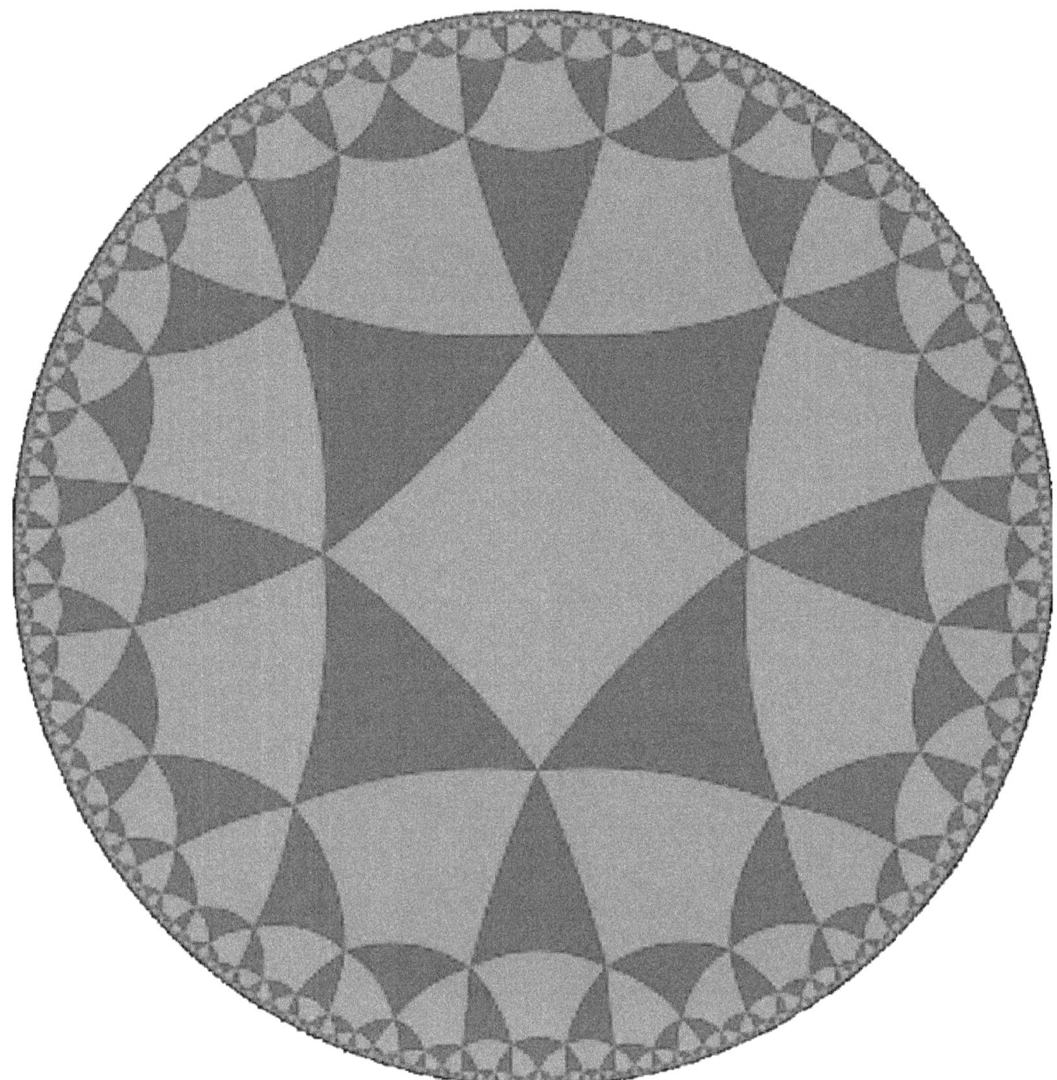

A tessellation of the hyperbolic plane by triangles and squares.

One can therefore consider an auxiliary theory in which "spacetime" is given by the boundary of anti-de Sitter space. This observation is the starting point for AdS/CFT correspondence, which states that the boundary of anti-de Sitter space can be regarded as the "spacetime" for a conformal field theory. The claim is that this conformal field theory is equivalent to the gravitational theory on the bulk anti-de Sitter space in the sense that there is a "dictionary" for translating calculations in one theory into calculations in the other. Every entity in one theory has a counterpart in the other theory. For example, a single particle in the gravitational theory might correspond to some collection of particles in the boundary theory. In addition, the predictions in the two theories are quantitatively identical so that if two particles have a 40 percent chance of colliding in the gravitational theory, then the corresponding collections in the boundary theory would also have a 40 percent chance of colliding.[22]

Notice that the boundary of anti-de Sitter space has fewer dimensions than anti-de Sitter space itself. For instance, in the three-dimensional example illustrated above, the boundary is a two-dimensional surface. The AdS/CFT correspondence is often described as a "holographic duality" because this relationship between the two theories is similar to the relationship between a three-dimensional object and its image as a hologram.[23] Although a hologram is two-dimensional, it encodes information about all three dimensions of the object it represents. In the same way, theories which are related by the AdS/CFT correspondence are conjectured to be *exactly* equivalent, despite living in different numbers of dimensions. The conformal field theory is like a hologram which captures information about the higher-dimensional quantum gravity theory.[19]

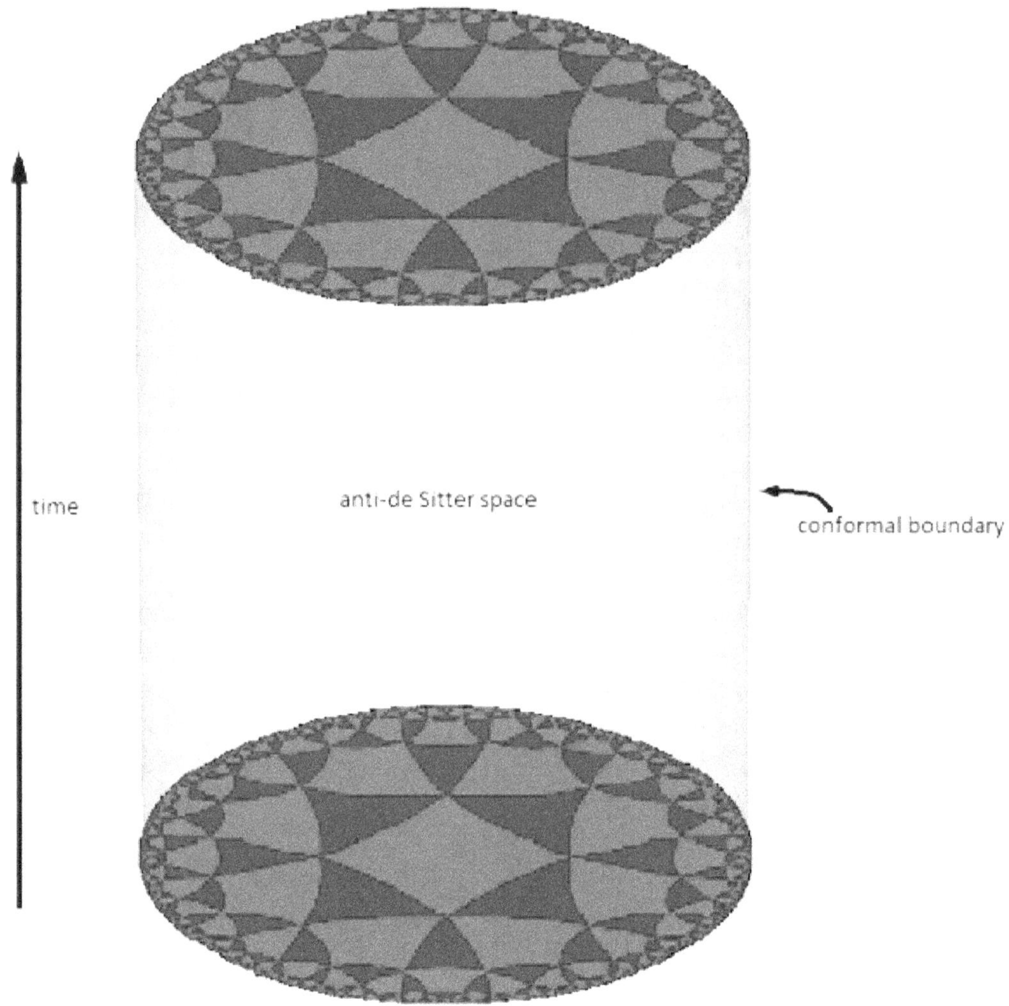

Three-dimensional anti-de Sitter space is like a stack of hyperbolic disks, each one representing the state of the universe at a given time. The resulting spacetime looks like a solid cylinder.

25.2.3 Examples of the correspondence

Following Maldacena's insight in 1997, theorists have discovered many different realizations of the AdS/CFT correspondence. These relate various conformal field theories to compactifications of string theory and M-theory in various numbers of dimensions. The theories involved are generally not viable models of the real world, but they have certain features, such as their particle content or high degree of symmetry, which make them useful for solving problems in quantum field theory and quantum gravity.[24]

The most famous example of the AdS/CFT correspondence states that type IIB string theory on the product space $AdS_5 \times S^5$ is equivalent to N = 4 supersymmetric Yang–Mills theory on the four-dimensional boundary.[25] In this example, the spacetime on which the gravitational theory lives is effectively five-dimensional (hence the notation AdS_5), and there are five additional "compact" dimensions (encoded by the S^5 factor). In the real world, spacetime is four-dimensional, at least macroscopically, so this version of the correspondence does not provide a realistic model of gravity. Likewise, the dual theory is not a viable model of any real-world system as it assumes a large amount of supersymmetry. Nevertheless, as explained below, this boundary theory shares some features in common with quantum chromodynamics, the fundamental theory of the strong force. It describes particles similar to the gluons of quantum chromodynamics together with certain fermions.[7] As a result, it has found applications in nuclear physics, particularly in the study of the quark–gluon plasma.[26]

Another realization of the correspondence states that M-theory on $AdS_7 \times S^4$ is equivalent to the so-called (2,0)-theory in six dimensions.[27] In this example, the spacetime of the gravitational theory is effectively seven-dimensional.

A hologram is a two-dimensional image which stores information about all three dimensions of the object it represents. The two images here are photographs of a single hologram taken from different angles.

The existence of the (2,0)-theory that appears on one side of the duality is predicted by the classification of superconformal field theories. It is still poorly understood because it is a quantum mechanical theory without a classical limit.[28] Despite the inherent difficulty in studying this theory, it is considered to be an interesting object for a variety of reasons, both physical and mathematical.[29]

Yet another realization of the correspondence states that M-theory on $AdS_4 \times S^7$ is equivalent to the ABJM superconformal field theory in three dimensions.[30] Here the gravitational theory has four noncompact dimensions, so this version of the correspondence provides a somewhat more realistic description of gravity.[31]

25.3 Applications to quantum gravity

25.3.1 A non-perturbative formulation of string theory

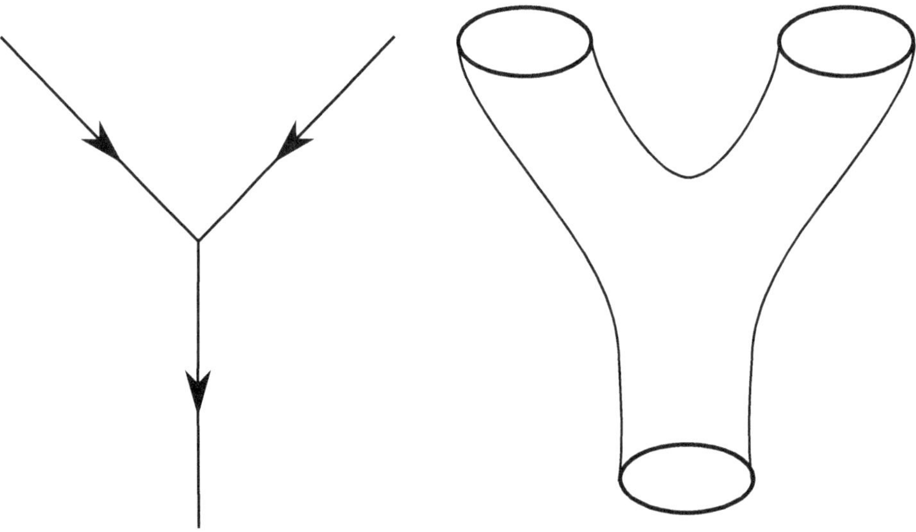

Interaction in the quantum world: world lines of point-like particles or a world sheet swept up by closed strings in string theory.

In quantum field theory, one typically computes the probabilities of various physical events using the techniques of perturbation theory. Developed by Richard Feynman and others in the first half of the twentieth century, perturbative quantum field theory uses special diagrams called Feynman diagrams to organize computations. One imagines that these diagrams depict the paths of point-like particles and their interactions.[32] Although this formalism is extremely useful for making predictions, these predictions are only possible when the strength of the interactions, the coupling constant, is small enough to reliably describe the theory as being close to a theory without interactions.[33]

The starting point for string theory is the idea that the point-like particles of quantum field theory can also be modeled as one-dimensional objects called strings. The interaction of strings is most straightforwardly defined by generalizing the perturbation theory used in ordinary quantum field theory. At the level of Feynman diagrams, this means replacing the one-dimensional diagram representing the path of a point particle by a two-dimensional surface representing the motion of a string. Unlike in quantum field theory, string theory does not yet have a full non-perturbative definition, so many of the theoretical questions that physicists would like to answer remain out of reach.[34]

The problem of developing a non-perturbative formulation of string theory was one of the original motivations for studying the AdS/CFT correspondence.[35] As explained above, the correspondence provides several examples of quantum field theories which are equivalent to string theory on anti-de Sitter space. One can alternatively view this correspondence as providing a *definition* of string theory in the special case where the gravitational field is asymptotically anti-de Sitter (that is, when the gravitational field resembles that of anti-de Sitter space at spatial infinity). Physically interesting quantities in string theory are defined in terms of quantities in the dual quantum field theory.[19]

25.3.2 Black hole information paradox

Main article: Black hole information paradox

In 1975, Stephen Hawking published a calculation which suggested that black holes are not completely black but emit a dim radiation due to quantum effects near the event horizon.[36] At first, Hawking's result posed a problem for theorists because it suggested that black holes destroy information. More precisely, Hawking's calculation seemed to conflict with one of the basic postulates of quantum mechanics, which states that physical systems evolve in time according to the Schrödinger equation. This property is usually referred to as unitarity of time evolution. The apparent contradiction between Hawking's calculation and the unitarity postulate of quantum mechanics came to be known as the black hole information paradox.[37]

The AdS/CFT correspondence resolves the black hole information paradox, at least to some extent, because it shows how a black hole can evolve in a manner consistent with quantum mechanics in some contexts. Indeed, one can consider black holes in the context of the AdS/CFT correspondence, and any such black hole corresponds to a configuration of particles on the boundary of anti-de Sitter space.[38] These particles obey the usual rules of quantum mechanics and in particular evolve in a unitary fashion, so the black hole must also evolve in a unitary fashion, respecting the principles of quantum mechanics.[39] In 2005, Hawking announced that the paradox had been settled in favor of information conservation by the AdS/CFT correspondence, and he suggested a concrete mechanism by which black holes might preserve information.[40]

25.4 Applications to quantum field theory

25.4.1 Nuclear physics

Main article: AdS/QCD

One physical system which has been studied using the AdS/CFT correspondence is the quark–gluon plasma, an exotic state of matter produced in particle accelerators. This state of matter arises for brief instants when heavy ions such as gold or lead nuclei are collided at high energies. Such collisions cause the quarks that make up atomic nuclei to deconfine at temperatures of approximately two trillion kelvins, conditions similar to those present at around 10^{-11} seconds after the Big Bang.[41]

The physics of the quark–gluon plasma is governed by quantum chromodynamics, but this theory is mathematically intractable in problems involving the quark–gluon plasma.[42] In an article appearing in 2005, Đàm Thanh Sơn and his collaborators showed that the AdS/CFT correspondence could be used to understand some aspects of the quark–gluon plasma by describing it in the language of string theory.[26] By applying the AdS/CFT correspondence, Sơn and his collaborators were able to describe the quark gluon plasma in terms of black holes in five-dimensional spacetime. The calculation showed that the ratio of two quantities associated with the quark–gluon plasma, the shear viscosity η and volume density of entropy s , should be approximately equal to a certain universal constant:

$$\frac{\eta}{s} \approx \frac{\hbar}{4\pi k}$$

where $\hbar$ denotes the reduced Planck's constant and k is Boltzmann's constant.[43] In addition, the authors conjectured that this universal constant provides a lower bound for η/s in a large class of systems. In 2008, the predicted value of this ratio for the quark–gluon plasma was confirmed at the Relativistic Heavy Ion Collider at Brookhaven National Laboratory.[44]

Another important property of the quark–gluon plasma is that very high energy quarks moving through the plasma are stopped or "quenched" after traveling only a few femtometers. This phenomenon is characterized by a number $\widehat{q}$ called the jet quenching parameter, which relates the energy loss of such a quark to the squared distance traveled through the plasma. Calculations based on the AdS/CFT correspondence have allowed theorists to estimate $\widehat{q}$, and the results agree roughly with the measured value of this parameter, suggesting that the AdS/CFT correspondence will be useful for developing a deeper understanding of this phenomenon.[45]

25.4.2 Condensed matter physics

A magnet levitating above a high-temperature superconductor. Today some physicists are working to understand high-temperature superconductivity using the AdS/CFT correspondence.[46]

Main article: AdS/CMT

Over the decades, experimental condensed matter physicists have discovered a number of exotic states of matter, including superconductors and superfluids. These states are described using the formalism of quantum field theory, but some phenomena are difficult to explain using standard field theoretic techniques. Some condensed matter theorists including Subir Sachdev hope that the AdS/CFT correspondence will make it possible to describe these systems in the language of string theory and learn more about their behavior.[47]

So far some success has been achieved in using string theory methods to describe the transition of a superfluid to an insulator. A superfluid is a system of electrically neutral atoms that flows without any friction. Such systems are often produced in the laboratory using liquid helium, but recently experimentalists have developed new ways of producing artificial superfluids by pouring trillions of cold atoms into a lattice of criss-crossing lasers. These atoms initially behave as a superfluid, but as experimentalists increase the intensity of the lasers, they become less mobile and then suddenly transition to an insulating state. During the transition, the atoms behave in an unusual way. For example, the atoms slow to a halt at a rate that depends on the temperature and on Planck's constant, the fundamental parameter of quantum mechanics, which does not enter into the description of the other phases. This behavior has recently been understood by considering a dual description where properties of the fluid are described in terms of a higher dimensional black hole.[48]

25.4.3 Criticism

With many physicists turning towards string-based methods to attack problems in nuclear and condensed matter physics, some theorists working in these areas have expressed doubts about whether the AdS/CFT correspondence can provide the tools needed to realistically model real-world systems. In a talk at the Quark Matter conference in 2006,[49] Larry McLerran pointed out that the N=4 super Yang–Mills theory that appears in the AdS/CFT correspondence differs significantly from quantum chromodynamics, making it difficult to apply these methods to nuclear physics. According to McLerran,

> $N = 4$ supersymmetric Yang–Mills is not QCD ... It has no mass scale and is conformally invariant. It has no confinement and no running coupling constant. It is supersymmetric. It has no chiral symmetry breaking or mass generation. It has six scalar and fermions in the adjoint representation ... It may be possible to correct some or all of the above problems, or, for various physical problems, some of the objections may not be relevant. As yet there is not consensus nor compelling arguments for the conjectured fixes or phenomena which would insure that the $N = 4$ supersymmetric Yang Mills results would reliably reflect QCD.[49]

In a letter to Physics Today, Nobel laureate Philip W. Anderson voiced similar concerns about applications of AdS/CFT to condensed matter physics, stating

> As a very general problem with the AdS/CFT approach in condensed-matter theory, we can point to those telltale initials "CFT"—conformal field theory. Condensed-matter problems are, in general, neither relativistic nor conformal. Near a quantum critical point, both time and space may be scaling, but even there we still have a preferred coordinate system and, usually, a lattice. There is some evidence of other linear-T phases to the left of the strange metal about which they are welcome to speculate, but again in this case the condensed-matter problem is overdetermined by experimental facts.[50]

25.5 History and development

25.5.1 String theory and nuclear physics

Main articles: History of string theory and 1/N expansion

The discovery of the AdS/CFT correspondence in late 1997 was the culmination of a long history of efforts to relate string theory to nuclear physics.[51] In fact, string theory was originally developed during the late 1960s and early 1970s as a theory of hadrons, the subatomic particles like the proton and neutron that are held together by the strong nuclear force. The idea was that each of these particles could be viewed as a different oscillation mode of a string. In the late 1960s, experimentalists had found that hadrons fall into families called Regge trajectories with squared energy proportional to angular momentum, and theorists showed that this relationship emerges naturally from the physics of a rotating relativistic string.[52]

On the other hand, attempts to model hadrons as strings faced serious problems. One problem was that string theory includes a massless spin-2 particle whereas no such particle appears in the physics of hadrons.[51] Such a particle would mediate a force with the properties of gravity. In 1974, Joel Scherk and John Schwarz suggested that string theory was therefore not a theory of nuclear physics as many theorists had thought but instead a theory of quantum gravity.[53] At the same time, it was realized that hadrons are actually made of quarks, and the string theory approach was abandoned in favor of quantum chromodynamics.[51]

In quantum chromodynamics, quarks have a kind of charge that comes in three varieties called colors. In a paper from 1974, Gerard 't Hooft studied the relationship between string theory and nuclear physics from another point of view by considering theories similar to quantum chromodynamics, where the number of colors is some arbitrary number N , rather than three. In this article, 't Hooft considered a certain limit where N tends to infinity and argued that in this limit certain calculations in quantum field theory resemble calculations in string theory.[54]

25.5.2 Black holes and holography

Main articles: Black hole information paradox, Thorne–Hawking–Preskill bet and Holographic principle

In 1975, Stephen Hawking published a calculation which suggested that black holes are not completely black but emit a dim radiation due to quantum effects near the event horizon.[36] This work extended previous results of Jacob Bekenstein who had suggested that black holes have a well defined entropy.[55] At first, Hawking's result appeared to contradict one of the main postulates of quantum mechanics, namely the unitarity of time evolution. Intuitively, the

Gerard 't Hooft obtained results related to the AdS/CFT correspondence in the 1970s by studying analogies between string theory and nuclear physics.

unitarity postulate says that quantum mechanical systems do not destroy information as they evolve from one state to another. For this reason, the apparent contradiction came to be known as the black hole information paradox.[56]

Later, in 1993, Gerard 't Hooft wrote a speculative paper on quantum gravity in which he revisited Hawking's work on black hole thermodynamics, concluding that the total number of degrees of freedom in a region of spacetime surrounding a black hole is proportional to the surface area of the horizon.[57] This idea was promoted by Leonard Susskind and is now known as the holographic principle.[58] The holographic principle and its realization in string the-

Stephen Hawking predicted in 1975 that black holes emit radiation due to quantum effects.

ory through the AdS/CFT correspondence have helped elucidate the mysteries of black holes suggested by Hawking's work and are believed to provide a resolution of the black hole information paradox.[39] In 2004, Hawking conceded that black holes do not violate quantum mechanics,[59] and he suggested a concrete mechanism by which they might

Leonard Susskind made early contributions to the idea of holography in quantum gravity.

preserve information.[40]

25.5.3 Maldacena's paper

In late 1997, Juan Maldacena published a landmark paper that initiated the study of AdS/CFT.[27] According to Alexander Markovich Polyakov, "[Maldacena's] work opened the flood gates."[60] The conjecture immediately excited great interest in the string theory community[39] and was considered in articles by Steven Gubser, Igor Klebanov and Polyakov,[61] and by Edward Witten.[62] These papers made Maldacena's conjecture more precise and showed that the conformal field theory appearing in the correspondence lives on the boundary of anti-de Sitter space.[60]

Juan Maldacena first proposed the AdS/CFT correspondence in late 1997.

One special case of Maldacena's proposal says that N=4 super Yang–Mills theory, a gauge theory similar in some ways to quantum chromodynamics, is equivalent to string theory in five-dimensional anti-de Sitter space.[30] This result helped clarify the earlier work of 't Hooft on the relationship between string theory and quantum chromodynamics, taking string theory back to its roots as a theory of nuclear physics.[52] Maldacena's results also provided a concrete realization of the holographic principle with important implications for quantum gravity and black hole physics.[1] By the year 2010, Maldacena's paper had become the most highly cited paper in high energy physics with over 7000 citations.[3] These subsequent articles have provided considerable evidence that the correspondence is correct, although so far it has not been rigorously proved.[63]

25.5.4 AdS/CFT finds applications

Main articles: AdS/QCD and AdS/CMT

In 1999, after taking a job at Columbia University, nuclear physicist Đàm Thanh Sơn paid a visit to Andrei Starinets, a friend from Sơn's undergraduate days who happened to be doing a Ph.D. in string theory at New York University.[64] Although the two men had no intention of collaborating, Sơn soon realized that the AdS/CFT calculations Starinets was doing could shed light on some aspects of the quark–gluon plasma, an exotic state of matter produced when heavy ions are collided at high energies. In collaboration with Starinets and Pavel Kovtun, Sơn was able to use the AdS/CFT correspondence to calculate a key parameter of the plasma.[26] As Sơn later recalled, "We turned the calculation on its head to give us a prediction for the value of the shear viscosity of a plasma ... A friend of mine in nuclear physics joked that ours was the first useful paper to come out of string theory."[47]

Today physicists continue to look for applications of the AdS/CFT correspondence in quantum field theory.[65] In addition to the applications to nuclear physics advocated by Đàm Thanh Sơn and his collaborators, condensed matter physicists such as Subir Sachdev have used string theory methods to understand some aspects of condensed matter physics. A notable result in this direction was the description, via the AdS/CFT correspondence, of the transition of a superfluid to an insulator.[48] Another emerging subject is the fluid/gravity correspondence, which uses the AdS/CFT correspondence to translate problems in fluid dynamics into problems in general relativity.[66]

25.6 Generalizations

25.6.1 Three-dimensional gravity

Main article: (2+1)-dimensional topological gravity

In order to better understand the quantum aspects of gravity in our four-dimensional universe, some physicists have considered a lower-dimensional mathematical model in which spacetime has only two spatial dimensions and one time dimension.[67] In this setting, the mathematics describing the gravitational field simplifies drastically, and one can study quantum gravity using familiar methods from quantum field theory, eliminating the need for string theory or other more radical approaches to quantum gravity in four dimensions.[68]

Beginning with the work of J. D. Brown and Marc Henneaux in 1986,[69] physicists have noticed that quantum gravity in a three-dimensional spacetime is closely related to two-dimensional conformal field theory. In 1995, Henneaux and his coworkers explored this relationship in more detail, suggesting that three-dimensional gravity in anti-de Sitter space is equivalent to the conformal field theory known as Liouville field theory.[70] Another conjecture formulated by Edward Witten states that three-dimensional gravity in anti-de Sitter space is equivalent to a conformal field theory with monster group symmetry.[71] These conjectures provide examples of the AdS/CFT correspondence that do not require the full apparatus of string or M-theory.[72]

25.6.2 dS/CFT correspondence

Main article: dS/CFT correspondence

Unlike our universe, which is now known to be expanding at an accelerating rate, anti-de Sitter space is neither expanding nor contracting. Instead it looks the same at all times.[18] In more technical language, one says that anti-de Sitter space corresponds to a universe with negative cosmological constant, whereas the real universe has a small positive cosmological constant.[73]

Although the properties of gravity at short distances should be somewhat independent of the value of the cosmological constant,[74] it is desirable to have a version of the AdS/CFT correspondence for positive cosmological constant. In 2001, Andrew Strominger introduced a version of the duality called the dS/CFT correspondence.[75] This duality involves a model of spacetime called de Sitter space with a positive cosmological constant. Such a duality is interesting from the point of view of cosmology since many cosmologists believe that the very early universe was close to being de Sitter space.[18] Our universe may also resemble de Sitter space in the distant future.[18]

25.6.3 Kerr/CFT correspondence

Main article: Kerr/CFT correspondence

Although the AdS/CFT correspondence is often useful for studying the properties of black holes,[76] most of the black holes considered in the context of AdS/CFT are physically unrealistic. Indeed, as explained above, most versions of the AdS/CFT correspondence involve higher-dimensional models of spacetime with unphysical supersymmetry.

In 2009, Monica Guica, Thomas Hartman, Wei Song, and Andrew Strominger showed that the ideas of AdS/CFT could nevertheless be used to understand certain astrophysical black holes. More precisely, their results apply to black holes that are approximated by extremal Kerr black holes, which have the largest possible angular momentum compatible with a given mass.[77] They showed that such black holes have an equivalent description in terms of conformal field theory. The Kerr/CFT correspondence was later extended to black holes with lower angular momentum.[78]

25.6.4 Higher spin gauge theories

The AdS/CFT correspondence is closely related to another duality conjectured by Igor Klebanov and Alexander Markovich Polyakov in 2002.[79] This duality states that certain "higher spin gauge theories" on anti-de Sitter space are equivalent to conformal field theories with O(N) symmetry. Here the theory in the bulk is a type of gauge theory describing particles of arbitrarily high spin. It is similar to string theory, where the excited modes of vibrating strings correspond to particles with higher spin, and it may help to better understand the string theoretic versions of AdS/CFT and possibly even prove the correspondence.[80] In 2010, Simone Giombi and Xi Yin obtained further evidence for this duality by computing quantities called three-point functions.[81]

25.7 See also

- Algebraic holography
- Ambient construction
- Randall–Sundrum model

25.8 Notes

[1] de Haro et al. 2013, p. 2

[2] Klebanov and Maldacena 2009

[3] "Top Cited Articles during 2010 in hep-th". Retrieved 25 July 2013.

[4] A standard textbook on general relativity is Wald 1984.

[5] Maldacena 2005, p. 58

[6] Griffiths 2004

[7] Maldacena 2005, p. 62

[8] See the subsection entitled "Examples of the correspondence". For examples which do not involve string theory or M-theory, see the section entitled "Generalizations".

[9] Wald 1984, p. 4

[10] Zwiebach 2009, p. 8

[11] Zwiebach 2009, pp. 7–8

[12] This analogy is used for example in Greene 2000, p. 186.

[13] A standard text is Peskin and Schroeder 1995.

[14] For an introduction to the applications of quantum field theory to condensed matter physics, see Zee 2010.

[15] Conformal field theories are characterized by their invariance under conformal transformations.

[16] For an introduction to conformal field theory emphasizing its applications to perturbative string theory, see Volume II of Deligne et al. 1999.

[17] Klebanov and Maldacena 2009, p. 28

[18] Maldacena 2005, p. 60

[19] Maldacena 2005, p. 61

[20] The mathematical relationship between the interior and boundary of anti-de Sitter space is related to the ambient construction of Charles Fefferman and Robin Graham. For details see Fefferman and Graham 1985, Fefferman and Graham 2011.

[21] Zwiebach 2009, p. 552

[22] Maldacena 2005, pp. 61–62

[23] Maldacena 2005, p. 57

[24] The known realizations of AdS/CFT typically involve unphysical numbers of spacetime dimensions and unphysical supersymmetries.

[25] This example is the main subject of the three pioneering articles on AdS/CFT: Maldacena 1998; Gubser, Klebanov, and Polyakov 1998; and Witten 1998.

[26] Merali 2011, p. 303; Kovtun, Son, and Starinets 2001

[27] Maldacena 1998

[28] For a review of the (2,0)-theory, see Moore 2012.

[29] See Moore 2012 and Alday, Gaiotto, and Tachikawa 2010.

[30] Aharony et al. 2008

[31] Aharony et al. 2008, sec. 1

[32] A standard textbook introducing the formalism of Feynman diagrams is Peskin and Schroeder 1995.

[33] Zee 2010, p. 43

[34] Zwiebach 2009, p. 12

[35] Maldacena 1998, sec. 6

[36] Hawking 1975

[37] For an accessible introduction to the black hole information paradox, and the related scientific dispute between Hawking and Leonard Susskind, see Susskind 2008.

[38] Zwiebach 2009, p. 554

[39] Maldacena 2005, p. 63

[40] Hawking 2005

[41] Zwiebach 2009, p. 559

[42] More precisely, one cannot apply the methods of perturbative quantum field theory.

[43] Zwiebach 2009, p. 561; Kovtun, Son, and Starinets 2001

[44] Merali 2011, p. 303; Luzum and Romatschke 2008

[45] Zwiebach 2009, p. 561

[46] Merali 2011

[47] Merali 2011, p. 303

[48] Sachdev 2013, p. 51

[49] McLerran 2007

[50] Anderson, Philip. "Strange connections to strange metals". Physics Today. Retrieved 14 August 2013.

[51] Zwiebach 2009, p. 525

[52] Aharony et al. 2008, sec. 1.1

[53] Scherk and Schwarz 1974

[54] 't Hooft 1974

[55] Bekenstein 1973

[56] Susskind 2008

[57] 't Hooft 1993

[58] Susskind 1995

[59] Susskind 2008, p. 444

[60] Polyakov 2008, p. 6

[61] Gubser, Klebanov, and Polyakov 1998

[62] Witten 1998

[63] Maldacena 2005, p. 63; Cowen 2013

[64] Merali 2011, pp. 302–303

[65] Merali 2011; Sachdev 2013

[66] Rangamani 2009

[67] For a review, see Carlip 2003.

[68] According to the results of Witten 1988, three-dimensional quantum gravity can be understood by relating it to Chern–Simons theory.

[69] Brown and Henneaux 1986

[70] Coussaert, Henneaux, and van Driel 1995

[71] Witten 2007

[72] Guica et al. 2009, p. 1

[73] Perlmutter 2003

[74] Biquard 2005, p. 33

[75] Strominger 2001

[76] See the subsection entitled "Black hole information paradox".

[77] Guica et al. 2009

[78] Castro, Maloney, and Strominger 2010

[79] Klebanov and Polyakov 2002

[80] See the Introduction in Klebanov and Polyakov 2002.

[81] Giombi and Yin 2010

25.9 References

- Aharony, Ofer; Bergman, Oren; Jafferis, Daniel Louis; Maldacena, Juan (2008). "N = 6 superconformal Chern-Simons-matter theories, M2-branes and their gravity duals”. *Journal of High Energy Physics* **2008** (10): 091. arXiv:0806.1218. Bibcode:2008JHEP...10..091A. doi:10.1088/1126-6708/2008/10/091.

- Aharony, Ofer; Gubser, Steven; Maldacena, Juan; Ooguri, Hirosi; Oz, Yaron (2000). “Large N Field Theories, String Theory and Gravity”. *Phys. Rept.* **323** (3–4): 183–386. arXiv:hep-th/9905111. Bibcode:1999PhR...323..183A. doi:10.1016/S0370-1573(99)00083-6.

- Alday, Luis; Gaiotto, Davide; Tachikawa, Yuji (2010). “Liouville correlation functions from four-dimensional gauge theories”. *Letters in Mathematical Physics* **91** (2): 167–197. arXiv:0906.3219. Bibcode:2010LMaPh..91..167A. doi:10.1007/s11005-010-0369-5.

- Bekenstein, Jacob (1973). “Black holes and entropy”. *Physical Review D* **7** (8): 2333. Bibcode:1973PhRvD...7.2333B. doi:10.1103/PhysRevD.7.2333.

- Biquard, Olivier (2005). *AdS/CFT Correspondence: Einstein Metrics and Their Conformal Boundaries*. European Mathematical Society. ISBN 978-3-03719-013-5.

- Brown, J. David; Henneaux, Marc (1986). “Central charges in the canonical realization of asymptotic symmetries: an example from three dimensional gravity”. *Communications in Mathematical Physics* **104** (2): 207–226. Bibcode:1986CMaPh.104..207B. doi:10.1007/BF01211590.

- Carlip, Steven (2003). *Quantum Gravity in 2+1 Dimensions*. Cambridge Monographs on Mathematical Physics. ISBN 978-0-521-54588-4.

- Castro, Alejandra; Maloney, Alexander; Strominger, Andrew (2010). “Hidden conformal symmetry of the Kerr black hole”. *Physical Review D* **82** (2). arXiv:1004.0996. Bibcode:2010PhRvD..82b4008C. doi:10.1103/PhysRevD.82.024

- Coussaert, Oliver; Henneaux, Marc; van Driel, Peter (1995). “The asymptotic dynamics of three-dimensional Einstein gravity with a negative cosmological constant”. *Classical and Quantum Gravity* **12** (12): 2961. arXiv:gr-qc/9506019. Bibcode:1995CQGra..12.2961C. doi:10.1088/0264-9381/12/12/012.

- Cowen, Ron (2013). “Simulations back up theory that Universe is a hologram”. *Nature News & Comment*. doi:10.1038/nature.2013.14328. Retrieved 21 December 2013.

- de Haro, Sebastian; Dieks, Dennis; 't Hooft, Gerard; Verlinde, Erik (2013). “Forty Years of String Theory Reflecting on the Foundations”. *Foundations of Physics* **43** (1): 1–7. Bibcode:2013FoPh...43....1D. doi:10.1007/s10701-012-9691-3.

- Deligne, Pierre; Etingof, Pavel; Freed, Daniel; Jeffery, Lisa; Kazhdan, David; Morgan, John; Morrison, David; Witten, Edward, eds. (1999). *Quantum Fields and Strings: A Course for Mathematicians*. American Mathematical Society. ISBN 978-0-8218-2014-8.

- Fefferman, Charles; Graham, Robin (1985). “Conformal invariants”. *Asterisque*: 95–116.

- Fefferman, Charles; Graham, Robin (2011). *The Ambient Metric*. Princeton University Press. ISBN 978-1-4008-4058-8.

- Giombi, Simone; Yin, Xi (2010). “Higher spin gauge theory and holography: the three-point functions”. *Journal of High Energy Physics* **2010** (9): 1–80. arXiv:0912.3462. Bibcode:2010JHEP...09..115G. doi:10.1007/JHEP09(2010)115.

- Greene, Brian (2000). *The Elegant Universe: Superstrings, Hidden Dimensions, and the Quest for the Ultimate Theory*. Random House. ISBN 978-0-9650888-0-0.

- Griffiths, David (2004). *Introduction to Quantum Mechanics*. Pearson Prentice Hall. ISBN 978-0-13-111892-8.

- Gubser, Steven; Klebanov, Igor; Polyakov, Alexander (1998). "Gauge theory correlators from non-critical string theory". *Physics Letters B* **428**: 105–114. arXiv:hep-th/9802109. Bibcode:1998PhLB..428..105G. doi:10.1016/S0370-2693(98)00377-3.

- Guica, Monica; Hartman, Thomas; Song, Wei; Strominger, Andrew (2009). "The Kerr/CFT Correspondence". *Physical Review D* **80** (12). arXiv:0809.4266. Bibcode:2009PhRvD..80l4008G. doi:10.1103/PhysRevD.80.124008.

- Hawking, Stephen (1975). "Particle creation by black holes". *Communications in mathematical physics* **43** (3): 199–220. Bibcode:1975CMaPh..43..199H. doi:10.1007/BF02345020.

- Hawking, Stephen (2005). "Information loss in black holes". *Physical Review D* **72** (8). arXiv:hep-th/0507171. Bibcode:2005PhRvD..72h4013H. doi:10.1103/PhysRevD.72.084013.

- Klebanov, Igor; Maldacena, Juan (2009). "Solving Quantum Field Theories via Curved Spacetimes" (PDF). *Physics Today* **62**: 28. Bibcode:2009PhT....62a..28K. doi:10.1063/1.3074260. Retrieved May 2013.

- Klebanov, Igor; Polyakov, Alexander (2002). "The AdS dual of the critical O(N) vector model". *Physics Letters B* **550** (3–4): 213–219. arXiv:hep-th/0210114. Bibcode:2002PhLB..550..213K. doi:10.1016/S0370-2693(02)02980-5.

- Kovtun, P. K.; Son, Dam T.; Starinets, A. O. (2001). "Viscosity in strongly interacting quantum field theories from black hole physics". *Physical review letters* **94** (11): 111601. arXiv:hep-th/0405231. Bibcode:2005PhRvL..94k1601K. doi:10.1103/PhysRevLett.94.111601. PMID 15903845.

- Luzum, Matthew; Romatschke, Paul (2008). "Conformal relativistic viscous hydrodynamics: Applications to RHIC results at $\sqrt{s_{NN}} = 200$ GeV". *Physical Review C* **78** (3). arXiv:0804.4015. doi:10.1103/PhysRevC.78.034915.

- Maldacena, Juan (1998). "The Large N limit of superconformal field theories and supergravity". *Advances in Theoretical and Mathematical Physics* **2**: 231–252. arXiv:hep-th/9711200. Bibcode:1998AdTMP...2..231M. doi:10.1063/1.59653.

- Maldacena, Juan (2005). "The Illusion of Gravity" (PDF). *Scientific American* **293** (5): 56–63. Bibcode:2005SciAm.293e..56M. doi:10.1038/scientificamerican1105-56. PMID 16318027. Retrieved July 2013.

- McLerran, Larry (2007). "Theory Summary : Quark Matter 2006". *Journal of Physics G: Nuclear and Particle Physics* **34** (8): S583. arXiv:hep-ph/0702004. Bibcode:2007JPhG...34..583M. doi:10.1088/0954-3899/34/8/S50.

- Merali, Zeeya (2011). "Collaborative physics: string theory finds a bench mate". *Nature* **478** (7369): 302–304. Bibcode:2011Natur.478..302M. doi:10.1038/478302a. PMID 22012369.

- Moore, Gregory (2012). "Lecture Notes for Felix Klein Lectures" (PDF). Retrieved 14 August 2013.

- Perlmutter, Saul (2003). "Supernovae, dark energy, and the accelerating universe". *Physics Today* **56** (4): 53–62. Bibcode:2003PhT....56d..53P. doi:10.1063/1.1580050.

- Peskin, Michael; Schroeder, Daniel (1995). *An Introduction to Quantum Field Theory*. Westview Press. ISBN 978-0-201-50397-5.

- Polyakov, Alexander (2008). "From Quarks to Strings". arXiv:0812.0183 [hep-th].

- Rangamani, Mukund (2009). "Gravity and Hydrodynamics: Lectures on the fluid-gravity correspondence". *Classical and quantum gravity* **26** (22): 4003. arXiv:0905.4352. Bibcode:2009CQGra..26v4003R. doi:10.1088/0264-9381/26/22/224003.

- Sachdev, Subir (2013). "Strange and stringy". *Scientific American* **308** (44): 44. Bibcode:2012SciAm.308a..44S. doi:10.1038/scientificamerican0113-44.

- Scherk, Joel; Schwarz, John (1974). "Dual models for non-hadrons". *Nuclear Physics B* **81** (1): 118–144. Bibcode:1974NuPhB..81..118S. doi:10.1016/0550-3213(74)90010-8.

- Strominger, Andrew (2001). "The dS/CFT correspondence". *Journal of High Energy Physics* **2001** (10): 034. arXiv:hep-th/0106113. Bibcode:2001JHEP...10..034S. doi:10.1088/1126-6708/2001/10/034.

- Susskind, Leonard (1995). "The World as a Hologram". *Journal of Mathematical Physics* **36** (11): 6377–6396. arXiv:hep-th/9409089. Bibcode:1995JMP....36.6377S. doi:10.1063/1.531249.

- Susskind, Leonard (2008). *The Black Hole War: My Battle with Stephen Hawking to Make the World Safe for Quantum Mechanics.* Little, Brown and Company. ISBN 978-0-316-01641-4.

- 't Hooft, Gerard (1974). "A planar diagram theory for strong interactions". *Nuclear Physics B* **72** (3): 461–473. Bibcode:1974NuPhB..72..461T. doi:10.1016/0550-3213(74)90154-0.

- 't Hooft, Gerard (1993). "Dimensional Reduction in Quantum Gravity". arXiv:gr-qc/9310026.

- Wald, Robert (1984). *General Relativity.* University of Chicago Press. ISBN 978-0-226-87033-5.

- Witten, Edward (1988). "2+1 dimensional gravity as an exactly soluble system". *Nuclear Physics B* **311** (1): 46–78. Bibcode:1988NuPhB.311...46W. doi:10.1016/0550-3213(88)90143-5.

- Witten, Edward (1998). "Anti-de Sitter space and holography". *Advances in Theoretical and Mathematical Physics* **2**: 253–291. arXiv:hep-th/9802150. Bibcode:1998AdTMP...2..253W.

- Witten, Edward (2007). "Three-dimensional gravity revisited". arXiv:0706.3359 [hep-th].

- Zee, Anthony (2010). *Quantum Field Theory in a Nutshell* (2nd ed.). Princeton University Press. ISBN 978-0-691-14034-6.

- Zwiebach, Barton (2009). *A First Course in String Theory.* Cambridge University Press. ISBN 978-0-521-88032-9.

Chapter 26

Multiverse

For other uses, see Multiverse (disambiguation).
See also: Many-worlds interpretation

The **multiverse** (or **meta-universe**) is the hypothetical set of infinite or finite possible universes (including the Universe we consistently experience) that together comprise everything that exists: the entirety of space, time, matter, and energy as well as the physical laws and constants that describe them. The various universes within the multiverse are sometimes called "**parallel universes**" or "alternate universes".

The structure of the multiverse, the nature of each universe within it and the relationships among the various constituent universes, depend on the specific multiverse hypothesis considered. Multiple universes have been hypothesized in cosmology, physics, astronomy, religion, philosophy, transpersonal psychology, and fiction, particularly in science fiction and fantasy. In these contexts, parallel universes are also called "alternate universes", "quantum universes", "interpenetrating dimensions", "parallel dimensions", "parallel worlds", "alternate realities", "alternate timelines", and "dimensional planes", among other names. The American philosopher and psychologist William James coined the term *multiverse* in 1895, but in a different context.[1]

The physics community continues to fiercely debate the multiverse hypothesis. Prominent physicists disagree about whether the multiverse may exist, and whether it is even a legitimate topic of scientific inquiry.[2] Serious concerns have been raised about whether attempts to exempt the multiverse from experimental verification may erode public confidence in science and ultimately damage the nature of fundamental physics [3] Some have argued that the multiverse question is philosophical rather than scientific because it lacks falsifiability; the ability to disprove a theory by means of scientific experiment has always been part of the accepted scientific method. [4] Paul Steinhardt has famously argued that no experiment can rule out a theory if it provides for all possible outcomes.[5]

Supporters of one of the multiverse hypotheses include Stephen Hawking,[6] Brian Greene,[7][8] Max Tegmark,[9] Alan Guth,[10] Andrei Linde,[11] Michio Kaku,[12] David Deutsch,[13] Leonard Susskind,[14] Raj Pathria,[15] Alexander Vilenkin,[16] Laura Mersini-Houghton,[17][18] Neil deGrasse Tyson[19] and Sean Carroll.[20]

Scientists who are not proponents of the multiverse include: Nobel laureate Steven Weinberg,[21] Nobel laureate David Gross,[22] Paul Steinhardt,[23] Neil Turok,[24] Viatcheslav Mukhanov,[25] Michael S. Turner,[26] Roger Penrose,[27]George Ellis,[28][29] Joe Silk, [30] Adam Frank, [31] Marcelo Gleiser, [31] Jim Baggott,[32] and Paul Davies.[33]

26.1 Multiverse hypotheses in physics

26.1.1 Categories

Max Tegmark and Brian Greene have devised classification schemes that categorize the various theoretical types of multiverse, or types of universe that might theoretically comprise a multiverse ensemble.

Max Tegmark's four levels

Cosmologist Max Tegmark has provided a taxonomy of universes beyond the familiar observable universe. The levels according to Tegmark's classification are arranged such that subsequent levels can be understood to encompass and expand upon previous levels, and they are briefly described below.[34][35]

Level I: Beyond our cosmological horizon A generic prediction of chaotic inflation is an infinite ergodic universe, which, being infinite, must contain Hubble volumes realizing all initial conditions.

Accordingly, an infinite universe will contain an infinite number of Hubble volumes, all having the same physical laws and physical constants. In regard to configurations such as the distribution of matter, almost all will differ from our Hubble volume. However, because there are infinitely many, far beyond the cosmological horizon, there will eventually be Hubble volumes with similar, and even identical, configurations. Tegmark estimates that an identical volume to ours should be about $10^{10^{115}}$ meters away from us.[9] Given infinite space, there would, in fact, be an infinite number of Hubble volumes identical to ours in the Universe.[36] This follows directly from the cosmological principle, wherein it is assumed our Hubble volume is not special or unique.

"Bubble universes": every disk is a bubble universe (Universe 1 to Universe 6 are different bubbles; they have physical constants that are different from our universe); our universe is just one of the bubbles.

Level II: Universes with different physical constants In the chaotic inflation theory, a variant of the cosmic inflation theory, the multiverse as a whole is stretching and will continue doing so forever,[37] but some regions of space stop stretching and form distinct bubbles, like gas pockets in a loaf of rising bread. Such bubbles are embryonic level I multiverses. Linde and Vanchurin calculated the number of these universes to be on the scale of $10^{10^{10,000,000}}$.[38]

Different bubbles may experience different spontaneous symmetry breaking resulting in different properties such as different physical constants.[36]

This level also includes John Archibald Wheeler's oscillatory universe theory and Lee Smolin's fecund universes theory.

Level III: Many-worlds interpretation of quantum mechanics Hugh Everett's many-worlds interpretation (MWI) is one of several mainstream interpretations of quantum mechanics. In brief, one aspect of quantum mechanics is that certain observations cannot be predicted absolutely. Instead, there is a range of possible observations, each with a different probability. According to the MWI, each of these possible observations corresponds to a different universe. Suppose a six-sided die is thrown and that the result of the throw corresponds to a quantum mechanics observable. All six possible ways the die can fall correspond to six different universes.

Tegmark argues that a level III multiverse does not contain more possibilities in the Hubble volume than a level I-II multiverse. In effect, all the different "worlds" created by "splits" in a level III multiverse with the same physical constants can be found in some Hubble volume in a level I multiverse. Tegmark writes that "The only difference between Level I and Level III is where your doppelgängers reside. In Level I they live elsewhere in good old three-dimensional space. In Level III they live on another quantum branch in infinite-dimensional Hilbert space." Similarly, all level II bubble universes with different physical constants can in effect be found as "worlds" created by "splits" at the moment of spontaneous symmetry breaking in a level III multiverse.[36] According to Yasunori Nomura[39] and Raphael Bousso and Leonard Susskind,[14] this is because global spacetime appearing in the (eternally) inflating multiverse is a redundant concept. This implies that the multiverses of Level I, II, and III are, in fact, the same thing. This hypothesis is referred to as "Multiverse = Quantum Many Worlds".

Related to the *many-worlds* idea are Richard Feynman's *multiple histories* interpretation and H. Dieter Zeh's *many-minds* interpretation.

Level IV: Ultimate ensemble The ultimate ensemble or mathematical universe hypothesis is the hypothesis of Tegmark himself.[40] This level considers equally real all universes that can be described by different mathematical structures. Tegmark writes that "abstract mathematics is so general that any Theory Of Everything (TOE) that is definable in purely formal terms (independent of vague human terminology) is also a mathematical structure. For instance, a TOE involving a set of different types of entities (denoted by words, say) and relations between them (denoted by additional words) is nothing but what mathematicians call a set-theoretical model, and one can generally find a formal system that it is a model of." He argues this "implies that any conceivable parallel universe theory can be described at Level IV" and "subsumes all other ensembles, therefore brings closure to the hierarchy of multiverses, and there cannot be say a Level V."[9]

Jürgen Schmidhuber, however, says the "set of mathematical structures" is not even well-defined, and admits only universe representations describable by constructive mathematics, that is, computer programs. He explicitly includes universe representations describable by non-halting programs whose output bits converge after finite time, although the convergence time itself may not be predictable by a halting program, due to Kurt Gödel's limitations.[41][42][43] He also explicitly discusses the more restricted ensemble of quickly computable universes.[44]

Brian Greene's nine types

American theoretical physicist and string theorist Brian Greene discussed nine types of parallel universes:[45]

Quilted The quilted multiverse works only in an infinite universe. With an infinite amount of space, every possible event will occur an infinite number of times. However, the speed of light prevents us from being aware of these other identical areas.

Inflationary The inflationary multiverse is composed of various pockets where inflation fields collapse and form new universes.

Brane The brane multiverse follows from M-theory and states that our universe is a 3-dimensional brane that exists with many others on a higher-dimensional brane or "bulk". Particles are bound to their respective branes except for gravity.

Cyclic The cyclic multiverse (via the ekpyrotic scenario) has multiple branes (each a universe) that collided, causing Big Bangs. The universes bounce back and pass through time, until they are pulled back together and again collide, destroying the old contents and creating them anew.

Landscape The landscape multiverse relies on string theory's Calabi–Yau shapes. Quantum fluctuations drop the shapes to a lower energy level, creating a pocket with a different set of laws from the surrounding space.

Quantum The quantum multiverse creates a new universe when a diversion in events occurs, as in the many-worlds interpretation of quantum mechanics.

Holographic The holographic multiverse is derived from the theory that the surface area of a space can simulate the volume of the region.

Simulated The simulated multiverse exists on complex computer systems that simulate entire universes.

Ultimate The ultimate multiverse contains every mathematically possible universe under different laws of physics.

26.1.2 Cyclic theories

Main article: Cyclic model

In several theories there is a series of infinite, self-sustaining cycles (for example: an eternity of Big Bang-Big crunches).

26.1.3 M-theory

See also: Introduction to M-theory, M-theory, Brane cosmology and String theory landscape

A multiverse of a somewhat different kind has been envisaged within string theory and its higher-dimensional extension, M-theory.[46] These theories require the presence of 10 or 11 spacetime dimensions respectively. The extra 6 or 7 dimensions may either be compactified on a very small scale, or our universe may simply be localized on a dynamical (3+1)-dimensional object, a D-brane. This opens up the possibility that there are other branes which could support “other universes”.[47][48] This is unlike the universes in the "quantum multiverse", but both concepts can operate at the same time.

Some scenarios postulate that our big bang was created, along with our universe, by the collision of two branes.[47][48]

26.1.4 Black-hole cosmology

Main article: Black-hole cosmology

A black-hole cosmology is a cosmological model in which the observable universe is the interior of a black hole existing as one of possibly many inside a larger universe. This includes the theory of white holes of which are on the opposite side of space time. While a black hole sucks everything in including light, a white hole releases matter and light, hence the name “white hole”.

26.1.5 Anthropic principle

Main article: Anthropic principle

The concept of other universes has been proposed to explain how our own universe appears to be fine-tuned for conscious life as we experience it. If there were a large (possibly infinite) number of universes, each with possibly different physical laws (or different fundamental physical constants), some of these universes, even if very few, would have the combination of laws and fundamental parameters that are suitable for the development of matter, astronomical structures, elemental diversity, stars, and planets that can exist long enough for life to emerge and evolve. The weak anthropic principle could then be applied to conclude that we (as conscious beings) would only exist in one of those few universes that happened to be finely tuned, permitting the existence of life with developed consciousness. Thus, while the probability might be extremely small that any particular universe would have the requisite conditions for life (as we understand life) to emerge and evolve, this does not require intelligent design as an explanation for the conditions in the Universe that promote our existence in it.

26.1.6 Search for evidence

Around 2010, scientists such as Stephen M. Feeney analyzed Wilkinson Microwave Anisotropy Probe (WMAP) data and claimed to find preliminary evidence suggesting that our universe collided with other (parallel) universes in the distant past.[49][50][51][52] However, a more thorough analysis of data from the WMAP and from the Planck satellite, which has a resolution 3 times higher than WMAP, failed to find any statistically significant evidence of such a bubble universe collision.[53][54] In addition, there is no evidence of any gravitational pull of other universes on ours.[55][56]

26.1.7 Criticism

Non-scientific claims

In his 2003 NY Times opinion piece, *A Brief History of the Multiverse,* author and cosmologist, Paul Davies, offers a variety of arguments that multiverse theories are non-scientific :[57]

> For a start, how is the existence of the other universes to be tested? To be sure, all cosmologists accept that there are some regions of the universe that lie beyond the reach of our telescopes, but somewhere on the slippery slope between that and the idea that there are an infinite number of universes, credibility reaches a limit. As one slips down that slope, more and more must be accepted on faith, and less and less is open to scientific verification. Extreme multiverse explanations are therefore reminiscent of theological discussions. Indeed, invoking an infinity of unseen universes to explain the unusual features of the one we do see is just as ad hoc as invoking an unseen Creator. The multiverse theory may be dressed up in scientific language, but in essence it requires the same leap of faith.
> — Paul Davies, *A Brief History of the Multiverse*

Taking cosmic inflation as a popular case in point, George Ellis, writing in August 2011, provides a balanced criticism of not only the science, but as he suggests, the scientific philosophy, by which multiverse theories are generally substantiated. He, like most cosmologists, accepts Tegmark's level I "domains", even though they lie far beyond the cosmological horizon. Likewise, the multiverse of cosmic inflation is said to exist very far away. It would be so far away, however, that it's very unlikely any evidence of an early interaction will be found. He argues that for many theorists, the lack of empirical testability or falsifiability is not a major concern. "Many physicists who talk about the multiverse, especially advocates of the string landscape, do not care much about parallel universes per se. For them, objections to the multiverse as a concept are unimportant. Their theories live or die based on internal consistency and, one hopes, eventual laboratory testing." Although he believes there's little hope that will ever be possible, he grants that the theories on which the speculation is based, are not without scientific merit. He concludes that multiverse theory is a "productive research program":[58]

> As skeptical as I am, I think the contemplation of the multiverse is an excellent opportunity to reflect on the nature of science and on the ultimate nature of existence: why we are here... In looking at this concept, we need an open mind, though not too open. It is a delicate path to tread. Parallel universes may or may not exist; the case is unproved. We are going to have to live with that uncertainty. Nothing is wrong with scientifically based philosophical speculation, which is what multiverse proposals are. But we should name it for what it is.
> — George Ellis, *Scientific American*, Does the Multiverse Really Exist?

Occam's razor

Proponents and critics disagree about how to apply Occam's razor. Critics argue that to postulate a practically infinite number of unobservable universes just to explain our own seems contrary to Occam's razor.[59] In contrast, proponents argue that, in terms of Kolmogorov complexity, the proposed multiverse is simpler than a single idiosyncratic universe.[36]

For example, multiverse proponent Max Tegmark argues:

> [A]n entire ensemble is often much simpler than one of its members. This principle can be stated more formally using the notion of algorithmic information content. The algorithmic information content in a number is, roughly speaking, the length of the shortest computer program that will produce that

> number as output. For example, consider the set of all integers. Which is simpler, the whole set or just one number? Naively, you might think that a single number is simpler, but the entire set can be generated by quite a trivial computer program, whereas a single number can be hugely long. Therefore, the whole set is actually simpler... (Similarly), the higher-level multiverses are simpler. Going from our universe to the Level I multiverse eliminates the need to specify initial conditions, upgrading to Level II eliminates the need to specify physical constants, and the Level IV multiverse eliminates the need to specify anything at all.... A common feature of all four multiverse levels is that the simplest and arguably most elegant theory involves parallel universes by default. To deny the existence of those universes, one needs to complicate the theory by adding experimentally unsupported processes and ad hoc postulates: finite space, wave function collapse and ontological asymmetry. Our judgment therefore comes down to which we find more wasteful and inelegant: many worlds or many words. Perhaps we will gradually get used to the weird ways of our cosmos and find its strangeness to be part of its charm.[36]
>
> — Max Tegmark, *"Parallel universes. Not just a staple of science fiction, other universes are a direct implication of cosmological observations." Scientific American 2003 May;288(5):40–51*

Princeton cosmologist Paul Steinhardt used the 2014 Annual Edge Question to voice his opposition to multiverse theorizing:

> A pervasive idea in fundamental physics and cosmology that should be retired: the notion that we live in a multiverse in which the laws of physics and the properties of the cosmos vary randomly from one patch of space to another. According to this view, the laws and properties within our observable universe cannot be explained or predicted because they are set by chance. Different regions of space too distant to ever be observed have different laws and properties, according to this picture. Over the entire multiverse, there are infinitely many distinct patches. Among these patches, in the words of Alan Guth, "anything that can happen will happen—and it will happen infinitely many times". Hence, I refer to this concept as a Theory of Anything. Any observation or combination of observations is consistent with a Theory of Anything. No observation or combination of observations can disprove it. Proponents seem to revel in the fact that the Theory cannot be falsified. The rest of the scientific community should be up in arms since an unfalsifiable idea lies beyond the bounds of normal science. Yet, except for a few voices, there has been surprising complacency and, in some cases, grudging acceptance of a Theory of Anything as a logical possibility. The scientific journals are full of papers treating the Theory of Anything seriously. What is going on?[23]
>
> — Paul Steinhardt, *"Theories of Anything"* edge.com'

Steinhardt claims that multiverse theories have gained currency mostly because too much has been invested in theories that have failed, e.g. inflation or string theory. He tends to see in them an attempt to redefine the values of science to which he objects even more strongly:

> A Theory of Anything is useless because it does not rule out any possibility and worthless because it submits to no do-or-die tests. (Many papers discuss potential observable consequences, but these are only possibilities, not certainties, so the Theory is never really put at risk.)[23]
>
> — Paul Steinhardt, *"Theories of Anything"* edge.com'

26.2 Multiverse hypotheses in philosophy and logic

26.2.1 Modal realism

Possible worlds are a way of explaining probability, hypothetical statements and the like, and some philosophers such as David Lewis believe that all possible worlds exist, and are just as real as the actual world (a position known as modal realism).[60]

26.2.2 Trans-world identity

A metaphysical issue that crops up in multiverse schema that posit infinite identical copies of any given universe is that of the notion that there can be identical objects in different possible worlds. According to the counterpart theory of David Lewis, the objects should be regarded as similar rather than identical.[61][62]

26.2.3 Fictional realism

The view that because fictions exist, fictional characters exist as well. There are fictional entities, in the same sense in which, setting aside philosophical disputes, there are people, Mondays, numbers and planets.[63][64]

26.3 See also

- Holographic principle
- Hugh Everett
- Impossible world
- Laura Mersini-Houghton
- Martin Rees, Astronomer Royal
- Modal realism
- Multiverse (religion)
- Parallel universe (fiction)
- Philosophy of physics
- Philosophy of space and time
- Reductionism
- Roger Penrose
- Simulated reality
- *The Fabric of Reality*

26.4 References

26.4.1 Notes

[1] James, William, *The Will to Believe*, 1895; and earlier in 1895, as cited in OED's new 2003 entry for "multiverse": James, William (October 1895), " "Is Life Worth Living?", *Internat. Jrnl. Ethics* **6**: 10, Visible nature is all plasticity and indifference, a multiverse, as one might call it, and not a universe.

[2] Kragh, H. (2009). "Contemporary History of Cosmology and the Controversy over the Multiverse". *Annals of Science* **66** (4): 529. doi:10.1080/00033790903047725.

[3] Ellis, George; Silk, Joe (December 16, 2014). "Scientific Method: Defend the Integrity of Physics". *Nature*.

[4] "Feynman on Scientific Method". *YouTube*. Retrieved July 28, 2012.

[5] Steinhardt, Paul (June 3, 2014). "Big Bang blunder bursts the Multiverse bubble". *Nature*.

[6] *Universe or Multiverse*. p. 19. ISBN 9780521848411. Some physicists would prefer to believe that string theory, or M-theory, will answer these questions and uniquely predict the features of the Universe. Others adopt the view that the initial state of the Universe is prescribed by an outside agency, code-named God, or that there are many universes, with ours being picked out by the anthropic principle. Hawking argues that string theory is unlikely to predict the distinctive features of the Universe. But neither is he is an advocate of God. He therefore opts for the last approach, favouring the type of multiverse which arises naturally within the context of his own work in quantum cosmology.

[7] Greene, Brian (January 24, 2011). *A Physicist Explains Why Parallel Universes May Exist*. *npr.org*. Interview with Terry Gross. Archived from the original on September 12, 2014. Retrieved September 12, 2014.

[8] Greene, Brian (January 24, 2011). *Transcript:A Physicist Explains Why Parallel Universes May Exist*. *npr.org*. Interview with Terry Gross. Archived from the original on September 12, 2014. Retrieved September 12, 2014.

[9] Tegmark, Max (2003). "Parallel Universes". *In "Science and Ultimate Reality: from Quantum to Cosmos", honoring John Wheeler's th birthday. J. D. Barrow, P.C.W. Davies, & C.L. Harper eds. Cambridge University Press ().* v1 **90** (2003). arXiv:astro-ph/0302131. Bibcode:2003SciAm.288e..40T. doi:10.1038/scientificamerican0503-40.

[10] "Alan Guth: Inflationary Cosmology: Is Our Universe Part of a Multiverse?". *YouTube*. Retrieved 6 October 2014.

[11] Linde, Andrei (January 27, 2012). "Inflation in Supergravity and String Theory: Brief History of the Multiverse" (PDF). *ctc.cam.ac.uk*. Archived (PDF) from the original on September 13, 2014. Retrieved September 13, 2014.

[12] Parallel Worlds: A Journey Through Creation, Higher Dimensions, and the Future of the Cosmos

[13] David Deutsch (1997). "The Ends of the Universe". The Fabric of Reality: The Science of Parallel Universes—and Its Implications. London: Penguin Press. ISBN 0-7139-9061-9.

[14] Bousso, R.; Susskind, L. (2012). "Multiverse interpretation of quantum mechanics". *Physical Review D* **85** (4). arXiv:1105.3796. doi:10.1103/PhysRevD.85.045007.

[15] Pathria, R. K. (1972). "The Universe as a Black Hole". *Nature* **240** (5379): 298. Bibcode:1972Natur.240..298P. doi:10.1038/240298a0.

[16] Vilenkin, Alex (2007). *Many Worlds in One: The Search for Other Universes*. ISBN 9780374707149.

[17] Catchpole, Heather (November 24, 2009). "Weird data suggests something big beyond the edge of the universe". *Cosmos (magazine)*. Retrieved July 27, 2014.

[18] Moon, Timur (May 19, 2013). "Planck Space Data Yields Evidence of Universes Beyond Our Own". *International Business Times*. Retrieved July 27, 2014.

[19] Freeman, David (March 4, 2014). "Why Revive 'Cosmos?' Neil DeGrasse Tyson Says Just About Everything We Know Has Changed". *huffingtonpost.com*. Archived from the original on September 12, 2014. Retrieved September 12, 2014.

[20] Sean Carroll (October 18, 2011). "Welcome to the Multiverse". *Discover (magazine)*. Retrieved May 5, 2015.

[21] Falk, Dan (March 17, 2015). "Science's Path from Myth to Multiverse". *Quanta Magazine* (New York: Simons Foundation).

[22] Davies, Paul (2008). "Many Scientists Hate the Multiverse Idea". *The Goldilocks Enigma: Why Is the Universe Just Right for Life?*. Houghton Mifflin Harcourt. p. 207. ISBN 9780547348469.

[23] Steinhardt, Paul (March 9, 2014). "Theories of Anything". *edge.org*. 2014 : WHAT SCIENTIFIC IDEA IS READY FOR RETIREMENT?. Archived from the original on March 9, 2014. Retrieved March 9, 2014.

[24] Gibbons, G.W.; Turok, Neil (2008). "The Measure Problem in Cosmology". *Phys.Rev.D* **77** (6): 063516. arXiv:hep-th/0609095. Bibcode:2008PhRvD..77f3516G. doi:10.1103/PhysRevD.77.063516.

[25] Mukhanov, Viatcheslav (2014). "Inflation without Selfreproduction". *Fortschritte der Physik* **63** (1): 36–41. doi:10.1002/prop.201400074.

[26] Woit, Peter (June 9, 2015). "A Crisis at the (Western) Edge of Physics". *Not Even Wrong*.

[27] Woit, Peter (June 14, 2015). "CMB @ 50". *Not Even Wrong*.

[28] Ellis, George F. R. (August 1, 2011). "Does the Multiverse Really Exist?". *Scientific American* (New York: Nature Publishing Group) **305** (2): 38–43. doi:10.1038/scientificamerican0811-38. ISSN 0036-8733. LCCN 04017574. OCLC 828582568. Retrieved September 12, 2014. (subscription required (help)).

[29] Ellis, George (2012). "The Multiverse: Conjecture, Proof, and Science" (PDF). *Slides for a talk at Nicolai Fest Golm 2012*. Archived (PDF) from the original on September 12, 2014. Retrieved September 12, 2014.

[30] Ellis, George; Silk, Joe (December 16, 2014), "Scientific Method: Defend the Integrity of Physics", *Nature*

[31] Frank, Adam; Gleiser, Marcelo (June 5, 2015). "A Crisis at the Edge of Physics". *New York Times*.

[32] Baggott, Jim (August 1, 2013). *Farewell to Reality: How Modern Physics Has Betrayed the Search for Scientific Truth*. Pegasus. ISBN 978-1-60598-472-8. ISBN 978-1-60598-574-9.

[33] Davies, Paul (April 12, 2003). "A Brief History of the Multiverse". *New York Times*.

[34] Tegmark, Max (May 2003). "Parallel Universes". *Scientific American*.

[35] Tegmark, Max (23 January 2003). *Parallel Universes* (PDF). Retrieved 7 February 2006.

[36] "Parallel universes. Not just a staple of science fiction, other universes are a direct implication of cosmological observations.", Tegmark M., Sci Am. 2003 May;288(5):40–51.

[37] "First Second of the Big Bang". *How The Universe Works 3*. 2014. Discovery Science.

[38] Zyga, Lisa "Physicists Calculate Number of Parallel Universes", *PhysOrg*, 16 October 2009.

[39] Nomura, Y. (2011). "Physical theories, eternal inflation, and the quantum universe". *Journal of High Energy Physics* **2011** (11). arXiv:1104.2324. doi:10.1007/JHEP11(2011)063.

[40] Tegmark, Max (2014). *Our Mathematical Universe: My Quest for the Ultimate Nature of Reality*. Knopf Doubleday Publishing Group. ISBN 9780307599803.

[41] J. Schmidhuber (1997): A Computer Scientist's View of Life, the Universe, and Everything. Lecture Notes in Computer Science, pp. 201–208, Springer: IDSIA – Dalle Molle Institute for Artificial Intelligence

[42] Schmidhuber, Juergen (2000). "Algorithmic Theories of Everything". *Sections in: Hierarchies of generalized Kolmogorov complexities and nonenumerable universal measures computable in the limit. International Journal of Foundations of Computer Science ():587-612 (2002). Section 6 in: the Speed Prior: A New Simplicity Measure Yielding Near-Optimal Computable Predictions. in J. Kivinen and R. H. Sloan, editors, Proceedings of the 15th Annual Conference on Computational Learning Theory (COLT 2002), Sydney, Australia, Lecture Notes in Artificial Intelligence, pages 216-–228. Springer, 2002* **13** (4): 1–5. arXiv:quant-ph/0011122. Bibcode:2000quant.ph.11122S.

[43] J. Schmidhuber (2002): Hierarchies of generalized Kolmogorov complexities and nonenumerable universal measures computable in the limit. International Journal of Foundations of Computer Science 13(4):587–612 IDSIA – Dalle Molle Institute for Artificial Intelligence

[44] J. Schmidhuber (2002): The Speed Prior: A New Simplicity Measure Yielding Near-Optimal Computable Predictions. Proc. 15th Annual Conference on Computational Learning Theory (COLT 2002), Sydney, Australia, Lecture Notes in Artificial Intelligence, pp. 216–228. Springer: IDSIA – Dalle Molle Institute for Artificial Intelligence

[45] In The Hidden Reality: Parallel Universes and the Deep Laws of the Cosmos, 2011

[46] Weinberg, Steven (2005). "Living in the Multiverse". arXiv:hep-th/0511037v1.

[47] Richard J Szabo, *An introduction to string theory and D-brane dynamics* (2004)

[48] Maurizio Gasperini, *Elements of String Cosmology* (2007)

[49] Lisa Zyga (December 17, 2010). "Scientists find first evidence that many universes exist". *PhysOrg.com*. phys.org. Retrieved 12 October 2013.

[50] "Astronomers Find First Evidence Of Other Universe". technologyreview.com. December 13, 2010. Retrieved 12 October 2013.

[51] Max Tegmark; Alexander Vilenkin (July 19, 2011). "The Case for Parallel Universes". Retrieved 12 October 2013.

[52] "Is Our Universe Inside a Bubble? First Observational Test of the 'Multiverse'". *Science Daily*. sciencedaily.com. Aug 3, 2011. Retrieved 12 October 2013.

[53] Feeney, Stephen M. et al. (2011). "First observational tests of eternal inflation: Analysis methods and WMAP 7-year results". *Physical Review D* **84** (4): 43507. arXiv:1012.3667. Bibcode:2011PhRvD..84d3507F. doi:10.1103/PhysRevD.84.043507.

[54] Feeney et al. (2011). "First observational tests of eternal inflation". *Physical review letters* **107** (7). arXiv:1012.1995. Bibcode:2011PhRvL.107g1301F. doi:10.1103/PhysRevLett.107.071301. . Bousso, Raphael; Harlow, Daniel; Senatore, Leonardo (2013). "Inflation after False Vacuum Decay: Observational Prospects after Planck". *Physical Review D* **91** (8). arXiv:1309.4060. Bibcode:2015PhRvD..91h3527B. doi:10.1103/PhysRevD.91.083527.

[55] Collaboration, Planck; Ade, P. A. R.; Aghanim, N.; Arnaud, M.; Ashdown, M.; Aumont, J.; Baccigalupi, C.; Balbi, A.; Banday, A. J.; Barreiro, R. B.; Battaner, E.; Benabed, K.; Benoit-Levy, A.; Bernard, J. -P.; Bersanelli, M.; Bielewicz, P.; Bikmaev, I.; Bobin, J.; Bock, J. J.; Bonaldi, A.; Bond, J. R.; Borrill, J.; Bouchet, F. R.; Burigana, C.; Butler, R. C.; Cabella, P.; Cardoso, J. -F.; Catalano, A.; Chamballu, A. et al. (2013-03-20). "[1303.5090] Planck intermediate results. XIII. Constraints on peculiar velocities". arXiv:[//arxiv.org/abs/1303.5090 1303.5090] [astro-ph.CO].

[56] "Blow for 'dark flow' in Planck's new view of the cosmos". *New Scientist*. 3 April 2013. Retrieved 10 March 2014.

[57] Davies, Paul (12 April 2003). "A Brief History of the Multiverse". *New York Times*. Retrieved 16 August 2011.

[58] Ellis, George F. R. (August 1, 2011). "Does the Multiverse Really Exist?". *Scientific American* (New York: Nature Publishing Group) **305** (2): 38–43. doi:10.1038/scientificamerican0811-38. ISSN 0036-8733. LCCN 04017574. OCLC 828582568. Retrieved August 16, 2011. (subscription required (help)).

[59] Trinh, Xuan Thuan (2006). Staune, Jean, ed. *Science & the Search for Meaning: Perspectives from International Scientists*. West Conshohocken, PA: Templeton Foundation. p. 186. ISBN 1-59947-102-7.

[60] Lewis, David (1986). *On the Plurality of Worlds*. Basil Blackwell. ISBN 0-631-22426-2.

[61] Deutsch, Harry (Summer 2002). Edward N. Zalta, ed. "Relative Identity". *The Stanford Encyclopedia of Philosophy*. Retrieved 6 October 2014.

[62] "Paul B. Kantor "The Interpretation of Cultures and Possible Worlds", 1 October 2002". Retrieved 6 October 2014.

[63] Schnieder, Benjamin; von Solodkoff, Tatjana (2009). "In Defence of Fictional Realism". *The Philosophical Quarterly* **59** (234): 138. doi:10.1111/j.1467-9213.2008.583.x.

[64] Thomasson, Amie L. (2009), "Fictional Entities", in Kim, Jaegwon; Sosa, Ernest; Rosenkrantz, Gary, *A Companion to Metaphysics* (PDF) (2nd ed.), Blackwell, pp. 10–18

26.4.2 Bibliography

- Bernard Carr, ed. (2007) *Universe or Multiverse?* Cambridge Univ. Press.
- Deutsch, David (1985). "Quantum theory, the Church–Turing principle and the universal quantum computer" (PDF). *Proceedings of the Royal Society of London A* (400): 97–117.
- Ellis, George F.R.; William R. Stoeger; Stoeger, W. R. (2004). "Multiverses and physical cosmology". *Monthly Notices of the Royal Astronomical Society* **347** (3): 921–936. arXiv:astro-ph/0305292. Bibcode:2004MNRAS.347..921E. doi:10.1111/j.1365-2966.2004.07261.x.
- Surya-Siddhanta: A Text Book of Hindu Astronomy by Ebenezer Burgess, ed. Phanindralal Gangooly (1989/1997) with a 45-page commentary by P. C. Sengupta (1935).

26.5 External links

- Interview with Tufts cosmologist Alex Vilenkin on his new book, "Many Worlds in One: The Search for Other Universes" on the podcast and public radio interview program ThoughtCast.
- Joseph Pine II about Multiverse, Presentation at Mobile Monday Amsterdam, 2008
- Multiverse – Radio-discussion on BBC Four with Melvyn Bragg

26.6 Text and image sources, contributors, and licenses

26.6.1 Text

- **String theory** *Source:* https://en.wikipedia.org/wiki/String_theory?oldid=670289441 *Contributors:* AxelBoldt, Sodium, Mav, Bryan Derksen, Zundark, The Anome, Tarquin, Taw, Eean, Malcolm Farmer, Hephaestos, Olivier, Drseudo, Stevertigo, Spiff~enwiki, Edward, PhilipMW, Michael Hardy, Bewildebeast, Dante Alighieri, Gabbe, Graue, Tgeorgescu, Mcarling, CesarB, Looxix~enwiki, Ahoerstemeier, Theresa knott, Suisui, Angela, Den fjättrade ankan~enwiki, Jdforrester, Julesd, Salsa Shark, Schneelocke, Charles Matthews, Timwi, Bemoeial, Jitse Niesen, 4lex, Greenrd, ErikStewart, Furrykef, Saltine, Phys, Omegatron, Bevo, Topbanana, Trent, Nufy8, Robbot, Craig Stuntz, Fredrik, Chris 73, R3m0t, COGDEN, Mirv, Wjhonson, Sverdrup, Academic Challenger, DHN, Hadal, Khlo, El-Benevolente, HaeB, Tobias Bergemann, Giftlite, DocWatson42, Christopher Parham, Awolf002, Mporter, Amorim Parga, Mikez, Harp, Kim Bruning, Tom harrison, Ferkelparade, Leflyman, Fropuff, No Guru, Anville, Moyogo, Curps, Pashute, Nomad~enwiki, Mboverload, Solipsist, SWAdair, DemonThing, Wmahan, Btphelps, MSTCrow, Decoy, Chowbok, Gadfium, Steuard, Pgan002, Quadell, Carandol~enwiki, Antandrus, Beland, JoJan, Khaosworks, Tothebarricades.tk, Thincat, Tomruen, Shidobu, Icairns, Lumidek, NoPetrol, Avihu, Fanghong~enwiki, Trevor MacInnis, Lacrimosus, Zro, D6, Urvabara, Felix Wan, Jkl, Discospinster, ElTyrant, Rich Farmbrough, Rhobite, Pjacobi, Alien life form, Vapour, Silence, Kzzl, LindsayH, Mani1, Pavel Vozenilek, Paul August, Bender235, Kjoonlee, Mashford, Kelvinc, Perlman10s, Panu~enwiki, Brian0918, Dpotter, Livajo, El C, Laurascudder, Shanes, Zegoma beach, RoyBoy, Causa sui, Bobo192, Directorstratton, Janna Isabot, Smalljim, John Vandenberg, Flxmghvgvk, I9Q79oL78KiL0QTFHgyc, Physicistjedi, Bongoo, 4v4l0n42, Merope, Geschichte, Linuxlad, Phils, Merenta, Alansohn, Gary, JYolkowski, Enirac Sum, Ryanmcdaniel, Arthena, Borisblue, Rd232, Plumbago, Axl, R Calvete, Lightdarkness, Kocio, Bart133, Wtmitchell, Isaac, Tycho, Cal 1234, Fadereu, CloudNine, Sciurinæ, Computerjoe, Kusma, DV8 2XL, Pwqn, Gene Nygaard, Ringbang, Ceyockey, Falcorian, Bobrayner, Joriki, Mel Etitis, Linas, BillC, Jacobolus, HFarmer, Before My Ken, Netdragon, MONGO, GeorgeOrr, Mpatel, Bbatsell, GregorB, ??????, Joke137, Christopher Thomas, Dysepsion, GSlicer, Jan.bannister, Graham87, Magister Mathematicae, Hillbrand, BD2412, Elvey, Galwhaa, Raymond Hill, JIP, RxS, Athelwulf, Edison, Sjakkalle, Rjwilmsi, Xgamer4, Jake Wartenberg, Arabani, MarSch, TheRingess, Jmcc150, Aero66, Crazynas, Juan Marquez, R.e.b., Bubba73, DoubleBlue, Zelos, AlisonW, Asafavi, Lionelbrits, Conorific, Zunz, Mathbot, Crazycomputers, RexNL, Gurch, Algri, TeaDrinker, Zifnabxar, XAXISx, Erik4, Phoenix2~enwiki, Antimatter15, Ggb667, Chobot, Visor, DVdm, Mhking, VolatileChemical, Bgwhite, Algebraist, Ben Tibbetts, YurikBot, Ugha, Wavelength, Borgx, NuclearFusion~enwiki, Angus Lepper, Hairy Dude, Jimp, Hillman, Cyferx, Wolfmankurd, Pip2andahalf, RussBot, Morononman, Crazytales, Pippo2001, Bhny, Pigman, SpuriousQ, Branman515, Stephenb, Gaius Cornelius, Eleassar, Bovineone, Cheesus, Shanel, NawlinWiki, Tong~enwiki, Mike18xx, SCZenz, Cleared as filed, Bdiah, Pym98, SColombo, Haemo, FF2010, Closedmouth, Reyk, Brina700, Chris Brennan, Vicarious, Brianlucas, Geoffrey.landis, Hitchhiker89, Spliffy, Pred, ArielGold, Roy Fultun, Ilmari Karonen, Katieh5584, Pentasyllabic, Lunch, DVD R W, WikiFew, That Guy, From That Show!, Street Scholar, AndrewWTaylor, QSquared, Sardanaphalus, Vanka5, MacsBug, Hvitlys, SmackBot, Kurochka, Zazaban, Tom Lougheed, Prodego, KnowledgeOfSelf, Hydrogen Iodide, Melchoir, Vald, Skrewtape, Atomota, Canthusus, GaeusOctavius, Cool3, Andyvn22, Skizzik, RobertM525, Dauto, Bluebot, SSJ 5, Keegan, Aidan Croft, Thumperward, Oli Filth, Silly rabbit, Timneu22, SchfiftyThree, Moshe Constantine Hassan Al-Silverburg, Complexica, Rediahs, RayAYang, Aero77, Adamstevenson, Ikiroid, Epastore, Baronnet, Ned Scott, Sbharris, Colonies Chris, Konstable, Sct72, Scwlong, Can't sleep, clown will eat me, Timothy Clemans, Onorem, Neilanderson, EvelinaB, TKD, KerathFreeman, Addshore, UU, The tooth, Pepsidrinka, Somebody2292, --=The Doctor=--, Fuhghettaboutit, Cybercobra, Irish Souffle, Nakon, Jdlambert, James McNally, MichaelBillington, Lostart, Insineratehymn, Drphilharmonic, SpiderJon, DMacks, Ihatetoregister, Where, Michael IFA, Yevgeny Kats, Vasiliy Faronov, Byelf2007, Angela26, Visium, Rory096, Zymurgy, Harryboyles, Mdl53711, T-dot, Titus III, Ergative rlt, MagnaMopus, UberCryxic, Vgy7ujm, Linnell, Mgiganteus1, Nonsuch, IronGargoyle, Ckatz, DoItAgain, AstroGod, Kirbytime, Jimbo Mahoney, FredrickS, Invisifan, Ryulong, Ryanjunk, MathStuf, Mike Doughney, Norm mit, Hindol, Dan Gluck, Huntscorpio, Iridescent, K, Sunoco, You? Me? Us?, CzarB, Rabinzkaman, JoeBot, Lottamiata, Tony Fox, Vrkaul, Torrazzo, Gil Gamesh, Areldyb, Courcelles, Tawkerbot2, Gebrah, Shamvil, DKqwerty, Lbr123, Harold f, Heqs, Devourer09, Duduong, Sarvagnya, Dewayne76, JForget, Cg-realms, InvisibleK, CRGreathouse, CmdrObot, Earthlyreason, Van helsing, Olaf Davis, CBM, Rawling, Jibal, Witten Is God, Nunquam Dormio, Giko, KnightLago, Thubsch, Leujohn, SlashDot, TheTito, Karenjc, Myasuda, Emarv, Cydebot, Gmusser, Gogo Dodo, Jkokavec, Kahananite, Quajafrie, Michael C Price, Doug Weller, DumbBOT, Narayanese, AlphaNumeric, SRoughsedge, Vanished User jdksfajlasd, Woland37, Zalgo, Daniel Olsen, UberScienceNerd, Bkazaz, DJBullfish, Thijs!bot, Epbr123, Rwmnau, Babemachine, Pimpin101, Mbell, O, Faigl.ladislav, Ucanlookitup, Andyjsmith, Headbomb, Tcturner2002, Marek69, Brahmajnani, Arthurcprado~enwiki, Y.t., D3gtrd, Babemonkey, Dark dude, Duncan McB, EdJohnston, MichaelMaggs, Ancientanubis, Natalie Erin, Hempfel, Jomoal99, Mmortal03, Mentifisto, Geekdom04, AntiVandalBot, Luna Santin, Seaphoto, Ed270791, Opelio, Doc Tropics, David136a, NithinBekal, Dotdotdotdash, Helicoptor, Poshzombie, MontanNito, Dylan Lake, Maximilian77, Shlomi Hillel, Db63376, SamIAmNot, Knotwork, Res2216firestar, Superior IQ Genius, MER-C, Andonic, Sitethief, 100110100, TallulahBelle, Nestamachine, Savant13, Daynightrader, Goldenglove, Charibdis, Acroterion, Ophion, Aigisthos, Editmyhandman, Aruben537, Magioladitis, WolfmanSF, Bongwarrior, VoABot II, Yandman, JamesBWatson, بساريم, Qutt, Jespinos, Kevinmon, Aka042, Froid, DAGwyn, Catgut, Panser Born, Ensign beedrill, Perspectival, JJ Harrison, Dirac66, Justanother, Aziz1005, Cpl Syx, ChazBeckett, Teardrop onthefire, WLU, Stephen shenker, Robin S, SkepticVK, Joshua Davis, Mkroh, B9 hummingbird hovering, S3000, Hdt83, MartinBot, FlieGerFaUstMe262, Ytomem, Shimwell, Arjun01, KrishSundaresan, Anaxial, Jay Litman, Alexcalamaro, Andrej.westermann, Smokizzy, LedgendGamer, Cyrus Andiron, Peteryoung144, Tgeairn, Artaxiad, HEL, AlphaEta, J.delanoy, AstroHurricane001, Maurice Carbonaro, Yonidebot, Morris729, M C Y 1008, 69gangsta420, It Is Me Here, Shawn in Montreal, Janus Shadowsong, Bailo26, Fredsie, Madagaskar07, Duchesserin, AntiSpamBot, CHIAGEHYANG, Chiswick Chap, Watsup1313, Belovedfreak, HaloInverse, NewEnglandYankee, Scott1329m, Thesis4Eva, Policron, Jrcla2, WJBscribe, Rnricklefs, Jamesofur, Eyelidlessness, Jonnyk aus, Kvdveer, JavierMC, Izno, Xiahou, CardinalDan, Sheliak, HamatoKameko, Malik Shabazz, Concertmusic, JohnBlackburne, JustinHagstrom, Fences and windows, Wooba doob, Philip Trueman, DoorsAjar, HowardFrampton, TXiKiBoT, Zidonuke, Red Act, Kriak, Calwiki, Technopat, Hqb, Andrius.v, Anonymous Dissident, Crohnie, AlysTarr, Qxz, Vanished user ikijeirw34iuaeolaseriffic, Impunv, Seraphim, Martin451, Don4of4, ABigGreenHippo, Huperphuff, LeaveSleaves, Kaenneth, StringyGuy, Maxim, Erth64net, Meters, Rickstauduhar, Enviroboy, Turgan, Anna512, PhysPhD, Northfox, NPguy, Matthew Sanders, Luke Walkerson, Newbyguesses, MissMJ, SieBot, Escher26, J.A.Ireland, BA (IHPST), 4wajzkd02, Robdunst, Dreamafter, Pallab1234, Dbelange, MTHarden, Lemonflash, Kylemew, Yintan, GlassCobra, Wpegden, Likebox, Flyer22, Exert, ProGeek314, Arbor to SJ, Babawhitemoose, Caidh, Dhatfield, Audree, Oxymoron83, Pretty Green, Weaselstomp, Manway, Alex.muller, Taco Manipulator, Tschach, Manheat84, Anchor Link Bot, Mikebernstein, ImperialismGo, Nergaal, Ionfield, Ayleuss, Sh4wz0r, Naturespace, Martarius, Phyte, ClueBot, The Thing That Should Not Be, String4d, Illusion96, Polyamorph, Mpd1989, Alexdeburca18, Wiggl3sLimited, Excirial, Kjramesh, Jusdafax, Resoru, WikiZorro, Eeekster, Verum~enwiki, Tamaratrouts, Gtstricky, Humanino, Brews ohare, NuclearWarfare, Cenarium, Razorflame, Scoobey, BOTarate, Sideswiper, Thingg, Capudo, BVBede, Versus22, Introductory adverb clause, MelonBot, SoxBot III, Egmontaz, Notpayingthepsychiatrist, DumZiBoT, BahTab, TimothyRias, Aj00200, Reaperfromhell, Dunkaroo207, XLinkBot, AlexGWU, Impshum, Saeed.Veradi, Little Mountain 5, Guy392, David424, Truthnlove, Qweeveen, Tayste, Addbot, Steven66s, Denali134, Elemented9, Varrey280303, Eric Drexler, Some jerk on the Internet, Fizzycyst, Uruk2008, DOI bot, Jojhutton, AngryBacon, Captain-tucker, Auspex1729, Kongr43gpen, Fgnievinski, Rhetoric Of A Sophist, Ronhjones, CanadianLinuxUser, Cst17, Download, Glane23, Bassbonerocks, Chzz, Favonian, Kronix35, LinkFA-Bot, Udugunit, Aktsu, Tassedethe, Numbo3-bot, Anpecota, Tide rolls, HerpesVirus, SDJ, OlEnglish, Scourge of God, Davidmedlar, Couldbenoway66, Yobot, Maxdamantus, Terrisknickers, Kartano, TaBOT-zerem, Julia W, Unique and proud of it, FireMouseHQ, Terrifictriffid, ArchonMagnus, CinchBug, Synchronism, AnomieBOT, Cleeseheb, 1exec1, Charlesvi, Bigdaddy4x4, Gitman4, Jim1138, IRP, Mintrick, Drweetmola, Ornamentalone, M00npirate, Gautam10, Csigabi, Poli-Psy, Materialscientist, 90 Auto, Citation bot, Teleprinter Sleuth, Vuerqex, Twri, Frankenpuppy, Fuzzy Bob Saget, DirlBot, Georgepowell2008, Heidisql, Cureden, Ekwos, Capricorn42, Gensanders, NFD9001, Anna Frodesiak, Tomwsulcer, A23649, Pra1998, Coretheapple, Ruy Pugliesi, Jagbag2, Vandalism destroyer, Ab1, Omnipaedista, Bandit5005, Shirik, RibotBOT, Waleswatcher, Saalstin, Amaury, Aaron35510, Caz34, Doulos Christos, Sewblon, Born Gay, Capricorn24, SchnitzelMannGreek, A. di M., SpacePyjamas, Kierkkadon, A.amitkumar, Dougofborg, StringLove, Nobelprizewinner, Astiburg, FrescoBot, Fortdj33, Paine Ellsworth, Goodbye Galaxy, HJ Mitchell, Steve Quinn, Vhann, Kwiki, Xhaoz, Citation bot 1, Batong, Gil987, Pinethicket, I dream of horses, Tallboyhoops1991, Three887, Steveo27five, RedBot, Sardinita, Vhsatheeshkumar, Swisstingle, DeletionUK, Reconsider the static, IVAN3MAN, Remingtonhill1, Orenburg1, Coltonhs, Willy Weazley, Smamaret, Bethovenn, Dinamik-bot, Dc987, Oswaldo Zapata, Egemont, Syebo, Alaithiran, Reaper Eternal, Seahorseruler, Ybungalobill, Quaker phil, Specs112, Dr. Aakash Patel, Tbhotch, StormbringerUK, Minimac, Mathgenius3141592, Keegscee, Omgwaffels, Mick le pick, Solancel, Aznhero3793, Dwielark, Afteread, Enauspeaker, EmausBot, MaooaM, Immunize, Az29, Milkocookie, Faolin42, Fotoni, RA0808, RenamedUser01302013, 8digits, Yukiseaside, Slightsmile, Tommy2010, Winner 42, Wikipelli, JonezyKiDx, Joe Gazz84, ZéroBot, Timeitsways, John Cline, Cogiati, Quaqa, Chrispaps2413, Nasulikid, Vollrath2323, Benjamin1414141414141414, Arbnos, Green Lane, A930913, Bamyers99, Azeraphale, H3llBot, Encyclopadia, Danga1988, Ollainen, PoisonGM, Wayne Slam, OnePt618, Knome335, L Kensington, Lulzprotuns, Kranix, Rpcappello, Vastly~enwiki, Donner60, CatFiggy, CountMacula, Orange Suede Sofa, Etov, M1k3 101, Bill william compton, Wakabaloola, TERBAFAN, Nickslspride34, NeuralLotus, Isocliff, Brechbill123, Xanchester, ClueBot NG, Martti Muukkonen, KagakuKyouju, Jeff Song, This lousy T-shirt, Satellizer, Name Omitted, Marcdean123, Wiki incorp, Frietjes, O.Koslowski, Alexdamaino9, Dream of Nyx, Blackhall616, Widr, Sashhere, WikiPuppies, Stu181, T00g00d96, Pluma, Storm.sarup, Helpful Pixie Bot, Manzeet, Waffleboy36, HMSSolent, Mikeshelton1, Bibcode Bot, 2001:db8, Phillip.phillipson, Hoaxinator, Lowercase sigmabot, Thor cherubim, Mrshabam, Nishch, Flowerhat15, AvocatoBot, Housegeek224, MahRanch, Benzband, Altaïr, Benhenchdickthomas, Shreyakstring, Sweaty maori sphincter, DaFalk, Dsabo74, Ratanmaitra, MM4EVAH, Steven.w.kowalski, Minsbot, JGallardo2600, Dylanlatham, Myfriendganesha, OCCullens, Likeaboss189, Sean271293, LinusE8, BattyBot, Several Pending, Aldrich2122, CommanderMoka, The Illusive Man, ChrisGualtieri, KoalamaN2, Trevorkid45, Catsloveit07, Alex Modzz, Rustyjamsen, Goh ryangoh, Dexbot, Exolius, Hilander316, Alman1234321, SuperCalzer, LightandDark2000, MeekMelange, BQND, Cdarrai1, Kephir, TheMonkeyboy524, Michael Anon, Mattfat8, Lugia2453, Anruy, Rachel weld, Jamesx12345, AHusain314, BossEditors, Hillbillyholiday, Joeinwiki, Mattninja, Theshadow444, Asaa82, Jakemarz197, Kzhang1025, Epicgenius, Spongbob456789, ?, TestMaster, Ianreisterariola, GrapperJ, Makeitnasty, Moemajdi, I am One of Many, Nualalvy, BAZINGASS, St3fanPC, Eyesnore, Isaac grozd, Jordanissexyaf1999, Baruch6525, Mosbruckercj, Ihatedirac2k13, Jonamithy121314, 123physicsquantum, Jt198, RaphaelQS, HeyJude70, AParker628, DimReg, A.k.blaze1, Joshuk, Zenibus, Nianoobasik, Ihelpapplen, Gamo To Apoel, SacredLabyrinth, Ginsuloft, Vampre1122, Dimension10, Howard Wolowitz, AddWittyNameHere, Polytope24, Elysion, Tutun12$, Longerboats5, SimonWombat8, Konveyor Belt, Vtank54, Micheal545, Hck24, Caliae19, Hexafish, Simpick, TheRealTheKoi, Bballbro62, Monkbot, ArmyPath, TheQ Editor, Jtsmith098, Joshmiller1, Hanseer360, XXvPIEvXx, Dbennett 24, Ghikpenos, Nick65633, Saundra03, Thehippothatknows, Sewwgers, Teelaskeletor, Cirksena, Balockaye1234, PloppyDoo, Yesufu29, Lumpy2k14, Podayeruma, Abstract92, Sbenfiel, Monkman2k4, Swegwegdgfyetkfoffkkfkfkv, John95541234, Poopman224, ScrapIronIV, Tetra quark, KasparBot, SHUCKYLUCKY, Fabiotheoto, FartGoblin and Anonymous: 1538

- **String (physics)** *Source:* https://en.wikipedia.org/wiki/String_(physics)?oldid=667292955 *Contributors:* Andres, Wereon, Fropuff, El C, Mpatel, Gwernol, Roboto de Ajvol, KnightRider~enwiki, Scwlong, Fiziker, EPM, ServAce85, Drewbarfield, TriTertButoxy, Astrobradley, Dan Gluck, Markjoseph125, Epbr123, Headbomb, The Radio Star, Hempfel, B-80, Qwerty Binary, Maurice Carbonaro, Andrius.v, PhysPhD, Anchor Link Bot, ClueBot, Addbot, Allowgolf~enwiki, Pcb95, FrescoBot, Fisuaq, Mathewmathewmeixnermeixner, Frietjes, Widr, Hansan29, Davida98, Polytope24 and Anonymous: 23

- **Compactification (physics)** *Source:* https://en.wikipedia.org/wiki/Compactification_(physics)?oldid=540832915 *Contributors:* The Anome, Michael Hardy, Charles Matthews, Mpatel, Eyu100, Eubot, Salsb, SmackBot, Ben Jos, Noah Salzman, JarahE, Dan Gluck, Headbomb, Isilanes, AlleborgoBot, Ozooxo, AnonyScientist, AlexGWU, Addbot, Luckas-bot, Wireader, Dogbert66, EmausBot, ZéroBot and Anonymous: 9

- **Supersymmetry** *Source:* https://en.wikipedia.org/wiki/Supersymmetry?oldid=670511651 *Contributors:* Bryan Derksen, Taw, Andre Engels, Roadrunner, Maury Markowitz, Ewen, Stevertigo, Edward, Michael Hardy, Arpingstone, Theresa knott, IMSoP, Jeandré du Toit, Samw, Smack, Charles Matthews, Maximus Rex, Phys, Raul654, BenRG, Rursus, Mor~enwiki, Ancheta Wis, Giftlite, Mporter, Ferkelparade, Monedula, Fropuff, Xerxes314, Anville, Gus Polly, Moyogo, Unconcerned, DO'Neil, Maarten van Vliet, Pharotic, LiDaobing, Sam Hocevar, Lumidek, Deglr6328, Arivero, Rich Farmbrough, Roybb95~enwiki, Bender235, El C, Nornagon~enwiki, Duk, Tweet Tweet, LostLeviathan, Pearle, Gary, Francescog~enwiki, Wtmitchell, RJFJR, Reaverdrop, Blaxthos, Killing Vector, Jordan14, Ted BJ, MONGO, Mpatel, MFH, SeventyThree, Bodera, VermillionBird, Drbogdan, Rjwilmsi, Josiah Rowe, R.e.b., Bubba73, Maxim Razin, Drrngrvy, FlaBot, Cless Alvein, Nowhither, Itinerant1, Gparker, KFP, Lmatt, Chobot, Vyroglyph, YurikBot, Wavelength, RussBot, Ohwilleke, Bhny, Epolk, Maxim Leyenson, Chaos, Romanc19s, Bota47, Mgnbar, Closedmouth, Arthur Rubin, RG2, That Guy, From That Show!, A bit iffy, SmackBot, Mira, Kurochka, Wangjiaji, Gilliam, Bluebot, Cadmasteradam, Complexica, Bazonka, Colonies Chris, Can't sleep, clown will eat me, QFT, Ruff ilb, Robma, Solarapex, Radagast83, Jgwacker, TheMaster42, Martijn Hoekstra, Ligulembot, Acjohnson55, Yevgeny Kats, Charleswestbrook, TriTertButoxy, Lambiam, Tktktk, Xiaphias, JarahE, Mdanziger, Dan Gluck, Newone, Marysunshine, Tawkerbot2, Cydebot, Hydraton31, Bazzargh, David edwards, Michael C Price, Crum375, Koeplinger, Headbomb, J.christianson, Escarbot, Salgueiro~enwiki, Kborland, Jpod2, Cgingold, Maliz, TimidGuy, C9, Kostisl, R'n'B, Zentropa77, Natsirtguy, Maurice Carbonaro, Kevin Hickerson, Shawn in Montreal, Idioma-bot, Sheliak, Cuzkatzimhut, Nxavar, Kawakameha, Cuboidal, Ptrslv72, PhysPhD, Kbrose, SieBot, Nn123645, ClueBot, Jcpilman, Chessmaster7m, Kitsunegami, Rhododendrites, Mastertek, Mishas42, Scrabby~enwiki, TimothyRias, WikHead, MystBot, Addbot, DOI bot, Zahd, Barak Sh, F Notebook, Lightbot, Luckas-bot, Yobot, Ibayn, TaBOT-zerem, Amirobot, Nonnormalizable, AnomieBOT, Girl Scout cookie, Citation bot, ArthurBot, Plumpurple, Tomwsulcer, Omnipaedista, Gsard, CES1596, FrescoBot, HaloStereo1, Paine Ellsworth, Xmikywayx, Citation bot 1, Gil987, Kikeku, Jonesey95, Eddie Nixon, MondalorBot, Aknochel, Tom1661, Gagoga ju, TobeBot, Puzl bustr, Andraas, EmausBot, Djloststylez, Ddimensões, Arbnos, Susy is it, ChuispastonBot, Isocliff, ClueBot NG, KagakuKyouju, IJVin, Frietjes, Helpful Pixie Bot, Bibcode Bot, BG19bot, Teika kazura, JayBeeEye, Ninmacer20,

ChrisGualtieri, Logosun, AHusain314, NA48, Rfassbind, Katherine Pendleton, Lioinnisfree, Liquidityinsta, TaiSakuma, Stamptrader, Kdmeaney, Qxxxxxq, Almaionescu, Monkbot, Janhaithabu, Mammoth2011, Jwill530, Stacie Croquet, Cuttlas1 and Anonymous: 171

- **Anthropic principle** *Source:* https://en.wikipedia.org/wiki/Anthropic_principle?oldid=663175679 *Contributors:* The Epopt, Derek Ross, Mav, Wesley, Bryan Derksen, The Anome, Malcolm Farmer, RK, XJaM, Roadrunner, SimonP, B4hand, DrRetard, Boud, Stormwriter, DopefishJustin, Menchi, Cyde, TakuyaMurata, GTBacchus, Alfio, Looxix~enwiki, Snoyes, Angela, Timwi, Pablo Mayrgundter, Reddi, Timc, Fairandbalanced, Samsara, AaronSw, Banno, Phil Boswell, Robbot, Fredrik, Goethean, Peak, Gandalf61, Tim Ivorson, Mirv, Tualha, Sverdrup, Academic Challenger, Desmay, Wikibot, Robinh, Johnstone, Xanzzibar, Paul Richter, Gene Ward Smith, Barbara Shack, Tom harrison, Snowdog, Alibaba, Highlander~enwiki, Gracefool, Golbez, Toby Woodwark, Andycjp, Pcarbonn, Karol Langner, Lumidek, Robin klein, Klemen Kocjancic, D6, Rfl, Rich Farmbrough, Rhobite, FT2, Vsmith, Lulu of the Lotus-Eaters, Edgarde, RJHall, Carlon, TheMile, Rbj, I9Q79oL78KiL0QTFHgyc, Timl, Tritium6, KarlHallowell, QuantumEleven, Orangemarlin, Lycanthrope, Nurban, Plumbago, Mc6809e, Swift, Deacon of Pndapetzim, Deathphoenix, DV8 2XL, Mattbrundage, Ringbang, Euphrosyne, Tomato~enwiki, Japanese Searobin, Pseudovector, Siafu, WilliamKF, Dandv, JFG, BlaiseFEgan, Joke137, Btyner, DaveApter, Marudubshinki, Aarghdvaark, Graham87, Drbogdan, Rjwilmsi, Zbxgscqf, Staecker, A ghost, Bubba73, Reinis, Cassowary, Billjefferys, Fragglet, Sderose, Diza, YurikBot, RussBot, Gaius Cornelius, Joncolvin, Ptcamn, Thiseye, JulesH, Number 57, Mattgrommes, Crumley, Georgewilliamherbert, Closedmouth, Asterion, Nekura, Robertd, SmackBot, Island1, Mitteldorf, 1dragon, InverseHypercube, McGeddon, Huhnra, Edgar181, Portillo, Rmosler2100, Jjalexand, Concerned cynic, Thumperward, Goldfinger820, Can't sleep, clown will eat me, Cybercobra, Infovoria, Localzuk, Richard001, Lpgeffen, Monoape, Luís Felipe Braga, Ligulembot, Vina-iwbot~enwiki, Bejnar, Byelf2007, John, Loodog, Jaganath, Wickethewok, Danburke, Ocatecir, RomanSpa, Hypnosifl, Ryulong, BranStark, Jlrobertson, DedalusJMMR~enwiki, Az1568, Albertod4, Friendly Neighbour, CRGreathouse, CmdrObot, Olaf Davis, JohnCD, Gregbard, Brianroemen, Cydebot, A876, Peterdjones, Joeseither, Fcn, Wexcan, Mbell, CSvBibra, Headbomb, Second Quantization, Peter Gulutzan, Kosmocentric, Mdriver1981, Mentifisto, WikiSlasher, EdgarCarpenter, Bm gub, FForeclosers, Dr. Submillimeter, Res2216firestar, MER-C, Andonic, Bpmullins, Magioladitis, VoABot II, Andrewthomas10, Swpb, Quark7, Caroldermoid, KConWiki, Epstewart, Dirac66, A3nm, Glen, JaGa, WLU, Info D, Gwern, Tirral, Sm8900, Richard Tierney, AstroHurricane001, Claus L. Rasmussen, Acalamari, SpigotMap, LittleHow, Richard D. LeCour, OriEri, Kenneth M Burke, Jarry1250, Michaelpremsrirat, Fences and windows, Mrbrownn, Rei-bot, Ask123, Charlesdrakew, Arioch7, Telecineguy, Insanity Incarnate, Monty845, PaddyLeahy, GirasoleDE, Paradoctor, METIfan, Crash Underride, Soler97, Lightmouse, Jwjdiamond, Miguel.mateo, OKBot, Iknowyourider, Firefly322, Cheesefondue, Myrvin, Epistemion, Martarius, ClueBot, Justin W Smith, J8079s, SuperHamster, Niceguyedc, ChandlerMapBot, Excirial, Alexbot, Dmyersturnbull, Tnxman307, Hans Adler, XLinkBot, Pgallert, WikHead, Aunt Entropy, Silylene, Addbot, DOI bot, Guoguo12, Discrepancy, TutterMouse, Cst17, ChenzwBot, West.andrew.g, Numbo3-bot, Jarble, Legobot, Yobot, Ht686rg90, AnomieBOT, ^musaz, IRP, Citation bot, ArthurBot, Xqbot, Nickkid5, Gap9551, GrouchoBot, Lukebarnesy, Dukejansen, Omnipaedista, RibotBOT, Shadowjams, WebCiteBOT, Mr Owl1234, Joso98, Nagualdesign, Wikianiki, Schnufflus, Machine Elf 1735, Citation bot 1, Anthony on Stilts, Momergil, Marsiancba, MarcelB612, Jordgette, Chico889, Cyanophycean314, Joehubris, Ammodramus, Ti-30X, RjwilmsiBot, Tesseract2, John of Reading, Bludsucker, Qrsdogg, Bettymnz4, Wikipelli, TeleComNasSprVen, GlacierSupremacy, Solomonfromfinland, Hhhippo, Fæ, StringTheory11, Dondervogel 2, Ppw0, G-13114, ClueBot NG, Wikiphysicsgr, Justlettersandnumbers, Plusorminuszero, Бертран, Jack Ponting, Helpful Pixie Bot, Bibcode Bot, Horn.imh, Island001, Joydeep, Trevayne08, CitationCleanerBot, MrBill3, Rho21111, Queen4thewin, Jgates104, Shaarang tenneti, YFdyh-bot, Ersober, Josophie, Shrikarsan, Andrey.a.mitin, Aubreybardo, NormDrez, Trackteur, Cirksena, Velvel2, I'm your Grandma., Tetra quark, AlanSkeptic, Isambard Kingdom, KasparBot and Anonymous: 350

- **Standard Model** *Source:* https://en.wikipedia.org/wiki/Standard_Model?oldid=667423128 *Contributors:* AxelBoldt, Derek Ross, CYD, Bryan Derksen, The Anome, Ed Poor, Andre Engels, Roadrunner, David spector, Isis~enwiki, Youandme, Ram-Man, Stevertigo, Edward, Patrick, Boud, Michael Hardy, SebastianHelm, Looxix~enwiki, Julesd, Glenn, AugPi, Mxn, Raven in Orbit, Reddi, Phr, Tpbradbury, Populus, Haoherb428, Phys, Floydian, Bevo, Pierre Boreal, AnonMoos, BenRG, Jeffq, Dmytro, Drxenocide, Robbot, Nurg, Securiger, Texture, Roscoe x, Fuelbottle, Mattflaschen, Tobias Bergemann, Alan Liefting, Ancheta Wis, Giftlite, Dbenbenn, Harp, Herbee, Monedula, LeYaYa, Xerxes314, Dratman, Alison, JeffBobFrank, Dmmaus, Pharotic, Brockert, Bodhitha, Andycjp, Sonjaaa, HorsePunchKid, APH, Icairns, AmarChandra, Gscshoyru, Kate, Arivero, FT2, Rama, Vsmith, David Schaich, Xezbeth, D-Notice, Dfan, Bender235, Pt, El C, Laurascudder, Shanes, Drhex, Fogger~enwiki, Brim, Rbj, Jeodesic, Jumbuck, Alansohn, Gary, ChristopherWillis, Guy Harris, Axl, Sligocki, Kocio, Stillnotelf, Alinor, Wtmitchell, Egg, TenOfAllTrades, H2g2bob, Killing Vector, Linas, Mindmatrix, Benbest, Dodiad, Mpatel, Faethon, TPickup, Faethon34, Palica, Dysepsion, Faethon36, Qwertyca, Drbogdan, Rjwilmsi, Zbxgscqf, Macumba, Strangethingintheland, Dstudent, R.e.b., Bubba73, Drrngrvy, Agasicles, FlaBot, Naraht, Agasides, DannyWilde, Dave1g, Itinerant1, Gparker, Jrtayloriv, Goudzovski, Chobot, Bgwhite, FrankTobia, YurikBot, Bambaiah, Ohwilleke, VoxMoose, Bhny, JabberWok, Bovineone, Krbabu, SCZenz, JulesH, Davemck, Lomn, E2mb0t~enwiki, Dna-webmaster, Jrf, Dv82matt, Tetracube, Hirak 99, Arthur Rubin, Netrapt, JLaTondre, Caco de vidro, RG2, GrinBot~enwiki, That Guy, From That Show!, Hal peridol, SmackBot, YellowMonkey, Tom Lougheed, Melchoir, Bazza 7, KocjoBot~enwiki, Jagged 85, Thunderboltz, Setanta747 (locked), Skizzik, Dauto, Chris the speller, Bluebot, TimBentley, Sirex98, Silly rabbit, Complexica, Metacomet, DHN-bot~enwiki, MovGP0, QFT, Kittybrewster, Addshore, Jmnbatista, Cybercobra, Jgwacker, BullRangifer, Soarhead77, Daniel.Cardenas, Yevgeny Kats, Byelf2007, TriTertButoxy, Craig Bolon, Ajnosek, Ekjon Lok, Bjankuloski06, Tarcieri, Waggers, JarahE, Michaelbusch, Lottamiata, Newone, Twas Now, IanOfNorwich, Srain, Patrickwooldridge, J Milburn, Mosaffa, Gatortpk, Vessels42, Geremia, Van helsing, Harrigan, Phatom87, Cydebot, David edwards, Verdy p, Michael C Price, Xantharius, Crum375, JamesAM, Thijs!bot, Epbr123, Headbomb, Phy1729, Stannered, Tariqhada, Seaphoto, Orionus, Voyaging, Gnixon, Jbaranao, Jrw@pobox.com, Len Raymond, Narssarssuaq, Bakken, CattleGirl, Davidoaf, Vanished user ty12kl89jq10, Lvwarren, Taborgate, Leyo, HEL, J.delanoy, Hans Dunkelberg, Stephanwehner, Wbellido, Aoosten, Jacksonwalters, The Transliterator, DadaNeem, Student7, Joshmt, WJBscribe, Jozwolf, Hexane2000, BernardZ, Awren, Sheliak, Physicist brazuca, Schucker, Goop Goop, Fences and windows, Dextrose, Mcewan, Swamy g, TXiKiBoT, Sharikkamur, Thrawn562, Voorlandt, Escalona, Setreset, PDFbot, Pleroma, UnitedStatesian, Piyush Sriva, Kacser, Billinghurst, Francis Flinch, Moose-32, Ptrslv72, David Barnard, SieBot, ShiftFn, Robdunst, Jim E. Black, SheepNotGoats, Gerakibot, Nozzer42, Mr swordfish, Wing gundam, Bamkin, Likebox, Arthur Smart, HungarianBarbarian, Commutator, KathrynLybarger, Iomesus, C0nanPayne, Crazz bug 5, ClueBot, Superwj5, Wwheaton, Garyzx, SuperHamster, Elsweyn, Maldmac, DragonBot, Djr32, Diagramma Della Verita, Nymf, Eeekster, Brews ohare, NuclearWarfare, PhySusie, Ordovico, Mastertek, DumZiBoT, BodhisattvaBot, Guarracino, Mitch Ames, Truthnlove, Stephen Poppitt, Tayste, Addbot, Deepmath, Eric Drexler, DWHalliday, Mjamja, Leszek Jańczuk, NjardarBot, Mwoldin, Bassbonerocks, Barak Sh, AgadaUrbanit, Lightbot, Smeagol 17, Abjiklam, Ve744, Luckas-bot, Yobot, Orion11M87, AnomieBOT, JackieBot, Icalanise, Citation bot, ArthurBot, Northryde, LilHelpa, Xqbot, Sionus, Professor J Lawrence, Tomwsulcer, Edsegal, GrouchoBot, Trongphu, QMarion II, Ernsts, A. di M., Bytbox, FrescoBot, Paine Ellsworth, Aliotra, Steve Quinn, Citation bot 1, Rameshngbot, MJ94, RedBot, MastiBot, Aknochel, Sijothankam, Puzl bustr, Beta Orionis, Physics therapist, Bj norge, Innotata, Jesse V., RjwilmsiBot, Mathewsyriac, Afteread, EmausBot, Bookalign, WikitanvirBot, Wilhelm-physiker, Bdijkstra, DerNeedle, Kenmint, Dbraize, Tanner Swett, HeptishHotik, ﺭﺍﻫﺐ ﻥﯼﺵﻥﻡﻩ, Suslindisambiguator, Quondum, Webbeh, UniversumExNihilo, Vanished user fijw983kjaslkekfhj45, RockMagnetist, Stormymountain, Ζετα ζ, Whoop whoop pull up, Isocliff, ClueBot NG, Smtchahal, Snotbot, Tonypak, O.Koslowski, CharleyQuinton, Dsperlich, Theopolisme, ZakMarksbury, Helpful Pixie Bot, Bibcode Bot, BG19bot, Tirebiter78, AvocatoBot, Lukys~enwiki, Stapletongrey, Ownedroad9, Chip123456, ChrisGualtieri, Khazar2, Billyfesh399, Rhlozier, JYBot, Dexbot, Doom636, Rongended, Cerabot~enwiki, Cjean42, Jayanta mallick, Joeinwiki, Kowtje, JPaestpreornJeolhlna, Eyesnore, Euan Richard, Nigstomper, Particle physicist, Prokaryotes, Jernahthern, Ginsuloft, Dimension10, JNrgbKLM, Krabaey, 1codesterS, FelixRosch, Delbert7, BradNorton1979, Lathamboyle, Tetra quark, KasparBot and Anonymous: 357

- **Quantum field theory** *Source:* https://en.wikipedia.org/wiki/Quantum_field_theory?oldid=667901100 *Contributors:* AxelBoldt, CYD, Mav, The Anome, XJaM, Roadrunner, Stevertigo, Michael Hardy, Tim Starling, IZAK, TakuyaMurata, SebastianHelm, Looxix~enwiki, Ahoerstemeier, Cyp, Glenn, Rotem Dan, Stupidmoron, Charles Matthews, Timwi, Jitse Niesen, Kbk, Rudminjd, Wik, Phys, Bevo, BenRG, Northgrove, Robbot, Bkalafut, Gandalf61, Rursus, Fuelbottle, Tobias Bergemann, Ancheta Wis, Giftlite, Lethe, Dratman, Alison, St3vo, Mboverload, DefLog~enwiki, ConradPino, Amarvc, Pcarbonn, Karol Langner, APH, AmarChandra, D6, CALR, Urvabara, Discospinster, Guanabot, Igorivanov~enwiki, Masudr, Pjacobi, Vsmith, Nvj, MuDavid, Bender235, Pt, El C, Shanes, Sietse Snel, Physicistjedi, KarlHallowell, PWilkinson, Helix84, Thialfi, Varuna, Gcbirzan, Docboat, Count Iblis, Egg, Mpatel, Marudubshinki, Graham87, Opie, Vanderdecken, Rjwilmsi, MarSch, Earin, R.e.b., RE, Strobilomyces, Arnero, Itinerant1, Alfred Centauri, Srleffler, Chobot, UkPaolo, Wavelength, Bambaiah, Hairy Dude, RussBot, TimNelson, Archelon, CambridgeBayWeather, SCZenz, Odddmonster, E2mb0t~enwiki, Semperf, Tetracube, Garion96, Erik J, Robert L, Banus, RG2, SmackBot, Stephan Schneider, Tom Lougheed, Melchoir, KocjoBot~enwiki, Mcld, Dauto, Chris the speller, Complexica, Threepounds, RuudVisser, QFT, Jmnbatista, Cybercobra, Rebooted, Victor Eremita, DJIndica, Lambiam, Mgiganteus1, Zarniwoot, Jim.belk, Stwalkerster, SirFozzie, Hu12, Dan Gluck, Iridescent, Joseph Solis in Australia, Albertod4, Van helsing, BeenAroundAWhile, Witten Is God, Cydebot, Jamie Lokier, Meno25, Michael C Price, The 80s chick, Mendicus~enwiki, AstroPig7, Msebast~enwiki, Mbell, Headbomb, Nick Number, Mentifisto, AntiVandalBot, Bt414, Bananan~enwiki, Martin Kostner, Moltrix, Kasimann, Kromatol, Puksik, Lerman, LLHolm, RogueNinja, Tlabshier, JEH, Nikolas Karalis, Storkk, JAnDbot, Igodard, Four Dog Night, N shaji, Bongwarrior, Andrea Allais, Soulbot, Etale, Maliz, Custos0, HEL, J.delanoy, Acalamari, Jeepday, Policron, Blckavnger, Juliancolton, Skou, Telecomtom, GrahamHardy, Sheliak, Cuzkatzimhut, VolkovBot, Bktennis2006, Marksr, HowardFrampton, The Original Wildbear, Dj thegreat, Markisgreen, TBond, Lejarrag, Moose-32, Raphtee, Sue Rangell, Neparis, Drschawrz, YohanN7, SieBot, TCO, Yintan, Likebox, Paolo.dL, Tugjob, Henry Delforn (old), Jecht (Final Fantasy X), OKBot, StewartMH, ClueBot, EoGuy, Wwheaton, The Wild West guy, Shvav~enwiki, Bob108, Brews ohare, Thingg, Count Truthstein, XLinkBot, PSimeon, SilvonenBot, Truthnlove, HexaChord, Addbot, ConCompS, Pinkgoanna, Leapold~enwiki, Dmhowarth26, Glane23, Hanish.polavarapu, Lightbot, Scientryst, R.ductor, Ettrig, Yndurain, Legobot, Luckas-bot, Yobot, Ht686rg90, Niout, Tamtamar, AnomieBOT, Ciphers, Palpher, IRP, Gjsreejith, Materialscientist, Citation bot, Bci2, ArthurBot, Northryde, LilHelpa, Caracolillo, Amareto2, MIRROR, Professor J Lawrence, Plasmon1248, Omnipaedista, RibotBOT, Spellage, JayJay, FrescoBot, Kenneth Dawson, D'ohBot, Knowandgive, N4tur4le, Hyqeom, Newt Winkler, Hickorybark, Lotje, Dinamik-bot, LilyKitty, Fortesque666, Reaper Eternal, Minimac, Marie Poise, Yaush, Dylan1946, EmausBot, Racerx11, GoingBatty, Carbosi, Thecheesykid, ZéroBot, Cogiati, Jjspinorfield1, Suslindisambiguator, Quondum, Maschen, Zueignung, Davidaedwards, Lom Konkreta, ClueBot NG, Gilderien, Iloveandrea, Vacation9, Heyheyheyhohoho, Fortune432, The ubik, Zak.estrada, Widr, Helpful Pixie Bot, Evanescent7, Ykentluo, Martin.uecker, Walterpfeifer, Pfeiferwalter, Klilidiplomus, W.D., CarrieVS, Khazar2, Momo1381, Dexbot, Cerabot~enwiki, Garuda0001, AHusain314, Thepalerider2012, A.entropy, Mark viking, Faizan, Aj7s6, संजीव कुमार, Lemnaminor, BerFinelli, Axel.P.Hedstrom, Kclongstocking, Mutley1989, I art a troler, Liquidityinsta, Prokaryotes, DemonThuum, Dingdong2680, Asherkirschbaum, Monkbot, Gjbayes, Thedarkcheese, BradNorton1979, UareNumber6, Teelaskeletor, YeOldeGentleman, Mret81, KasparBot and Anonymous: 292
- **Graviton** *Source:* https://en.wikipedia.org/wiki/Graviton?oldid=668468782 *Contributors:* CYD, Bryan Derksen, Timo Honkasalo, XJaM, Fubar Obfusco, Maury Markowitz, Kaczor~enwiki, Jketola, TakuyaMurata, Eric119, Looxix~enwiki, Glenn, Cyan, Wooster, Charles Matthews, Timwi, Wik, BenRG, Donarreiskoffer, Scott McNay, Stephan Schulz, Arkuat, Chris Roy, Merovingian, Davidl9999, Giftlite, Xerxes314, Jason Quinn, Matt Crypto, CryptoDerk, RetiredUser2, Icairns, Zfr, Lumidek, Ukexpat, Urvabara, Discospinster, Pjacobi, Vapour, Brian0918, El C, Joanjoc~enwiki, Dalf, Army1987, Mpvdm, La goutte de pluie, Physicistjedi, Daniel Arteaga~enwiki, Zenosparadox, Dethtron5000, Keenan Pepper, Viridian, Falcorian, Skeejay, Simetrical, Dr Archeville, Mpatel, Kyleca, Tmassey, Christopher Thomas, Tevatron~enwiki, Kbdank71, Nightscream, Koavf, Mike Peel, Ems57fcva, FlaBot, RexNL, Chobot, DVdm, Roboto de Ajvol, Spacepotato, Anonymous editor, SnoopY~enwiki, Salsb, Bachrach44, Hyperbrand, NickBush24, Pnrj, RL0919, EEMIV, IslandGyrl, Bota47, C h fleming, Petri Krohn, Mario23, Alias Flood, Tim314, Teply, GrinBot~enwiki, SmackBot, Amcbride, Melchoir, Eskimbot, Gilliam, Skizzik, Timneu22, Complexica, Villarinho, Colonies Chris, V1adis1av, Chlewbot, Xyzzyplugh, Jmnbatista, Fuhghettaboutit, Sadi Carnot, Yevgeny Kats, TenPoundHammer, Lambiam, Zaphraud, JorisvS, Mr Stephen, Ramuman, Quasar Jarosz, Lottamiata, Firewall62, Kurtan~enwiki, CmdrObot, BeenAroundAWhile, WeggeBot, Shultz IV, UncleBubba, Michael C Price, Anthmoo, Thijs!bot, Epbr123, Headbomb, KevinS06, Opelio, Spartaz, JAnDbot, Xoneca, SHCarter, Pikazilla, Robin S, STBot, Kostisl, J.delanoy, Tarotcards, Coppertwig, Wesino, Sava ankit2006, Tygrrr, Idioma-bot, Sheliak, JoAnneThrax, TXiKiBoT, WilliamSommerwerck, Hqb, Anonymous Dissident, Antixt, SieBot, Flyer22, Henry Delforn (old), ClueBot, Ergn, Darkicebot, DenverRedhead, Addbot, Eric Drexler, Uruk2008, DOI bot, BrianBop, PJonDevelopment, F Notebook, Legobot, Picturesofnothing, Dov Henis, Alfredschrader, Eric-Wester, AnomieBOT, VanishedUser sdu9aya9fasdsopa, Jim1138, Materialscientist, Citation bot, Tomflaherty, ProtectionTaggingBot, Waleswatcher, FrescoBot, Juto20, LucienBOT, Paine Ellsworth, I dream of horses, Tom.Reding, RedBot, Omar.tigereyes, IVAN3MAN, Ashish.kotwal, Michael9422, D0wnfalle, EmausBot, Octaazacubane, 8digits, Slightsmile, K6ka, Thecheesykid, User10 5, Rcsprinter123, Orbjeeples, Puffin, Herk1955, ClueBot NG, Raidr, Helpful Pixie Bot, Bibcode Bot, BG19bot, Shapoopy178, ServiceAT, PhnomPencil, Trevayne08, Brainssturm, Tjamcclain2, ChrisGualtieri, Ariscod, TheUyulala, LightandDark2000, Jessybun, Makecat-bot, Kryomaxim, JRYon, Andyhowlett, Mark viking, Yorsh07, CensoredScribe, WPratiwi, Monkbot, Bryan Paul Senior, Dr.Begich, Nompynuthead and Anonymous: 196
- **Quantum gravity** *Source:* https://en.wikipedia.org/wiki/Quantum_gravity?oldid=667958630 *Contributors:* AstroNomer~enwiki, Matusz, Miguel~enwiki, Roadrunner, Stevertigo, Ubiquity, Bobby D. Bryant, Mcarling, NuclearWinner, Anders Feder, Susurrus, Coren, Charles Matthews, Timwi, Reddi, Tpbradbury, Phys, Bevo, Raul654, BenRG, Frazzydee, Jeffq, Sdedeo, Rholton, Wereon, Ilya (usurped), Seth Ilys, Ancheta Wis, Giftlite, Herbee, Fropuff, Endlessnameless, Malyctenar, Jason Quinn, Finn-Zoltan, YapaTi~enwiki, Lumidek, Marcus2, Joyous!, TJSwoboda, Vitaleyes, Davidcllifford, JimJast, Guanabot, FT2, Masudr, Pjacobi, Pie4all88, David Schaich, Bender235, Clement Cherlin, El C, PhilHibbs, Army1987, Apyule, VBGFscJUn3, PWilkinson, Daniel Arteaga~enwiki, Keenan Pepper, Cjthellama, DonJStevens, Velella, Dabbler, Tycho, Cal 1234, RJFJR, Count Iblis, ThomasWinwood, Anarchimede, Scarykitty, Woohookitty, Igny, ToddFincannon, Mpatel, GregorB, Joke137, Christopher Thomas, Marudubshinki, Graham87, Yurik, Kroggz, Rjwilmsi, Eoghanacht, Jrasowsky, JHMM13, Smithfarm, Ems57fcva, FayssalF, Itinerant1, Lmatt, Chobot, Hmonroe, YurikBot, Hillman, ErkDemon, JocK, SCZenz, Roy Brumback, Bota47, Zunaid, JonathanD, 2over0, Arthur Rubin, Modify, LeonardoRob0t, Caco de vidro, RG2, KasugaHuang, Resolute, SmackBot, Samdutton, Vald, Eskimbot, Hbackman, Onebravemonkey, Chris the speller, Ben.c.roberts, Cthuljew, Silly rabbit, Complexica, Colonies Chris, QFT, Soosed, Theanphibian, Shushruth, Ck lostsword, Yevgeny Kats, DJIndica, Lambiam, Vampus,

Vincenzo.romano, Jaganath, JorisvS, RoboDick~enwiki, IronGargoyle, Dicklyon, SirFozzie, Treyp, Twunchy, Piccor, Kurtan~enwiki, Harold f, CalebNoble, Duduong, Paulmlieberman, TVC 15, UncleBubba, TAz69x, Sam Staton, ST47, B, Patrick O'Leary, Epbr123, Koeplinger, Klasovsky, Markus Pössel, Keraunos, Headbomb, Marek69, MichaelMaggs, Tim Shuba, MER-C, ParadiZio, Clementvidal, Perlygatekeeper, VoABot II, Alvatros~enwiki, Bdalevin, SHCarter, Jpod2, DAGwyn, Nucleophilic, LorenzoB, Rickard Vogelberg, DancingPenguin, Rettetast, Victor Blacus, AstroHurricane001, Yonidebot, Acalamari, Mstuomel, Fullmetal2887, NewEnglandYankee, DorganBot, CardinalDan, Idioma-bot, Sheliak, VolkovBot, Pleasantville, Seattle Skier, AlnoktaBOT, TXiKiBoT, Dllahr, Rdekleer, Saibod, Cyberchip, Wikiwikimoore, Carlorovelli, LoreMiles, StevenJohnston, SieBot, LeadSongDog, Bentogoa, Coldcreation, ReluctantPhilosopher, StaticG, GarbagEcol, ClueBot, The Thing That Should Not Be, EoGuy, Polyamorph, Andwor9, Notburnt, Tms9, Alexbot, Resoru, Eekster, Tamaratrouts, Brews ohare, SchreiberBike, Askahrc, BOTarate, Lambtron, DumZiBoT, XLinkBot, Rror, Facts707, SilvonenBot, Theonlydavewilliams, Mhsb, Truthnlove, Ttimespan, Trifonov~enwiki, Addbot, Mortense, Grayfell, Eric Drexler, Gravitophoton, DOI bot, AkhtaBot, CanadianLinuxUser, Frosty726, LaaknorBot, Delaszk, Tassedethe, Tide rolls, Taketa, Titan1129, Krano, Luckas-bot, Yobot, WikiDan61, Pigetrational, Wireader, Allowgolf~enwiki, Wiki Roxor, Jim1138, IRP, Sz-iwbot, Quantity, Materialscientist, Citation bot, ArthurBot, LilHelpa, Amareto2, Ekwos, KrisBogdanov, Rolfguthmann, StealthCopyEditor, ??, Dan6hell66, Rabsmith, Hep thinker, Paine Ellsworth, DrArthurRubinPHD, Lagelspeil, Nunc aut numquam, Vacuunaut, Van Speijk, Knowandgive, Craig Pemberton, Udifuchs, Citation bot 2, Citation bot 1, Citation bot 4, Jonesey95, Hirvenkürpa, Tom.Reding, Pmokeefe, Casimir9999, Dac04, Dude1818, Valeriy Pischenko, Follyland, TrueTeargem, N0814444, Earthandmoon, Korepin, DARTH SIDIOUS 2, Musictime4me, RjwilmsiBot, EmausBot, Francophile124, Octaazacubane, Fotoni, Slightsmile, Garfield Salazar, Hhhippo, JSquish, John Cline, Fæ, Brazmyth, Throwmeaway, Arbnos, Ebrambot, Kusername, DanielBurnstein, TonyMath, L Kensington, Maschen, Donner60, Parusaro, Apratim07, Terra Novus, Isocliff, Googledin!, ClueBot NG, SpikeTorontoRCP, Science writer, Preon, Raidr, Jhmmok, 336, Widr, Helpful Pixie Bot, Bibcode Bot, Bardsley Rides a Segway, Apelikedawg, FiveColourMap, Trevayne08, Mr.viktor.stepanov, Brainssturm, BattyBot, Jimw338, Ryanr666, Kryomaxim, Garuda0001, Saehry, Sanathdevalapurkar, Andyhowlett, GabeIglesia, Sanathlab, Roiwallace, Spencer.mccormick, Spencerfjase, MrShlongNo1, Marc D. Garrett, D00d00ballz, Gigantmozg, Polytope24, Frinthruit, Anrnusna, Dfyytj, Monkbot, Umut Alihan Dikel, Amortias, Klj1234, Pfpguy, KasparBot and Anonymous: 290

- **Quantum chromodynamics** *Source:* https://en.wikipedia.org/wiki/Quantum_chromodynamics?oldid=663173036 *Contributors:* AxelBoldt, CYD, Zundark, Youandme, Ewen, Stevertigo, Michael Hardy, Ahoerstemeier, Whkoh, Emperorbma, Jitse Niesen, Phys, Robbot, Fredrik, Ojigiri~enwiki, Seth Ilys, Alan Liefting, Giftlite, JamesMLane, Monedula, Xerxes314, JeffBobFrank, Jason Quinn, Elroch, Icairns, Sam Hocevar, Lumidek, Sctfn, Eep², David Schaich, JonL, Goplat, AdamSolomon, Pt, El C, CDN99, Robotje, Slicky, Physicistjedi, Azn king28, Fwb22, Guy Harris, Ricky81682, TenOfAllTrades, Skyring, Kusma, Alai, Mpatel, Betsythedevine, Mendaliv, VermillionBird, Rjwilmsi, Coemgenus, FlaBot, Thenewdeal87, Adoniscik, Algebraist, YurikBot, Wavelength, Bambaiah, Hairy Dude, Moto Perpetuo, Ohwilleke, JabberWok, Kirill Lokshin, Spike Wilbury, BlackAndy, Thiseye, CecilWard, Voidxor, Zzuuzz, Banus, Finell, SmackBot, Henriok, Vald, ProveIt, GaeusOctavius, Chris the speller, Bluebot, TimBentley, Complexica, Colonies Chris, Modest Genius, Berland, Grover cleveland, Garry Denke, TriTertButoxy, DJIndica, Jaganath, RoboDick~enwiki, NNemec, Slakr, Ryulong, Tawkerbot2, Memetics, Capefeather, Runningonbrains, Cydebot, DavidMcCabe, Headbomb, WVhybrid, Noclevername, Escarbot, Salgueiro~enwiki, Shambolic Entity, Andonic, Hut 8.5, Pkoppenb, .anacondabot, Robomojo, Corvidaecorvus, Maliz, Connor Behan, TechnoFaye, R'n'B, HEL, DrKiernan, Acalamari, Shomroni, Lseixas, Skullfunk, GrahamHardy, Idioma-bot, Sheliak, Cuzkatzimhut, VolkovBot, TXiKiBoT, Calwiki, Rei-bot, Saibod, KP-Adhikari, Ptrslv72, SieBot, Dawn Bard, Likebox, Anchor Link Bot, ClueBot, WDavis1911, Pechmerle, PixelBot, Brews ohare, Chrisarnesen, XLinkBot, SilvonenBot, SkyLined, Truthnlove, Addbot, DOI bot, AnnaFrance, SpBot, Lightbot, Zorrobot, Legobot, Luckas-bot, Yobot, Tamtamar, Nallimbot, Citation bot, LilHelpa, Info21, Chrisfox8, Pra1998, Petros000, FrescoBot, Ecuqkindler, Timmeken, Ganondolf, Meier99, Tarsilia, McSaks, Autumnalmonk, EmausBot, Mnkyman, Wikipelli, Brazmyth, Quondum, Aschwole, Rcsprinter123, Maschen, Fwilczek, RolteVolte, Neduard, QuantumSquirrel, Teaktl17, ClueBot NG, Helpful Pixie Bot, Bibcode Bot, Dalit Llama, PhnomPencil, Vkpd11, Snow Blizzard, Cjean42, Joeinwiki, Trompedo, KasparBot and Anonymous: 137
- **Brane** *Source:* https://en.wikipedia.org/wiki/Brane?oldid=647595087 *Contributors:* Bth, Michael Hardy, DIG~enwiki, Samuelsen, JWSchmidt, Silverfish, Wetman, Bcorr, Blainster, Fropuff, Just Another Dan, D3, Lumidek, Yuriz, Rhobite, H0riz0n, Ben Standeven, RoyBoy, Mairi, Constantine, GatesPlusPlus, Kocio, Agquarx, Mpatel, Liface, BD2412, Quiddity, R.e.b., Mathbot, BradBeattie, Metropolitan90, YurikBot, Wavelength, NawlinWiki, Wknight94, Closedmouth, SmackBot, Kurochka, Jwestbrook, Autarch, Seanor32, Silly rabbit, Colonies Chris, Nsmith4658, Mesons, Monotonehell, TheVikingRaider, Yevgeny Kats, Spiritia, PaddyM, Czoller, Calmargulis, BeenAroundAWhile, Adailton, Julius M-D, J. W. Love, Julia Rossi, Chrisjj3, MER-C, Steveprutz, Just H, N.Nahber, Urco, Alexrussell101, Cyborg Ninja, Idioma-bot, VolkovBot, Anonymous Dissident, Paucabot, Drschawrz, SieBot, Tresiden, OKBot, ClueBot, The Thing That Should Not Be, SilvonenBot, NonvocalScream, Addbot, Jujutsuka, Royote, LilHelpa, Patmethenyfan, Omnipaedista, Nagualdesign, Kgrad, Tbhotch, Tesseract2, EmausBot, MathMaven, ClueBot NG, Mikeflem, Gilderien, Baseball Watcher, Frietjes, DBigXray, Lowercase sigmabot, OCCullens, BattyBot, Brirush, E8xE8, Dimension10, Polytope24 and Anonymous: 50
- **D-brane** *Source:* https://en.wikipedia.org/wiki/D-brane?oldid=629000738 *Contributors:* Zundark, TakuyaMurata, Karada, JWSchmidt, AugPi, Smack, Schneelocke, Gandalf61, Michael Snow, Fropuff, Anville, Just Another Dan, Phe, Lumidek, Rgrg, H0riz0n, El C, Constantine, I9Q79oL78KiL0QTFHgyc, FlaBot, Bhny, Nick, Zwobot, Sardanaphalus, KnightRider~enwiki, Teemu Ruskeepää, Colonies Chris, Scwlong, QFT, Eric Olson, Fuhghettaboutit, Vampus, JarahE, Twyder, Eewild, 345Kai, Cydebot, Headbomb, J. W. Love, Nick Number, Magioladitis, Jpod2, STBot, HEL, VolkovBot, TXiKiBoT, PhysPhD, Jonathanrcoxhead, Excirial, Alexbot, ResidueOfDesign, Addbot, LaaknorBot, Tassedethe, Lightbot, Luckas-bot, Yobot, Amirobot, Azcolvin429, Royote, Citation bot, Twri, Omnipaedista, Galaktiker, Mentibot, Wakabaloola, Petrb, Frietjes, Luizpuodzius, OCCullens, Polytope24 and Anonymous: 29
- **Perturbation theory** *Source:* https://en.wikipedia.org/wiki/Perturbation_theory?oldid=663496066 *Contributors:* CYD, FlorianMarquardt, Stevertigo, Michael Hardy, Kku, AugPi, Ideyal, Jitse Niesen, Phys, Robbot, Lowellian, Tobias Bergemann, Giftlite, BenFrantzDale, Neilc, Karol Langner, The Land, Igorivanov~enwiki, MuDavid, Bender235, Cmdrjameson, Haham hanuka, Keenan Pepper, RJFJR, Count Iblis, Dirac1933, Mattbrundage, Djsasso, Oleg Alexandrov, Linas, Yansa, SeventyThree, Bubba73, Mathbot, ChrisChiasson, YurikBot, Piet Delport, Tong~enwiki, Joel7687, Dhollm, Tony1, DerHannes, Artemisfowl3rd, SmackBot, Mmernex, Tom Lougheed, Mcld, Chris the speller, Bduke, Complexica, Nbarth, Colonies Chris, MaxSem, Ohconfucius, Nishkid64, Harryboyles, Tomatoman, JorisvS, Frokor, Hiiiiiiiiiiiiiiiiii, Chetvorno, Khromegnome, CBM, Myasuda, Cydebot, Tawkerbot4, Roy W. Wright, Headbomb, Ben pcc, Engelbaet, David Eppstein, Alexei Kopylov, P.wormer, Cuzkatzimhut, Maghnus, Bphillab, Lechatjaune, EverGreg, Vsst, SieBot, JerroldPease-Atlanta, Nancy, Yhkhoo, ClueBot, Warbler271, PtolemyGalen, Mild Bill Hiccup, Zl1vette, CohesionBot, Guiermo, Lacce, Crowsnest, DumZiBoT, Terry0051, Queenmomcat, Download, Yobot, TaBOT-zerem, Pownuk, Nfr-Maat, J04n, Resident Mario, Pradameinhoff, FrescoBot, Lotje, Mhilferink, Yger, Qweilun, Mattedia, Zfeinst, ClueBot NG, Wcherowi, LPOG1, Helpful Pixie Bot, Vlos2008, PhnomPencil, Anylai, Andyhowlett, Pdecalculus, Hctrmyess and Anonymous: 74
- **Gauge theory** *Source:* https://en.wikipedia.org/wiki/Gauge_theory?oldid=663144153 *Contributors:* The Anome, Michael Hardy, Tobias Bergemann, Ancheta Wis, TedPavlic, Xezbeth, MuDavid, Bender235, Pt, Phils, BD2412, Rjwilmsi, JocK, Modify, Teply, SmackBot,

RDBury, Henning Makholm, Byelf2007, Michael C Price, Biblbroks, Headbomb, Nick Number, Fashionslide, VectorPosse, Magioladitis, Bakken, Email4mobile, JaGa, Policron, Squids and Chips, Cuzkatzimhut, VolkovBot, Red Act, Michael H 34, Setreset, Jwpitts, Tcamps42, Moonriddengirl, ClueBot, Mastertek, TimothyRias, XLinkBot, Addbot, Mortense, Eric Drexler, Bte99, Zorrobot, Luckas-bot, AnomieBOT, Christopher.Gordon3, Citation bot, Northryde, Xqbot, Pra1998, Gsard, A. di M., Erik9bot, FrescoBot, Fortdj33, Citation bot 1, Ganondolf, RedBot, RobinK, Mary at CERN, EmausBot, Brent Perreault, Slawekb, Cogiati, Maschen, Isocliff, ClueBot NG, Helpful Pixie Bot, Bibcode Bot, Dzustin, Brendan.Oz, ChrisGualtieri, SD5bot, Dexbot, Enyokoyama, Joeinwiki, Dath Thou Even Lift, Dhm4444, Dbw1976, KasparBot and Anonymous: 40

- **Group theory** *Source:* https://en.wikipedia.org/wiki/Group_theory?oldid=668380925 *Contributors:* AxelBoldt, Zundark, The Anome, KF, Cwitty, Edward, Michael Hardy, Wshun, Dcljr, Ellywa, JWSchmidt, Bogdangiusca, Poor Yorick, Rossami, Jordi Burguet Castell, Charles Matthews, Lfh, Dysprosia, Jitse Niesen, Hyacinth, Fibonacci, Phys, Bevo, Kwantus, Finlay McWalter, PuzzletChung, Gromlakh, Romanm, Mayooranathan, Gandalf61, MathMartin, Rursus, Papadopc, ComplexZeta, Giftlite, Graeme Bartlett, Recentchanges, Dratman, Doshell, LiDaobing, Alberto da Calvairate~enwiki, Karl-Henner, Rich Farmbrough, FT2, Luqui, ArnoldReinhold, H0Okwurm, Paul August, Tompw, Jaimedv, Adan, Obradovic Goran, Friviere, Ranveig, Masv~enwiki, HenryLi, Oleg Alexandrov, Tbsmith, Archie Paulson, OdedSchramm, Kmg90, PeterPearson, V8rik, BD2412, Chun-hian, Josh Parris, Rjwilmsi, Dennis Estenson II, Salix alba, Ligulem, R.e.b., Brighterorange, FlaBot, Chris Pressey, Mathbot, Margosbot~enwiki, Rune.welsh, MTC, Chobot, YurikBot, Hairy Dude, Hillman, Michael Slone, Grubber, Cate, Merlincooper, Petter Strandmark, DYLAN LENNON~enwiki, Crasshopper, Googl, Tigershrike, Willtron, GrinBot~enwiki, RonnieBrown, Palapa, SmackBot, Reedy, Melchoir, Scullin, Natebarney, Cessator, BiT, GBL, Bluebot, Pieter Kuiper, MalafayaBot, Ligulembot, Pilotguy, Davipo, Christopherodonovan, Lambiam, Richard L. Peterson, Utopianheaven, Mike Fikes, Tawkerbot2, Chetvorno, CRGreathouse, Ale jrb, Gregbard, Rifleman 82, Tyskis, Mungomba, Headbomb, WVhybrid, Nadav1, RobHar, NERIUM, Escarbot, Seaphoto, M cuffa, VictorAnyakin, JAnDbot, Bongwarrior, Jakob.scholbach, CountingPine, Baccyak4H, Gabriel Kielland, David Eppstein, MaEr, David Callan, J.delanoy, Cmbankester, Indeed123, Gombang, Treisijs, Useight, Lemonaftertaste, VolkovBot, JohnBlackburne, EchoBravo, Philip Trueman, Eakirkman, Magmi, Eubulides, ArzelaAscoli, Arcfrk, Andreas Carter, Peter Stalin, Drschawrz, SieBot, Ivan Štambuk, WereSpielChequers, Viskonsas, Messagetolove, Lightmouse, JackSchmidt, NobillyT, StaticGull, Alpha Beta Epsilon, Justin W Smith, Alksentrs, Padicgroup, Bhuna71, Mspraveen, Avouac, Watchduck, Edwinconnell, Xylthixlm, Hans Adler, Vegetator, Johnuniq, TimothyRias, XLinkBot, JinJian, CàlculIntegral, Addbot, Manuel Trujillo Berges, SpellingBot, Fluffernutter, Kristine8~enwiki, Favonian, Tide rolls, Luckas-bot, Yobot, TaBOT-zerem, Julia W, Eamonster, AnomieBOT, DemocraticLuntz, Rubinbot, Μυρμηγκάκι, WinoWeritas, Citation bot, Calcio33, Auclairde, FrescoBot, Lothar von Richthofen, Orhanghazi, Sławomir Biały, Citation bot 1, Boulaur, Hard Sin, Hamtechperson, Ngyikp, D stankov, Jauhienij, Debator of mathematics, Lightlowemon, FoxBot, Yger, SomeRandomPerson23, EmausBot, Fly by Night, Tommy2010, Shishir332, D15724C710N, Quondum, Kranix, Adgjdghjdety, Gottlob Gödel, ClueBot NG, Lord Roem, Ciro.santilli, HMSSolent, BG19bot, Ijgt, CimanyD, Meclee, Brad7777, Jochen Burghardt, Brirush, CsDix, Laxfan1977, Chetan bagora, Edmundthe, KasparBot and Anonymous: 136
- **Feynman diagram** *Source:* https://en.wikipedia.org/wiki/Feynman_diagram?oldid=669390851 *Contributors:* CYD, Bryan Derksen, XJaM, Nate Silva, Shii, Stevertigo, Edward, Michael Hardy, Looxix~enwiki, Ahoerstemeier, Ryan Cable, Nikai, Kaihsu, Charles Matthews, Ww, Doradus, Furrykef, Phys, Omegatron, Bevo, Bhiggs, BenRG, Peak, Rasmus Faber, Intangir, SC, Wikibot, Anthony, Tobias Bergemann, Danenberg, Ancheta Wis, Giftlite, Sj, Harp, Alison, Remy B, Vivektewary, Sigfpe, Beland, Pmanderson, Icairns, Sam Hocevar, Robin klein, AlexChurchill, Urvabara, Jkl, Pyrop, Guanabot, Pjacobi, Vsmith, Rspeer, Mal~enwiki, Bender235, Ben Standeven, Robert P. O'Shea, AnyFile, Jjk, Matt McIrvin, Scentoni, Tritium6, Mdd, Neonumbers, Rgclegg, Ferrierd, Mac Davis, Tony Sidaway, RJFJR, Drat, Dominic, Markko, Linas, LoopZilla, Daira Hopwood, Ketiltrout, Koavf, Zbxgscqf, Commander, Strait, Miserlou, Ligulem, RE, FlaBot, Moskvax, RobertG, Gnostic804, Acyso, DoomBringer, Chobot, YurikBot, JabberWok, Gaius Cornelius, Rodier, Anomalocaris, SEWilcoBot, Welsh, SCZenz, Ragesoss, Nubby, Larsobrien, Ejl, DRosenbach, Dna-webmaster, Tomj, Caco de vidro, GrinBot~enwiki, KasugaHuang, SmackBot, Tom Lougheed, GaeusOctavius, Chris the speller, Jjalexand, JustThisGuy, Pieter Kuiper, Complexica, Colonies Chris, Salmar, Wikipedia brown, Xiner, Huon, Tesseran, DJIndica, Eliyak, SilverStar, Xiphoris, JanBielawski, Bitwise, Beefyt, Paul venter, Joseph Solis in Australia, Wikifarzin, Patrickwooldridge, 8754865, CmdrObot, Van helsing, Jsmaye, Joelholdsworth, Myasuda, Cydebot, Xxanthippe, Michael C Price, Thijs!bot, Epbr123, Barticus88, Headbomb, Davidhorman, Shlomi Hillel, Leevclarke, Sluzzelin, AniRaptor2001, CosineKitty, Jameskeates, Bakken, SHCarter, Swpb, Warchef, Connor Behan, R'n'B, Choihei, Sefog, Chiswick Chap, Fylwind, Joshmt, Vyn, Brvman, Sheliak, Mulanhua, Quilbert, Rei-bot, Kevin Steinhardt, Mbusux, Richwil, Ptrslv72, Drschawrz, SieBot, Likebox, Taemyr, Staylor71, Martarius, WurmWoode, ChandlerMapBot, Sjdunn9, DragonBot, Chutsu, GlasGhost, DumZiBoT, TimothyRias, Mchaddock, PSimeon, SilvonenBot, Truthnlove, Out of Phase User, Metsavend, Download, Chamal N, Favonian, Chenzw-Bot, Barak Sh, Mikkim64, AgadaUrbanit, Conroy23, Alfie66, Legobot, Luckas-bot, Yobot, Amirobot, AnomieBOT, Piano non troppo, Xqbot, Srich32977, PhysicsR, Noamz, Seeleschneider, Createangelos, Kismalac, Ysyoon, Nurefsan, Meier99, Hickorybark, Earthandmoon, Aaivazis, EmausBot, John of Reading, Bookalign, WikitanvirBot, Bornerdogge, Maschen, Zueignung, Xanchester, Anagogist, Antiqueight, Smack the donkey, Pfeiferwalter, OCCullens, Bakkedal, ChrisGualtieri, Equatorbit, Mrmagikpants, Jamesx12345, Pjpeters, Mark viking, MutluMan, Someone not using his real name, UltraBird, Airwoz, Monkbot, Quogle, Cheweblaze, FivePillarPurist and Anonymous: 124
- **Bosonic string theory** *Source:* https://en.wikipedia.org/wiki/Bosonic_string_theory?oldid=666466612 *Contributors:* TakuyaMurata, Arpingstone, Phys, Fropuff, Anville, Sigfpe, ~Brain.Wav~, Keenan Pepper, Kocio, Mpatel, Joke137, BD2412, Chobot, Mario23, Sardanaphalus, SmackBot, Kurochka, Skizzik, Scwlong, Zapvet, JoeBot, Michael C Price, Thijs!bot, Headbomb, Dr. Blofeld, Maurice Carbonaro, WJBscribe, Idioma-bot, Sheliak, Venny85, Robdunst, Dombom, Estirabot, Trabelsiismail, Addbot, Jasper Deng, Quasar.dj, Materialscientist, Waleswatcher, FrescoBot, ClueBot NG, Sudip2118, Frietjes, Poopypoopyx2, Calabe1992, Bibcode Bot, BG19bot, Luizpuodzius, Ilikeediting555111, Hmainsbot1, Mogism, Andyhowlett, Polytope24, Monkbot, Rantonels and Anonymous: 34
- **Superstring theory** *Source:* https://en.wikipedia.org/wiki/Superstring_theory?oldid=669906748 *Contributors:* Mav, Bryan Derksen, Stevertigo, Michael Hardy, Erik Zachte, Minesweeper, Looxix~enwiki, Ahoerstemeier, JWSchmidt, Cyan, Palfrey, Evercat, Schneelocke, Hashar, Charles Matthews, Tpbradbury, Motor, David Shay, Omegatron, Bevo, Bcorr, Robbot, Fredrik, Hadal, Vuara, Giftlite, Barbara Shack, Herbee, Fropuff, Anville, Maarten van Vliet, WalkinDownThirtyThree, Christopherlin, Steuard, Karol Langner, Lumidek, Prestonmarkstone, Rich Farmbrough, Igorivanov~enwiki, Autiger, Pavel Vozenilek, El C, Rgdboer, Shadow demon, Causa sui, Billymac00, BM, Gary, Pion, Tycho, Cal 1234, Redvers, Postrach, Supercool Dude, Mindmatrix, Mpatel, Joke137, Mandarax, Bill37212, Yamamoto Ichiro, Bubbleboys, Chobot, Ben Tibbetts, Wavelength, RussBot, Chris Capoccia, Chensiyuan, Cate, Chaos, NawlinWiki, Astral, Voidxor, TheMadBaron, Zerodamage, Allens, SmackBot, Android 93, Kurochka, McGeddon, Kintetsubuffalo, Cesoid, Silly rabbit, Stevage, Baronnet, Colonies Chris, Scwlong, Mesons, Kurrupt3d, Bjankuloski06en~enwiki, Makyen, MathStuf, Hu12, Iridescent, Kahalachan, Gatortpk, Mattbr, Neelix, Gregbard, Nauticashades, Cydebot, Davidanzaldua, ChKa, Headbomb, Escarbot, Jj137, Shlomi Hillel, Dougher, JAnDbot, 100110100, 28421u2232nfenfcenc, Aziz1005, Jean-Pierre Petit~enwiki, Hans Dunkelberg, Maurice Carbonaro, Bot-Schafter, SmilesALot, Student7, Cmichael, Sheliak, JayCo777, Calwiki, Andrius.v, Molinogi, Billinghurst, Lamro, En-

viroboy, Seraphita~enwiki, Drschawrz, Henry Delforn (old), Lightmouse, Altzinn, Gratedparmesan, ClueBot, Arakunem, Vergil 577, Frdayeen, Vizzini101, Niceguyedc, Neverquick, Resoru, Mastertek, Kakofonous, Princess Janay, Alex123irish123, Madeinmexico567, Oldnoah, Madeinmexico566, Truthnlove, YeAaMsLtA, Addbot, Physicman123, CWatchman, Cuaxdon, Semdino, AnnaFrance, LinkFA-Bot, TaBOT-zerem, Evans1982, Gerixau, Eric-Wester, Magog the Ogre, AnomieBOT, DemocraticLuntz, ^musaz, Josh Guffin, Jim1138, Citation bot, Renaissancee, BLP-outrageous move logs, Omnipaedista, RibotBOT, Paine Ellsworth, Steve Quinn, Tom.Reding, Klavesin, Tkachyk, Dinamik-bot, Bj norge, Idh0854, Arbnos, Vramasub, L Kensington, Particle hep, Isocliff, ClueBot NG, Widr, Adminium, Delivernews, Bibcode Bot, Khanduras, Quarkgluonsoup, Flowerhat15, MythosMagic, OCCullens, Aldrich2122, Graphium, AHusain314, Jochen Burghardt, WorldWideJuan, Jakec, Liquidityinsta, E8xE8, Polytope24, FlaviusCorcoata, Cirksena, BakedLikaBiscuit, KasparBot, Rantonels and Anonymous: 171

- **Type I string theory** *Source:* https://en.wikipedia.org/wiki/Type_I_string_theory?oldid=629001015 *Contributors:* DJ Clayworth, Fropuff, Moyogo, Lumidek, Brian0918, Velella, Mpatel, EricCHill, Reyk, Sardanaphalus, SmackBot, Unyoyega, Colonies Chris, Jmnbatista, LordAnubisBOT, MarkJefferys, Idioma-bot, Sheliak, Thomas.W, Sagnotti, Addbot, Lightbot, Materialscientist, Omnipaedista, Erik9bot, Calmer Waters, EmausBot, ClueBot NG, BG19bot, Frze, Hmainsbot1, Zacht.carnevale, Dimension10, Polytope24 and Anonymous: 6

- **Type II string theory** *Source:* https://en.wikipedia.org/wiki/Type_II_string_theory?oldid=629086741 *Contributors:* Stevertigo, Fropuff, Christopherlin, Lumidek, Jag123, MarSch, Mgnbar, Reyk, Sardanaphalus, SmackBot, Unyoyega, Colonies Chris, Jmnbatista, Joshua Davis, LordAnubisBOT, Idioma-bot, Sheliak, OKBot, Jovianeye, MystBot, Addbot, AkhtaBot, Yobot, Omnipaedista, Erik9bot, FrescoBot, EmausBot, 336, Luizpuodzius, Dimension10, Polytope24, Luca.agozzino and Anonymous: 5

- **Heterotic string theory** *Source:* https://en.wikipedia.org/wiki/Heterotic_string_theory?oldid=640076972 *Contributors:* Charles Matthews, Denni, Hugo~enwiki, Giftlite, Fropuff, Lumidek, Rich Farmbrough, Pearle, NTK, Mpatel, Chobot, YurikBot, Bhny, Hwasungmars, 2over0, Sardanaphalus, KnightRider~enwiki, SmackBot, Schmiteye, Fplay, QFT, Jmnbatista, Headbomb, Lamontacranston, STBot, MarkJefferys, Sheliak, JohnBlackburne, Spiral5800, Legoktm, TimothyRias, MystBot, Addbot, Debresser, Tassedethe, Lightbot, OlEnglish, Jack who built the house, AnomieBOT, Citation bot, Omnipaedista, Erik9bot, Steve Quinn, Orenburg1, EmausBot, WaterfordPBR, Staszek Lem, ClueBot NG, BG19bot, AHusain314, I am One of Many, Dimension10, Polytope24 and Anonymous: 31

- **S-duality** *Source:* https://en.wikipedia.org/wiki/S-duality?oldid=665947180 *Contributors:* CYD, The Anome, Michael Hardy, Angela, Charles Matthews, Phys, Lumidek, Gauge, Linas, Mpatel, Rjwilmsi, CJLL Wright, Shell Kinney, Sardanaphalus, SmackBot, Reedy, Colonies Chris, Scwlong, QFT, Pjoef, MenoBot, AnonyScientist, Addbot, Yobot, JackieBot, Citation bot, Omnipaedista, Erik9bot, Tom.Reding, Vhsatheeshkumar, Lightlowemon, ZéroBot, ClueBot NG, Bibcode Bot, Enyokoyama, AHusain314, Polytope24 and Anonymous: 7

- **T-duality** *Source:* https://en.wikipedia.org/wiki/T-duality?oldid=665947380 *Contributors:* TakuyaMurata, Charles Matthews, Reddi, David Gerard, Alison, Lumidek, Lysdexia, BD2412, Roboto de Ajvol, WAS 4.250, Sardanaphalus, Unyoyega, MalafayaBot, Scwlong, QFT, Dan Gluck, Outriggr, Headbomb, RogueTeddy, Eujin16, Truthnlove, Addbot, MagnusA.Bot, Yobot, Citation bot, LilHelpa, Omnipaedista, HissingFauna, Patchy1, Tom.Reding, EmausBot, ZéroBot, Bibcode Bot, AHusain314, Polytope24, Anrnusna and Anonymous: 10

- **M-theory** *Source:* https://en.wikipedia.org/wiki/M-theory?oldid=668256489 *Contributors:* AxelBoldt, CYD, Eloquence, BF, Bryan Derksen, Zundark, The Anome, Ap, Tim Chambers, Hari, Maury Markowitz, Stevertigo, Michael Hardy, Tim Starling, Gabbe, Tompagenet, Ixfd64, CesarB, Looxix~enwiki, JWSchmidt, Darkwind, Marco Krohn, Jeandré du Toit, Evercat, Schneelocke, Charles Matthews, Timwi, Reddi, Malcohol, Bevo, Jusjih, Slawojarek, Sander123, Fredrik, R3m0t, RedWolf, Blainster, DHN, Hadal, HaeB, Tobias Bergemann, David Gerard, Giftlite, DocWatson42, Jmnbpt, Barbara Shack, Fropuff, Moyogo, Sigfpe, Daen, Antandrus, Lumidek, ChrisCostello, Mike Rosoft, Spiffy sperry, Urvabara, Noisy, Discospinster, H0riz0n, Vsmith, Loren36, El C, Momotaro, Shanes, RoyBoy, Triona, Constantine, Smalljim, I9Q79oL78KiL0QTFHgyc, Giraffedata, Wolfrider~enwiki, Physicistjedi, MPerel, Gsklee, ShardPhoenix, Axl, Mac Davis, Kocio, Burn, Hu, Wtmitchell, SidP, DV8 2XL, Ringbang, Kazvorpal, Omnist, Sharkie, Joelpt, Angr, Firsfron, FeanorStar7, Pol098, WadeSimMiser, Mpatel, GregorB, Jugger90, Paxsimius, Mandarax, Chun-hian, Grammarbot, Rjwilmsi, Nightscream, Koavf, Zbxgscqf, Oblivious, Yug, Lionelbrits, Ruidlopes, The ARK, Latka, Mathbot, Diza, Phoenix2~enwiki, DVdm, Eric B, Bomb319, Loom91, Zafiroblue05, Bhny, Stephenb, KSchutte, Bovineone, Salsb, Erielhonan, Bobak, Asarelah, Dna-webmaster, Sandstein, Superdude99, Zzuuzz, Imaninjapirate, Arthur Rubin, Ilmari Karonen, Caco de vidro, DVD R W, Hide&Reason, Jmeden2000, Teo64x, Sardanaphalus, MartinGugino, RupertMillard, SmackBot, Kurochka, K-UNIT, Rwp, Rlbates99, Ajt, Ian Rose, Gilliam, Wlmg, DividedByNegativeZero, Mirokado, Bluebot, Cush, SMP, Ben.c.roberts, MalafayaBot, Nbarth, DHN-bot~enwiki, Colonies Chris, Joemah, N.MacInnes, Xiner, Nunocordeiro, Mbertsch, Addshore, EPM, Nakon, Kiplantt, Bigmantonyd, Martijn Hoekstra, Kabain52, Brdforallseasons, Sayden, Doug Bell, Jaganath, Shadowlynk, IronGargoyle, Jochietoch, Hu12, Jxh2154, Tawkerbot2, Valoem, Gebrah, Albertod4, Kurtan~enwiki, Harold f, Devourer09, Cyrusc, CRGreathouse, Olaf Davis, Lambertian, Friendlystar, Rowellcf, Bmk, Myasuda, DepartedUser2, Ekajati, Cydebot, Fluence, Mike Christie, Meno25, Gagueci, Kahananite, Michael C Price, Alexnye, IComputerSaysNo, Lord Satorious, Krowe, Mrockman, Thijs!bot, Epbr123, Daniel, Headbomb, NeilHalfway, James086, KrakatoaKatie, AntiVandalBot, Blue Tie, Alphachimpbot, J rowley, Shambolic Entity, SuperLuigi31, Buchhemi, Fetchcomms, 100110100, WolfmanSF, VoABot II, Madevin314, SHCarter, Rami R, Jqshenker, Just H, War wizard90, Rickard Vogelberg, Stephen Shenker, Theoretic, MartinBot, Kostisl, R'n'B, Euku, Numbo3, Maurice Carbonaro, Nly8nchz, Thucydides411, LordAnubisBOT, Janus Shadowsong, Peskydan, Isoko, Belovedfreak, Antony-22, Wesino, WJBscribe, Thomas795135, Blood Oath Bot, Idioma-bot, Sheliak, Gogobera, Jeff G., Rei-bot, Ask123, Pennstatephil, JhsBot, Mazarin07, Peace keeper II, Antixt, Why Not A Duck, PhysPhD, Rknasc, Guystout, Drschawrz, YohanN7, SieBot, Robdunst, Paradoctor, Wing gundam, Holt27, Astroboyretro, Caidh, OKBot, Divinestuff, Wpac5, Ayleuss, Beofluff, Loren.wilton, ClueBot, Master Shake 9, The Thing That Should Not Be, Haemorrhage, Arakunem, Drmies, IMNTU, Yupjohnny, Huntthetroll, Patrik Andersson, Dank, Gardv, DumZiBoT, Jfosc, Maky, Truthnlove, Autocoast~enwiki, Albambot, Addbot, Uruk2008, Cuaxdon, CanadianLinuxUser, WikiUserPedia, Barak Sh, Tassedethe, Carapheonix, Togekiss101, Tide rolls, OlEnglish, Snaily, Legobot, Luckas-bot, Yobot, Fraggle81, Pcap, Foolo~enwiki, CinchBug, Tempodivalse, AnomieBOT, KDS4444, Götz, Charlesvi, Dalton h, Marcka, Alexzabbey, Jim1138, IRP, AdjustShift, Materialscientist, Citation bot, Quebec99, Ruike, TinucherianBot II, Ekwos, Techwiz2000, Omnipaedista, Peanuts4life, Pinethicket, Vicenarian, Tom.Reding, EDG161, Jusses2, Serols, ActivExpression, SkyMachine, Gerda Arendt, Tkachyk, 122589423KM, சஞ்சீவி சிவகுமார், Reaper Eternal, Apb91781, 786 zikhar, LcawteHuggle, Adam1217, EmausBot, GoingBatty, Pyschobbens, StringTheory11, Smiwi, Suslindisambiguator, SporkBot, PoisonGM, Besneatte, SBaker43, Denholm Reynholm, RockMagnetist, ClueBot NG, Blueshift333, Rgwkenyon, Helpful Pixie Bot, Bibcode Bot, BG19bot, SharkinthePool, Msaunier, MusikAnimal, Copernicus01, Elginfball10, Qed3, ShotmanMaslo, Zujua, Kooky2, Mediran, Chris5631, FEYKATD, Ecila3, Lugia2453, Frosty, AHusain314, Armanschwarz, Among Men, Faizan, Epicgenius, Diekilldie, EddieHugh, Dustin V. S., RaphaelQS, Beakr, DavidLeighEllis, Tedsanders, TFA Protector Bot, Vampre1122, Polytope24, Evandas, Oneidiotsavant, Pretickle, TheRealTheKoi, Shantsforeverandalways, QuantumMatt101, FACBot, Kh3368, Sizeofint, Jyhtgqwqsdfghjydwq, Mberkson12, KasparBot, Cmealo, Yadav.aakash.500 and Anonymous: 447

- **AdS/CFT correspondence** *Source:* https://en.wikipedia.org/wiki/AdS/CFT_correspondence?oldid=665115955 *Contributors:* Bryan Derksen, The Anome, Stevertigo, Michael Hardy, Charles Matthews, Grendelkhan, Phys, COGDEN, Giftlite, Mporter, LeYaYa, Fropuff, Michael Devore, Eequor, Aside, Quadell, Tomruen, Lumidek, Rich Farmbrough, Cacycle, Pjacobi, Schuetzm, David Schaich, Dmr2, Jag123, Physicistjedi, Merlinme, Mpatel, Jugger90, BD2412, Rjwilmsi, FlaBot, Bgwhite, YurikBot, Salsb, Grafen, Hwasungmars, Psu256, Knotnic, Sardanaphalus, Honza Záruba, GaeusOctavius, Silly rabbit, Colonies Chris, QFT, John, JorisvS, Mr Stephen, Dan Gluck, Metre01, Bobamnertiopsis, Gebrah, Thijs!bot, Mbell, Headbomb, Landscape~enwiki, JAnDbot, Magioladitis, WolfmanSF, CommonsDelinker, Sigmundur, TXiKiBoT, Hqb, Piperh, Drschawrz, Mnudelman, Elassint, Ozooxo, Piledhigheranddeeper, Kville105125, Mastertek, Another Believer, AnonyScientist, Addbot, Tassedethe, Lightbot, SPat, AnomieBOT, Citation bot, PowerUserPCDude, Eugeneelgato, Cannolis, Vhsatheeshkumar, Meier99, Nilock, RjwilmsiBot, GoingBatty, Riggr Mortis, Serketan, Werieth, StringTheory11, SporkBot, AManWithNoPlan, YtivarG, Isocliff, Raidr, Bibcode Bot, BG19bot, QuarkyPi, Hmainsbot1, AHusain314, BronzeRatio, Mark viking, VoxelBot, Liquidityinsta, TFA Protector Bot, Dimension10, Polytope24, DThung, Username123454321, DvtheDv, Sileby10, Barbcash, Kismit33 and Anonymous: 61

- **Multiverse** *Source:* https://en.wikipedia.org/wiki/Multiverse?oldid=669446959 *Contributors:* Derek Ross, Dan~enwiki, Bryan Derksen, The Anome, Css, BenBaker, Ant, B4hand, Modemac, Hephaestos, Olivier, Patrick, Boud, Michael Hardy, Tim Starling, Modster, Delirium, William M. Connolley, Angela, Ciphergoth, Poor Yorick, ^{212}, Emperorbma, Renamed user 4, Dcoetzee, Morwen, Fairandbalanced, BenRG, Sjorford, Tlogmer, Nurg, Rursus, Hadal, Wereon, Johnstone, Giftlite, Smjg, Graeme Bartlett, Mshonle~enwiki, Gtrmp, Mporter, Barbara Shack, ShaunMacPherson, Bfinn, Lefty, Anville, Beardo, Gracefool, Pascal666, Siroxo, Eequor, Pne, Hannes Karnoefel, Fishal, Gdm, Pcarbonn, MadIce, Khaosworks, CSTAR, Slivester, Sam Hocevar, JeffreyN, Soilguy5, Sam, Fermion, Waza, MakeRocketGoNow, Grunt, Mike Rosoft, Freakofnurture, Dbachmann, Paul August, Petersam, BlueNight, Zippedmartin, Livajo, Tverbeek, Haxwell, RoyBoy, Bobo192, Androo, K0hlrabi, Rbj, Reuben, Dlarmore, Thanos6, I9Q79oL78KiL0QTFHgyc, Martg76, Apostrophe, Zetawoof, Dillee1, Alansohn, SnowFire, Walter Görlitz, Arthena, Hu, Titanium Dragon, Knowledge Seeker, Pfahlstrom, DV8 2XL, Pwqn, Netkinetic, Kitch, Falcorian, Ott, Zntrip, Dmitrybrant, Stemonitis, Firsfron, Roboshed, Mel Etitis, FeanorStar7, ApLundell, Netdragon, Jeff3000, Meneth, Rchamberlain, CharlesC, Noetica, Joke137, Ictlogist, Teflon Don, Drbogdan, Rjwilmsi, Electricnet, Salix alba, Mike Peel, Bubba73, Dianelos, Brighterorange, Cfortunato, Cassowary, JeffStickney, FlaBot, Pitamakan, Old Moonraker, Latka, Saswann, Phoenix2~enwiki, Chobot, DVdm, Bgwhite, Whosasking, PointedEars, YurikBot, ThunderPeel2001, Fabartus, Bhny, Admiral Roo, Hellbus, Mpfrank, Eleassar, NawlinWiki, Joncolvin, Mike18xx, Spikehay, Inhighspeed, Taigei, Obey, RonCram, Shadowblade, Tomisti, WAS 4.250, Deville, Joecab, Moogsi, KGasso, Xaxafrad, Gating, JoanneB, Johnpseudo, Ilmari Karonen, Markbenecke, Hathaldir~enwiki, Johnbmckenna, NeilN, Philip Stevens, Eog1916, SmackBot, Reedy, K1234567890y, MJMyers2~enwiki, Vald, Jagged 85, Eskimbot, KelleyCook, Kintetsubuffalo, Alex earlier account, Gilliam, Portillo, Ohnoitsjamie, Kmarinas86, Armeria, Chris the speller, Glastohead, Apeloverage, Foosher, Jerome Charles Potts, Octahedron80, Scwlong, Vanished User 0001, Zaian, Xiner, Rrburke, Robma, Nakon, Geoffr, Adrigon, Metamagician3000, Kurrupt3d, Vina-iwbot~enwiki, Ollj, CHBMch05, Ohconfucius, Lambiam, Rory096, Amnuel~enwiki, Police officer, Vampus, JoshuaZ, IronGargoyle, JHunterJ, Makyen, Noah Salzman, BenRayfield, Ehheh, AxG, Z E U S, DCNanney, Clq, SandyGeorgia, Eridani, Doczilla, Dl2000, ShakingSpirit, Iridescent, Joseph Solis in Australia, Newone, Imad marie, Blehfu, Gco, Valoem, Xammer, CRGreathouse, Vyznev Xnebara, Rwflammang, Editorius, K00bine, Sanspeur, Hemlock Martinis, Mchmike, Peterdjones, Gogo Dodo, DarthSidious, Michael C Price, Kozuch, Danogo, Silicianer, UberScienceNerd, JamesAM, Thijs!bot, Daniel Newman, Wylfing, GentlemanGhost, Headbomb, Second Quantization, Parsiferon, Chunminghan, Riction, AntiVandalBot, AbstractClass, Gioto, Caledones, Christinedoby, EdgarCarpenter, QuiteUnusual, Handsaw, Darklilac, Tim Shuba, Ran4, LinkinPark, Jenattiyeh, Geshel, .anacondabot, Magioladitis, VoABot II, Hawkaris, Craze3, Hiplibrarianship, Winterus, Urco, Wikianon, Rickard Vogelberg, Greenguy1090, MartinBot, STBot, Anarchia, The Anonymous One, EdBever, Tgeairn, Trusilver, Multisport3, UBeR, MattB2, Arion 3x3, McSly, Tarotcards, RickardV, Richard D. LeCour, NewEnglandYankee, Mirage GSM, Dorftrottel, VolkovBot, Camrn86, Riraito, Sunwukongmonkeygod, TXiKiBoT, Mercurywoodrose, Kilo39, Kriak, AstroWiki, Albertmin, Anonymous Dissident, Riemann Zeta, Shiritai, Aaron Bowen, Masqline, BotKung, Jamelan, Mjkrol, Parsifal, Roy7, Sapphic, BrianY, Zx-man, Logan, Neparis, Hrafn, Peter.thelander, Silver Spoon, Spartan, Scarian, Malcolmxl5, Richardvanas1, Krawi, Curst Saden, ThisIsRealPuma, Alexbook, Yintan, Bernard Marx, Araignee, Hac13, Flyer22, Radon210, Dominik92, Wmpearl, Kriss12, Sedecrem, OKBot, Kurtilein, BigBang616, Escape Orbit, GeorgiaSLee, Vinay Jha, IomPsinso, Athenean, XDanielx, WikipedianMarlith, Martarius, ClueBot, Troublemane, Boing! said Zebedee, Nfenn, K a r n a, Undisputedloser, Chimesmonster, Masterpiece2000, Stupidjassi, SolomonFreer, BobKawanaka, Hans Adler, Rui Gabriel Correia, Aitias, Yboc2000, Stoned philosophaster, Miami33139, DumZiBoT, Wednesday Next, XLinkBot, Ajcheema, ErkinBatu, Thatguyflint, Cloudblazer, Santasa99, Mbroderick271, Addbot, Ito123456789, Moosester, Uruk2008, DOI bot, Tcncv, Discrepancy, Morriswa, Longhandle1137, ExternalGazer, Download, Bwikiroa, Richard-r-v, Smurfsider, Tide rolls, MuZemike, Bartledan, SeniorInt, LuK3, Athanatsius, Luckas-bot, Yobot, Ht686rg90, HieronymousCrowley, GateKeeper, SwisterTwister, Naroood, MrH3MinuteMile, Rickybaksh, Kookyunii, AnomieBOT, 1exec1, Kingpin13, Projectyugo, Materialscientist, Citation bot, Allen234, Xqbot, Addihockey10, Smk65536, J JMesserly, Coretheapple, J04n, Foreignshore, Artemka373, Sophus Bie, Keeganp, Another disinterested reader, Steven Avraham Rosten, KronicTOOL, Alxeedo, Steve Quinn, Machine Elf 1735, Citation bot 1, Anthony on Stilts, Osprey9713, Gil987, Pinethicket, Three887, Tom.Reding, Dazedbythebell, Chrisdann, Casimir9999, BlackHades, Tlhslobus, Zhonghuo~enwiki, SkyMachine, Trappist the monk, Lotje, Nickyus, Bacabed, Bluefist, TheGrimReaper NS, Jeffrd10, Skakkle, Jesse V., Twoshotted, Beyond My Ken, Djf2564, DASHBot, EmausBot, WikitanvirBot, RA0808, Annaalexandrae, Wesley J M, Slightsmile, Wikipelli, Slawekb, Solomonfromfinland, Rafi5749, Italia2006, ZéroBot, Naiveandsilly, Daonguyen95, Chasrob, RUBEN TESOLIN, Hyblackeagle22, Red Echidna, AnnSec, AManWithNoPlan, PoisonGM, 09dhummel, Tiiliskivi, Hanjunjade, Crux007, ChuispastonBot, Terraflorin, Spydra, 28bot, Isocliff, Petrb, ClueBot NG, Thanose, Satellizer, Kikichugirl, Movses-bot, Tabletrack, Primergrey, O.Koslowski, Dream of Nyx, Parthdu, North Atlanticist Usonian, Helpful Pixie Bot, Bibcode Bot, Jeraphine Gryphon, BG19bot, Akash pagla, MusikAnimal, Rnm102696, Future kosmos, In another mutiverse, Meabandit, Achowat, GreenUniverse, OCCullens, BattyBot, Ha33 blondechick, Mdann52, Nerudography, ChrisGualtieri, Mediran, EuroCarGT, BorealisXIV, Quafman, MarioCGR, Lscagle, Cwobeel, Sminthopsis84, Ddddurga, Jcardazzi, Lugia2453, Graphium, Vibhor.sen, Marvelous8, Asdf62872, Reatlas, Maniesansdelire, Martinsolis1970, Epicgenius, Debouch, Draviste, Sneazy, Jakec, Dustin V. S., Rolf h nelson, CensoredScribe, Tlönorbis, Matthewunthird, Christopherryan100, Jwratner1, Ginsuloft, Kogge, JosephSpiral, Aubreybardo, Cr.marthi, Charizard5904, MagicatthemovieS, Qwertyuiopasdfghjkl111, Romponu, RLight1, Tim001Jack, Bluesnickers, Samstan2100, Monkbot, DR177, Jamekam, Ffioninwonderland, Johnevella, Thundergodz, MarquisMartinez, DanBalance, Anunaki truth, Tetra quark, Smaleganderr5, Isambard Kingdom, Sleepy Geek, Anand2202, Oduduserra, Thomasberggren, Multiverse Guy, SocraticOath, Nøkkenbuer, Sweggysweg, Ksb1231, Daveptt, Phseek and Anonymous: 632

26.6.2 Images

- **File:AdS3.svg** *Source:* https://upload.wikimedia.org/wikipedia/commons/4/47/AdS3.svg *License:* CC BY-SA 3.0 *Contributors:* This file was derived from: AdS3 (new).png
Original artist:
- derivative work: Alex Dunkel (Maky)
- **File:AdS3_(new).png** *Source:* https://upload.wikimedia.org/wikipedia/commons/f/fe/AdS3_%28new%29.png *License:* CC BY-SA 3.0 *Contributors:* Own work *Original artist:* Polytope24
- **File:Ambox_important.svg** *Source:* https://upload.wikimedia.org/wikipedia/commons/b/b4/Ambox_important.svg *License:* Public domain *Contributors:* Own work, based off of Image:Ambox scales.svg *Original artist:* Dsmurat (talk · contribs)
- **File:Ambox_question.svg** *Source:* https://upload.wikimedia.org/wikipedia/commons/1/1b/Ambox_question.svg *License:* Public domain *Contributors:* Based on Image:Ambox important.svg *Original artist:* Mysid, Dsmurat, penubag
- **File:Ambox_rewrite.svg** *Source:* https://upload.wikimedia.org/wikipedia/commons/1/1c/Ambox_rewrite.svg *License:* Public domain *Contributors:* self-made in Inkscape *Original artist:* penubag
- **File:CERN_LHC_Tunnel1.jpg** *Source:* https://upload.wikimedia.org/wikipedia/commons/f/fc/CERN_LHC_Tunnel1.jpg *License:* CC BY-SA 3.0 *Contributors:* Own work *Original artist:* Julian Herzog (website)
- **File:Caesar3.svg** *Source:* https://upload.wikimedia.org/wikipedia/commons/2/2b/Caesar3.svg *License:* Public domain *Contributors:* Own work *Original artist:* Cepheus
- **File:Calabi-Yau.png** *Source:* https://upload.wikimedia.org/wikipedia/commons/d/d4/Calabi-Yau.png *License:* CC BY-SA 2.5 *Contributors:* own work by Lunch
http://en.wikipedia.org/wiki/Image:Calabi-Yau.png (english Wikipedia) *Original artist:* Lunch
- **File:Calabi_yau.jpg** *Source:* https://upload.wikimedia.org/wikipedia/commons/f/f3/Calabi_yau.jpg *License:* Public domain *Contributors:* Mathematica output, created by author *Original artist:* Jbourjai
- **File:Calabi_yau_formatted.svg** *Source:* https://upload.wikimedia.org/wikipedia/commons/8/8c/Calabi_yau_formatted.svg *License:* Public domain *Contributors:* This file was derived from: Calabi yau.jpg
Original artist:
- derivative work: Polytope24
- **File:Cayley_graph_of_F2.svg** *Source:* https://upload.wikimedia.org/wikipedia/commons/d/d2/Cayley_graph_of_F2.svg *License:* Public domain *Contributors:* ? *Original artist:* ?
- **File:Clebsch_Cublic.png** *Source:* https://upload.wikimedia.org/wikipedia/commons/7/7c/Clebsch_Cublic.png *License:* CC BY-SA 3.0 *Contributors:* I created this on my own computer using the free software Surfer *Original artist:* Fly by Night
- **File:Commons-logo.svg** *Source:* https://upload.wikimedia.org/wikipedia/en/4/4a/Commons-logo.svg *License:* ? *Contributors:* ? *Original artist:* ?
- **File:Compactification_example.svg** *Source:* https://upload.wikimedia.org/wikipedia/commons/f/f5/Compactification_example.svg *License:* CC BY-SA 4.0 *Contributors:* Brian Greene (2004). The Elegant Universe (DVD). Part II (String's the thing): WGBH Boston Video. Event occurs at 43:55. OCLC 54019786 *Original artist:* Alex Dunkel (Maky)
- **File:Crab_Nebula.jpg** *Source:* https://upload.wikimedia.org/wikipedia/commons/0/00/Crab_Nebula.jpg *License:* Public domain *Contributors:* HubbleSite: gallery, release. *Original artist:* NASA, ESA, J. Hester and A. Loll (Arizona State University)
- **File:D3-brane_et_D2-brane.PNG** *Source:* https://upload.wikimedia.org/wikipedia/commons/8/88/D3-brane_et_D2-brane.PNG *License:* Public domain *Contributors:* Image:D-brane.PNG, oeuvre personnelle. *Original artist:* <a href='//commons.wikimedia.org/w/index.php?title=User:Rogilbert,<span>,&,</span>,action=edit,<span>,&,</span>,redlink=1' class='new' title='User:Rogilbert (page does not exist)'>Rogilbert</a>
- **File:Dualities_of_string_and_M-theory.jpg** *Source:* https://upload.wikimedia.org/wikipedia/commons/b/bd/Dualities_of_string_and_M-theory.jpg *License:* CC BY-SA 3.0 *Contributors:* Own work *Original artist:* Polytope24
- **File:ECClines-3.svg** *Source:* https://upload.wikimedia.org/wikipedia/commons/d/d0/ECClines-3.svg *License:* CC BY-SA 3.0 *Contributors:* Own work based on Image:ECexamples01.png by en:User:Image:Dake and Image:ECClines-2.svg by SuperManu. *Original artist:* <a href='//commons.wikimedia.org/wiki/User:YassineMrabet/Gallery' title='User:YassineMrabet/Gallery'>**G**</a><a href='//commons.wikimedia.org/wiki/User:YassineMrabet' title='User:YassineMrabet'>YassineMrabet</a><a href='//commons.wikimedia.org/wiki/User_talk:YassineMrabet' title='User talk:YassineMrabet'>Talk</a><a class='external text' href='http://commons.wikipedia.org/w/index.php?title=User_talk:YassineMrabet,<span>,&,</span>,action=edit,<span>,&,</span>,section=new'></a>
- **File:Edward_Witten.jpg** *Source:* https://upload.wikimedia.org/wikipedia/commons/9/97/Edward_Witten.jpg *License:* Public domain *Contributors:* Own work *Original artist:* Ojan
- **File:Elementary_particle_interactions_in_the_Standard_Model.png** *Source:* https://upload.wikimedia.org/wikipedia/commons/a/a7/Elementary_particle_interactions_in_the_Standard_Model.png *License:* CC0 *Contributors:* Own work *Original artist:* Eric Drexler
- **File:Feynman-Diagram.svg** *Source:* https://upload.wikimedia.org/wikipedia/commons/e/e3/Feynman-Diagram.svg *License:* Public domain *Contributors:* own work, based on Image:Feynman-Diagram.jpg *Original artist:* helix84
- **File:Feynman-diagram-ee-scattering.png** *Source:* https://upload.wikimedia.org/wikipedia/en/f/fb/Feynman-diagram-ee-scattering.png *License:* Cc-by-sa-3.0 *Contributors:* ? *Original artist:* ?
- **File:Feynman_EP_Annihilation.svg** *Source:* https://upload.wikimedia.org/wikipedia/commons/b/ba/Feynman_EP_Annihilation.svg *License:* Public domain *Contributors:* My creation; reimplementation of PNG on Wikipedia *Original artist:* bitwise
- **File:Feynman_diagram_general_properties.svg** *Source:* https://upload.wikimedia.org/wikipedia/commons/7/7a/Feynman_diagram_general_properties.svg *License:* CC0 *Contributors:* Own work *Original artist:* Maschen

- **File:Feynmann_Diagram_Gluon_Radiation.svg** *Source:* https://upload.wikimedia.org/wikipedia/commons/1/1f/Feynmann_Diagram_Gluon_Radiation.svg *License:* CC BY 2.5 *Contributors:* Non-Derived SVG of Radiate_gluon.png, originally the work of SilverStar at Feynmann-diagram-gluon-radiation.svg, updated by joelholdsworth. *Original artist:* Joel Holdsworth (Joelholdsworth)
- **File:Fifths.png** *Source:* https://upload.wikimedia.org/wikipedia/commons/c/ce/Fifths.png *License:* CC-BY-SA-3.0 *Contributors:* ? *Original artist:* ?
- **File:Fluxtube_meson.png** *Source:* https://upload.wikimedia.org/wikipedia/commons/9/93/Fluxtube_meson.png *License:* Public domain *Contributors:* http://inspirehep.net/record/840296 , arXiv:0912.3181 [hep-lat] and M. Cardoso et al., *Lattice QCD computation of the colour fields for the static hybrid quark-gluon-antiquark system, and microscopic study of the Casimir scaling*, Physical Review D, **81** (2010)) *Original artist:* RolteVolte (<a href='//commons.wikimedia.org/wiki/User_talk:RolteVolte' title='User talk:RolteVolte'>talk</a>) 16:42, 23 October 2011 (UTC)
- **File:Folder_Hexagonal_Icon.svg** *Source:* https://upload.wikimedia.org/wikipedia/en/4/48/Folder_Hexagonal_Icon.svg *License:* Cc-by-sa-3.0 *Contributors:* ? *Original artist:* ?
- **File:GabrieleVeneziano.jpg** *Source:* https://upload.wikimedia.org/wikipedia/commons/9/95/GabrieleVeneziano.jpg *License:* CC BY-SA 2.5 *Contributors:* Taken by Betsythedevine *Original artist:* Original uploader was Betsythedevine at en.wikipedia
- **File:Gerard_'t_Hooft.jpg** *Source:* https://upload.wikimedia.org/wikipedia/commons/2/2c/Gerard_%27t_Hooft.jpg *License:* CC BY-SA 3.0 *Contributors:* Own work *Original artist:* Wammes Waggel
- **File:Gravity_Probe_B.jpg** *Source:* https://upload.wikimedia.org/wikipedia/commons/5/51/Gravity_Probe_B.jpg *License:* Public domain *Contributors:* ? *Original artist:* ?
- **File:Holomouse2.jpg** *Source:* https://upload.wikimedia.org/wikipedia/commons/5/5f/Holomouse2.jpg *License:* Public domain *Contributors:*
- Holo-Mouse.jpg *Original artist:* Holo-Mouse.jpg: Georg-Johann Lay
- **File:Hqmc-vector.svg** *Source:* https://upload.wikimedia.org/wikipedia/commons/6/68/Hqmc-vector.svg *License:* CC BY 3.0 *Contributors:* Own work *Original artist:* VermillionBird
- **File:Ilc_9yr_moll4096.png** *Source:* https://upload.wikimedia.org/wikipedia/commons/3/3c/Ilc_9yr_moll4096.png *License:* Public domain *Contributors:* http://map.gsfc.nasa.gov/media/121238/ilc_9yr_moll4096.png *Original artist:* NASA / WMAP Science Team
- **File:John_Schwarz_(Australia_1988).jpg** *Source:* https://upload.wikimedia.org/wikipedia/en/6/61/John_Schwarz_%28Australia_1988%29.jpg *License:* Public domain *Contributors:* ? *Original artist:* ?
- **File:Joseph_Polchinski.jpg** *Source:* https://upload.wikimedia.org/wikipedia/commons/c/cf/Joseph_Polchinski.jpg *License:* Public domain *Contributors:* Transferred from en.wikipedia; transfered to Commons by User:Magnus Manske using CommonsHelper. *Original artist:* Original uploader was Lumidek at en.wikipedia
- **File:JuanMaldacena.jpg** *Source:* https://upload.wikimedia.org/wikipedia/commons/b/bc/JuanMaldacena.jpg *License:* CC BY 3.0 *Contributors:* Own work by the original uploader *Original artist:* Lumidek at English Wikipedia
- **File:Kaluza_Klein_compactification.svg** *Source:* https://upload.wikimedia.org/wikipedia/commons/d/dd/Kaluza_Klein_compactification.svg *License:* Public domain *Contributors:* Own work *Original artist:* Isilanes
- **File:Kaon-Decay.svg** *Source:* https://upload.wikimedia.org/wikipedia/commons/7/70/Kaon-Decay.svg *License:* CC-BY-SA-3.0 *Contributors:* Own work *Original artist:* User JabberWok on en.wikipedia
- **File:KleinInvariantJ.jpg** *Source:* https://upload.wikimedia.org/wikipedia/commons/3/37/KleinInvariantJ.jpg *License:* Public domain *Contributors:* made with mathematica, own work *Original artist:* Jan Homann
- **File:Knot_table-blank_unknot.svg** *Source:* https://upload.wikimedia.org/wikipedia/commons/f/f0/Knot_table-blank_unknot.svg *License:* Public domain *Contributors:* This file was derived from: Knot table.svg

  A similar table appears in

  Original artist:
- derivative work: 84user (removed Unknot label)
- **File:Labeled_Triangle_Reflections.svg** *Source:* https://upload.wikimedia.org/wikipedia/commons/3/38/Labeled_Triangle_Reflections.svg *License:* Public domain *Contributors:* Own work *Original artist:* Jim.belk
- **File:LeonardSusskindStanford2009_cropped.jpg** *Source:* https://upload.wikimedia.org/wikipedia/commons/f/f8/LeonardSusskindStanford2009_cropped.jpg *License:* CC BY-SA 3.0 *Contributors:* File:LeonardSusskindStanford2009.jpg *Original artist:* Jonathan Maltz
- **File:Limits_of_M-theory.svg** *Source:* https://upload.wikimedia.org/wikipedia/commons/b/b8/Limits_of_M-theory.svg *License:* CC BY-SA 3.0 *Contributors:* This file was derived from: Limits of M-theory.png
  Original artist:
- derivative work: Alex Dunkel (Maky)
- **File:Meissner_effect_p1390048.jpg** *Source:* https://upload.wikimedia.org/wikipedia/commons/5/55/Meissner_effect_p1390048.jpg *License:* CC-BY-SA-3.0 *Contributors:* self photo *Original artist:* Mai-Linh Doan
- **File:Miri2.jpg** *Source:* https://upload.wikimedia.org/wikipedia/commons/d/db/Miri2.jpg *License:* CC BY-SA 3.0 *Contributors:* Own work *Original artist:* J:136401
- **File:ModularGroup-FundamentalDomain-01.png** *Source:* https://upload.wikimedia.org/wikipedia/commons/a/ad/ModularGroup-FundamentalDomain png *License:* CC-BY-SA-3.0 *Contributors:* from en wikipedia *Original artist:* Fropuff
- **File:Multiverse_-_level_II.svg** *Source:* https://upload.wikimedia.org/wikipedia/commons/3/34/Multiverse_-_level_II.svg *License:* Public domain *Contributors:* Vectorisation of Multiverse - level II.GIF (by K1234567890y), by Lokal_Profil *Original artist:*
- Original by K1234567890y

- **File:Nuvola_apps_edu_mathematics_blue-p.svg** *Source:* https://upload.wikimedia.org/wikipedia/commons/3/3e/Nuvola_apps_edu_mathematics_blue-p.svg *License:* GPL *Contributors:* Derivative work from Image:Nuvola apps edu mathematics.png and Image:Nuvola apps edu mathematics-p.svg *Original artist:* David Vignoni (original icon); Flamurai (SVG convertion); bayo (color)
- **File:Office-book.svg** *Source:* https://upload.wikimedia.org/wikipedia/commons/a/a8/Office-book.svg *License:* Public domain *Contributors:* This and myself. *Original artist:* Chris Down/Tango project
- **File:Open_and_closed_strings.svg** *Source:* https://upload.wikimedia.org/wikipedia/commons/5/56/Open_and_closed_strings.svg *License:* Public domain *Contributors:* Own work *Original artist:* Xoneca
- **File:Point&string.png** *Source:* https://upload.wikimedia.org/wikipedia/commons/4/47/Point%26string.png *License:* Public domain *Contributors:* ? *Original artist:* ?
- **File:Portal-puzzle.svg** *Source:* https://upload.wikimedia.org/wikipedia/en/f/fd/Portal-puzzle.svg *License:* Public domain *Contributors:* ? *Original artist:* ?
- **File:QCD.svg** *Source:* https://upload.wikimedia.org/wikipedia/commons/2/2b/QCD.svg *License:* CC BY-SA 3.0 *Contributors:* Own work *Original artist:* Cjean42
- **File:Quantum_gravity.png** *Source:* https://upload.wikimedia.org/wikipedia/commons/6/6a/Quantum_gravity.png *License:* Public domain *Contributors:* Own work *Original artist:* Raidr
- **File:Question_book-new.svg** *Source:* https://upload.wikimedia.org/wikipedia/en/9/99/Question_book-new.svg *License:* Cc-by-sa-3.0 *Contributors:*

  Created from scratch in Adobe Illustrator. Based on Image:Question book.png created by User:Equazcion *Original artist:*

  Tkgd2007
- **File:Rubik'{}s_cube.svg** *Source:* https://upload.wikimedia.org/wikipedia/commons/a/a6/Rubik%27s_cube.svg *License:* CC-BY-SA-3.0 *Contributors:* Based on Image:Rubiks cube.jpg *Original artist:* This image was created by me, Booyabazooka
- **File:Schrödinger'{}s_flower.png** *Source:* https://upload.wikimedia.org/wikipedia/commons/8/86/Schr%C3%B6dinger%27s_flower.png *License:* CC BY-SA 4.0 *Contributors:* Own work *Original artist:* Clementvidal
- **File:Science.jpg** *Source:* https://upload.wikimedia.org/wikipedia/commons/5/54/Science.jpg *License:* Public domain *Contributors:* ? *Original artist:* ?
- **File:Spacetime_curvature.png** *Source:* https://upload.wikimedia.org/wikipedia/commons/2/22/Spacetime_curvature.png *License:* CC-BY-SA-3.0 *Contributors:* ? *Original artist:* ?
- **File:Spacetime_dimensionality.svg** *Source:* https://upload.wikimedia.org/wikipedia/commons/5/56/Spacetime_dimensionality.svg *License:* CC BY-SA 3.0 *Contributors:* On the dimensionality of spacetime, by Max Tegmark. This version extracted directly from the PostScript file. *Original artist:* Max Tegmark
- **File:Spin_network.svg** *Source:* https://upload.wikimedia.org/wikipedia/commons/5/52/Spin_network.svg *License:* CC BY-SA 3.0 *Contributors:* Own work *Original artist:* Markus Poessel (Mapos)
- **File:Standard_Model.svg** *Source:* https://upload.wikimedia.org/wikipedia/commons/f/f4/Standard_Model.svg *License:* CC BY-SA 3.0 *Contributors:* Own work *Original artist:* Cjean42
- **File:Standard_Model_Feynman_Diagram_Vertices.png** *Source:* https://upload.wikimedia.org/wikipedia/commons/7/75/Standard_Model_Feynman_Diagram_Vertices.png *License:* CC BY-SA 3.0 *Contributors:* I made it in Adobe Illustrator *Original artist:* Garyzx
- **File:Standard_Model_Of_Particle_Physics,_Most_Complete_Diagram.jpg** *Source:* https://upload.wikimedia.org/wikipedia/commons/7/73/Standard_Model_Of_Particle_Physics%2C_Most_Complete_Diagram.jpg *License:* CC BY-SA 4.0 *Contributors:* Own work *Original artist:* Latham Boyle
- **File:Standard_Model_of_Elementary_Particles.svg** *Source:* https://upload.wikimedia.org/wikipedia/commons/0/00/Standard_Model_of_Elementary_Particles.svg *License:* CC BY 3.0 *Contributors:* Own work by uploader, PBS NOVA [1], Fermilab, Office of Science, United States Department of Energy, Particle Data Group *Original artist:* MissMJ
- **File:Stephen_Hawking.StarChild.jpg** *Source:* https://upload.wikimedia.org/wikipedia/commons/e/eb/Stephen_Hawking.StarChild.jpg *License:* Public domain *Contributors:* Original. Source (StarChild Learning Center). Directory listing. *Original artist:* NASA
- **File:StringTheoryDualities.svg** *Source:* https://upload.wikimedia.org/wikipedia/commons/8/8a/StringTheoryDualities.svg *License:* CC BY-SA 3.0 *Contributors:*
- StringTheoryDualities.jpg *Original artist:*
- derivative work: Alex Dunkel (Maky)
- **File:Stylised_Lithium_Atom.svg** *Source:* https://upload.wikimedia.org/wikipedia/commons/e/e1/Stylised_Lithium_Atom.svg *License:* CC-BY-SA-3.0 *Contributors:* based off of Image:Stylised Lithium Atom.png by Halfdan. *Original artist:* SVG by Indolences. Recoloring and ironing out some glitches done by Rainer Klute.
- **File:Sum_over_genera.png** *Source:* https://upload.wikimedia.org/wikipedia/commons/1/1e/Sum_over_genera.png *License:* CC BY-SA 4.0 *Contributors:* Own work *Original artist:* Rantonels
- **File:Torus.png** *Source:* https://upload.wikimedia.org/wikipedia/commons/1/17/Torus.png *License:* Public domain *Contributors:* ? *Original artist:* ?
- **File:Torus_cycles.png** *Source:* https://upload.wikimedia.org/wikipedia/commons/5/54/Torus_cycles.png *License:* CC-BY-SA-3.0 *Contributors:* ? *Original artist:* ?
- **File:Uniform_tiling_433-t0.png** *Source:* https://upload.wikimedia.org/wikipedia/commons/1/13/Uniform_tiling_433-t0.png *License:* Public domain *Contributors:* Transferred from en.wikipedia *Original artist:* Tomruen at en.wikipedia
- **File:Uniform_tiling_433-t0_(formatted).svg** *Source:* https://upload.wikimedia.org/wikipedia/commons/2/21/Uniform_tiling_433-t0_%28formatted%29.svg *License:* CC BY-SA 4.0 *Contributors:* This file was derived from: Uniform tiling 433-t0.png
  Original artist:

- derivative work: Polytope24
- **File:Wiki_letter_w_cropped.svg** *Source:* https://upload.wikimedia.org/wikipedia/commons/1/1c/Wiki_letter_w_cropped.svg *License:* CC-BY-SA-3.0 *Contributors:*
- Wiki_letter_w.svg *Original artist:* Wiki_letter_w.svg: Jarkko Piiroinen
- **File:Wiktionary-logo-en.svg** *Source:* https://upload.wikimedia.org/wikipedia/commons/f/f8/Wiktionary-logo-en.svg *License:* Public domain *Contributors:* Vector version of Image:Wiktionary-logo-en.png. *Original artist:* Vectorized by Fvasconcellos (talk · contribs), based on original logo tossed together by Brion Vibber
- **File:Winding_Number_-1.svg** *Source:* https://upload.wikimedia.org/wikipedia/commons/1/18/Winding_Number_-1.svg *License:* Public domain *Contributors:* Own work *Original artist:* Jim.belk
- **File:Winding_Number_-2.svg** *Source:* https://upload.wikimedia.org/wikipedia/commons/9/92/Winding_Number_-2.svg *License:* Public domain *Contributors:* Own work *Original artist:* Jim.belk
- **File:Winding_Number_0.svg** *Source:* https://upload.wikimedia.org/wikipedia/commons/6/64/Winding_Number_0.svg *License:* Public domain *Contributors:* Own work *Original artist:* Jim.belk
- **File:Winding_Number_1.svg** *Source:* https://upload.wikimedia.org/wikipedia/commons/0/0e/Winding_Number_1.svg *License:* Public domain *Contributors:* Own work *Original artist:* Jim.belk
- **File:Winding_Number_2.svg** *Source:* https://upload.wikimedia.org/wikipedia/commons/a/aa/Winding_Number_2.svg *License:* Public domain *Contributors:* Own work *Original artist:* Jim.belk
- **File:Winding_Number_3.svg** *Source:* https://upload.wikimedia.org/wikipedia/commons/5/54/Winding_Number_3.svg *License:* Public domain *Contributors:* Own work *Original artist:* Jim.belk
- **File:World_lines_and_world_sheet.svg** *Source:* https://upload.wikimedia.org/wikipedia/commons/2/25/World_lines_and_world_sheet.svg *License:* Public domain *Contributors:* Point&string.png *Original artist:* Kurochka, svg version by Actam

26.6.3 Content license

-

www.ingramcontent.com/pod-product-compliance
Lightning Source LLC
Chambersburg PA
CBHW081141180526
45170CB00006B/1883

* 9 7 8 1 5 2 3 2 2 4 1 0 4 *